Nanostructure Science and Technology

Series Editor:
David J. Lockwood, FRSC
National Research Council of Canada
Ottawa, Ontario, Canada

More information about this series at http://www.springer.com/series/6331

Anatoli Korkin • Stephen Goodnick
Robert Nemanich
Editors

Nanoscale Materials and Devices for Electronics, Photonics and Solar Energy

 Springer

Editors
Anatoli Korkin
Nano and Giga Solutions, Inc.
Gilbert, AZ, USA

Stephen Goodnick
Engineering Research Center
Arizona State University
Tempe, AZ, USA

Robert Nemanich
Department of Physics
Arizona State University
Tempe, AZ, USA

ISSN 1571-5744 ISSN 2197-7976 (electronic)
Nanostructure Science and Technology
ISBN 978-3-319-18632-0 ISBN 978-3-319-18633-7 (eBook)
DOI 10.1007/978-3-319-18633-7

Library of Congress Control Number: 2015945943

Springer Cham Heidelberg New York Dordrecht London

Printed on acid-free paper

Springer International Publishing AG Switzerland is part of Springer Science+Business Media (www.springer.com)

Preface

The *Nano and Giga Challenges* (NGC) conference series has had a long tradition of tutorial lectures given by world-renowned researchers. As early as the first forum in Moscow, Russia, in 2002, the organizers realized that publication of the lectures notes from NGC2002 would be a valuable legacy of the meeting and a significant educational resource and knowledge base for students, young researchers, and experts alike. Our first book was published by *Elsevier* and named after the meeting itself—*Nano and Giga Challenges in Microelectronics* [1]. Our subsequent books based on the tutorial lectures of the NGCM2004 [2], NGC2007 [3], NGC2009 [4], NGC2011 [5], and the current book derived from the NGC2014 conference have been published by Springer in the *Nanostructure Science and Technology* series.

Energy and information are essential elements for the development of human society, which are interconnected. Processing and storage of information requires energy consumption, while the efficient use and access to new energy sources requires new information (ideas and expertise) and the design of novel systems such as photovoltaic devices, fuel cells, and batteries. Semiconductor physics creates the knowledge base for the development of information (computers, cell phones, etc.) and energy (photovoltaics) technologies. The exchange of ideas and expertise between these two technologies is critical and expands beyond semiconductors. Efficient use of solar energy requires development of novel energy storage devices while biosystems provide new paradigms for the development of materials and devices for information (processing and storage) and energy (e.g., biofuel and artificial photosynthesis) technologies and biomedical applications (sensors and diagnostics).

Progress in information and renewable energy technologies requires miniaturization of devices and reduction of costs, energy, and material consumption. The latest generation of electronic devices is now approaching nanometer scale dimensions; new materials are being introduced into electronics manufacturing at an unprecedented rate; and alternative technologies to mainstream CMOS are evolving. The low cost of natural energy sources has created economic barriers to the development of alternative and more efficient solar energy systems, fuel cells, and batteries. However, there is emergent understanding that the sustainable development of human society requires use of new alternate sources of energy to natural gas and oil.

Nanotechnology is widely accepted as a source of potential solutions in securing future progress for information and energy technologies. Our conference series is an interdisciplinary forum in education, research, and innovations in the development of new materials, devices, and systems for these key technologies. The NGC2014 conference (the sixth Nano and Giga Forum) invited academic and industrial researchers to present tutorial and original research papers dedicated to solving scientific and technological problems in the following areas of electronics, photonics, and renewable energy: atomic scale materials design, bio- and molecular electronics, high frequency electronics, fabrication of nanodevices, magnetic materials and spintronics, materials and processes for integrated and subwave optoelectronics, nanoCMOS, new materials for FETs and other devices, nanoelectronics system architecture, nano-optics and lasers, non-silicon materials and devices, chemical and biosensors, quantum effects in devices, nanoscience and technology applications in the development of novel solar energy devices, and fuel cells and batteries. We also invited inventors, entrepreneurs, and business leaders to explore the unique opportunity provided by our interdisciplinary forum for technical due diligence and potential commercialization of emerging new technologies.

The success of the NGC2014 conference [6], which resulted in the publication of this book, would have not been possible without generous support from many sponsors and research institutions. We gratefully acknowledge contributions and support of Arizona State University (host of the conference), Springer Publisher, National Institute of Health (NIH), and many other local, national, and international organizations and individual supporters.

Gilbert, AZ, USA   Anatoli Korkin
Tempe, AZ, USA   Stephen Goodnick
Tempe, AZ, USA   Robert Nemanich

References

1. *Nano and Giga Challenges in Microelectronics*, ed. by J. Greer, A. Korkin, J. Labanowski (Elsevier, Amsterdam, Netherlands, 2003)
2. *Nanotechnology for Electronic Materials and Devices*, ed. by A. Korkin, E. Gusev, J. Labanowski, S. Luryi (Springer, New York, 2007)
3. *Nanoelectronics and Photonics: From Atoms to Materials, Devices, and Architectures*, ed. by A. Korkin, F. Rosei (Springer, New York, 2008)
4. *Nanotechnology for Electronics, Photonics, and Renewable Energy*, ed. by A. Korkin, P. Krstic, J. Wells (Springer, New York, 2010)

5. *Nanoscale Applications for Information and Energy Systems*, ed. by A. Korkin, D.J. Lockwood (Springer, New York, 2013)
6. Nano and Giga Challenges in Electronics, Photonics and Renewable Energy: From Materials to Devices to System Architecture, Symposium and Spring School (Tutorial Lectures), Phoenix, Arizona, 10–14 March 2014; http://www.nanoandgiga.com/ngc2014

Contents

Contributors

Francis Balestra IMEP-LAHC, Grenoble INP-Minatec, Grenoble, France

Sanjay Kumar Banerjee University of Texas at Austin, Austin, TX, USA

Oleg L. Berman Physics Department, New York City College of Technology, The City University of New York, Brooklyn, NY 11201, USA

Adam Blake Department of Physics, Arizona State University, Tempe AZ 85281, USA

Hassan Elsentriecy Department of Chemical and Environmental Engineering, University of Arizona, Tucson, AZ 85281, USA

Dominic F. Gervasio Department of Chemical and Environmental Engineering, University of Arizona, Tucson, AZ 85281, USA

A.M. Kannan Fulton School of Engineering, Arizona State University, Mesa, AZ, USA

Roman Ya. Kezerashvili Physics Department, New York City College of Technology, The City University of New York, Brooklyn, NY 11201, USA

Wojciech Knap LC2-Laboratory, CNRS-Universite Montpellier 2, Montpellier, France

Marek Korkusinski Quantum Theory Group, Security and Disruptive Technologies, National Research Council, Ottawa, ON, Canada

Stuart Lindsay Department of Physics, Biodesign Institute, Arizona State University, Tempe, AZ 85287, USA

Department of Chemistry and Biochemistry, Biodesign Institute, Arizona State University, Tempe, AZ 85287, USA

Yurii E. Lozovik Institute of Spectroscopy, Russian Academy of Sciences, 142190 Troitsk, Moscow Region, Russia

MIEM at National Research University HSE, 109028 Moscow, Russia

Vladimir Mitin Department of EE, University at Buffalo, SUNY, Buffalo, NY, USA

Xuehao Mou University of Texas at Austin, Austin, TX, USA

Taiichi Otsuji RIEC, Tohoku University, Sendai, Japan

Vyacheslav V. Popov Kotelnikov Institute of Radio Engineering and Electronics, Saratov, Russia

Leonard Franklin Register University of Texas at Austin, Austin, TX, USA

Maxim Ryzhii Department of CSC, University of Aizu, Aizu-Wakamatsu, Japan

Victor Ryzhii RIEC, Tohoku University, Sendai, Japan

Akira Satou RIEC, Tohoku University, Sendai, Japan

Michael Shur Department of ECSC, Rensselaer Polytechnic Institute, Troy, MI, USA

Luis Phillipi da Silva Department of Chemical and Environmental Engineering, University of Arizona, Tucson, AZ 85281, USA

Maxim Sukharev Science and Mathematics Faculty, College of Letters and Sciences, Arizona State University, Mesa AZ 85212, USA

Stephane Boubanga Tombet RIEC, Tohoku University, Sendai, Japan

K. Vignarooban Fulton School of Engineering, Arizona State University, Mesa, AZ, USA

Xinhai Xu Department of Chemical and Environmental Engineering, University of Arizona, Tucson, AZ 85281, USA

Chapter 1
Ultralow-Power Device Operation

Francis Balestra

Abstract The historic trend in micro/nano-electronics these last 40 years has been to increase both speed and density by scaling down the electronic devices, together with reduced energy dissipation per binary transition. We are facing today dramatic challenges dealing with the limits of energy consumption and heat removal, inducing fundamental tradeoffs for the future ICs. A substantial reduction of the static and dynamic power is strongly needed for the development of future high-performance/ultralow-power terascale integration and autonomous nanosystems.

This chapter of the tutorial book addresses the main trends, challenges, limits, and possible solutions for strongly reducing the energy per binary switching. Several paths are possible, the most promising one being the reduction of static and dynamic energy consumption using conventional logic with a reduction in the stored energy and therefore a decrease of device capacitance C (device integration) and applied bias V, together with a decrease of leakage currents of nanodevices.

The best potential solutions are ultrathin-film SOI (silicon-on-insulator) and multi-gate devices, nanowires, and small-slope switches (tunnel FETs, ferroelectric gate FETs, NEMS) using alternative channel, source/drain, and gate materials. We will present the main challenges to continue More's law, the novel materials and device architectures, and the possible combination of these boosters, needed for the development of future ultralow-power ICs.

1.1 Introduction

Silicon is the leading semiconductor in the electronics industry. More than 95 % of the integrated circuits are fabricated with Si-based devices. The basic component which is currently used for very large-scale integrated (VLSI) circuits, such as microprocessors and memories, is the metal oxide semiconductor field-effect transistor (MOSFET). The first structure with a thermally oxidized silicon was

F. Balestra (✉)
IMEP-LAHC, Grenoble INP-Minatec, BP 257, 38016 Grenoble, France
e-mail: balestra@minatec.inpg.fr

© Springer International Publishing Switzerland 2015
A. Korkin et al. (eds.), *Nanoscale Materials and Devices for Electronics,*
Photonics and Solar Energy, Nanostructure Science and Technology,
DOI 10.1007/978-3-319-18633-7_1

Fig. 1.1 Bulk silicon
CMOS structure

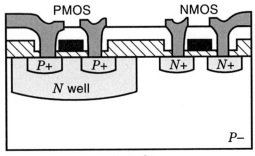

proposed and fabricated in the 1960s [1, 2] and hugely developed during the past 40 years. The main advantages of this device, compared with many others using the field-effect principle, are the quality of the Si/dielectric system and the possibility to develop high-performance and very low-power components. The silicon/Si dioxide interface and more recently Si/high-k interface have been improved over these years, leading to high-performance transistors with a very low defect density. Compared with their bipolar counterparts, in which both types of carriers are involved, the virtues of MOS transistors are low cost and high density. However, bipolar devices are also used for some applications needing high speed and/or low noise.

Complementary MOS (CMOS) has become the dominant MOS technology and uses both n-channel (electrons) and p-channel (holes) MOSFETs (Fig. 1.1). This structure has been chosen historically owing to its low-power consumption property compared with other ones (NMOS, n-channel MOSFET, etc.). More than 80 % of the recent VLSI circuits are fabricated with the CMOS technology. The general trends in micro/nano-electronics are the reduction of device geometry which allows one to obtain an improvement of device performance and integration density. Other high priorities in this field are the decrease of power consumption, the enhancement of reliability, and the reduction of the cost of electronic systems.

However, the device scaling down to deep sub-0.1 μm dimension faces formidable challenges. After 40 years of technology development, MOSFETs are pushed toward their fundamental and technological limits. Indeed, the scaling scenario put forth in widely accepted industry roadmaps calls for sub-10 nm gate length MOSFETs in the next decade. In this channel length range, the optimization of the transistors need in particular very thin source/drain junction depths and a good control of the electrostatic potential and charge in the active silicon layer in order to reduce parasitic effects (for instance, short-channel effects and leakage currents) and to improve device performance (e.g., drain current and transconductance).

However, the historic trend in micro/nano-electronics these last 40 years has been to increase both speed and density by scaling down the electronic devices, together with reduced energy dissipation per binary transition. We are facing today dramatic challenges dealing with the limits of energy consumption (static + dynamic) and heat removal, inducing fundamental tradeoffs for the future ICs.

The worldwide energy consumption has increased by 40 % these last 20 years. The ICT part represents 15 % of the electricity consumption and will increase by a factor 3 in the next 20 years. For instance, today, in 2 days the information generated in the world corresponds to the one generated between the beginning of the ICT era till 2003, and 150 billion e-mails are exchanged every day on our planet.

Therefore, the researches on ultimate reduction of computation dissipation are strongly needed for the development of future very low-power and high-performance terascale integration and autonomous (nano)systems.

This tutorial chapter addresses the main challenges, limits, and possible solutions for strongly reducing the energy per binary switching. Since the 90 nm node, V_{dd} scaling has been slowed leading to accelerated energy consumption and heating and a move from a constant field toward a constant voltage scaling. In 2005 the increase in microprocessor frequency abruptly ceased, but the integration level continues to increase and parallel processors were proposed. We are thus facing dramatic challenges and two main paths are possible for reducing the energy dissipated, which is the most critical limit for future ICs: the conventional logic with a reduction in the stored energy and the leakage currents and therefore a decrease of device capacitance C (device integration) or applied bias V as well as using novel materials and device architectures and the adiabatic logic using a slow clock.

The ultimate limit of energy dissipation in irreversible logical operations requires of at least $k.T.\ln(2)$ for each bit of information lost (corresponding to an increase of entropy [3]). Reversible operations can be performed without dissipation. In principle all computations could be performed without dissipation using only reversible operations [4]; however in practice any computer will dissipate energy (e.g., due to error-correcting codes to maintain reliable operation or due to residual resistance for non-superconducting wires and electrodes). It has recently been shown the possibility of sub-kT energy dissipation in charging a capacitor adiabatically, with charging energy of many kT delivered to the capacitor (by charging and discharging the capacitor gradually/adiabatically, the power dissipation is significantly lower—at the cost of lowering the switching speed) [5]. Therefore, it is possible to use charges, and not necessarily other state variables, for reversible computation. For irreversible computation (with the least energy $kT\ln(2)$), the ultimate limit of logic switching can be obtained using the Heisenberg uncertainty relations, leading to a minimum size of devices of 1.5 nm, a maximum density of devices of $5 \times 10^{13}/cm^2$, and a minimum switching time of 0.04 ps, inducing a power dissipation of 4×10^6 W/cm^2, which is far beyond the ultimate theoretical limit for heat removal from 2D Si surface (1,000 W/cm^2). These numbers illustrate well the challenges we are facing for the ultimate ICs.

The reduction of the stored energy in conventional logic can be done with a strong reduction in V using new physics and/or devices with 60 or sub-60 mV/dec subthreshold swing S, which is the limit of MOSFETs at 300 K. The main following concepts are possible: energy filtering (tunnel FET, with MOS-, NW-, CNT-, or graphene based, using band-to-band tunneling to filter energy distribution of electrons in the source), internal voltage step-up (ferroelectric gate FET, inducing a negative capacitance to amplify the change in channel potential induced by the gate), or nano-electromechanical structures.

All these topics will be presented in the next sections.

1.2 Novel Materials and Innovative Device Architectures for Very Low-Power CMOS

1.2.1 The SOI Structure

The SOI (silicon-on-insulator) CMOS is shown in Fig. 1.2. A buried insulator, which is typically an oxide layer, is fabricated in the silicon substrate using various methods. A number of advantages, suitable for many applications, are obtained with the SOI structure, which allows us to push back the technological and physical limits intrinsic to the bulk Si structure [6, 7]:

1. The latch-up, a parasitic n-p-n-p structure (thyristor) which can be triggered in bulk silicon CMOS structures, is suppressed by the dielectric isolation of the SOI technology.
2. Ultrathin SOI films, inducing a full depletion ("fully depleted" devices) or inversion ("fully inverted" devices) of the active silicon layer, can easily be fabricated and have been shown to lead to very interesting behaviors. In this respect, a low-threshold voltage together with small leakage currents can be obtained which is of major importance for low-power–low-voltage integrated circuits used in portable electronic systems. Another advantage of such thin films is related to higher drain current and device speed.

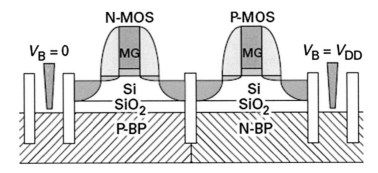

Fig. 1.2 Silicon on insulator (SOI) CMOS structure, with possible back-gate control for optimizing device performance

3. The parasitic source/drain junction capacitances are substantially reduced using thin-film SOI, leading to high-frequency operation (several tens of gigahertz have been demonstrated).
4. The leakage currents can be significantly smaller in SOI than that of bulk Si devices (several orders of magnitude) due to the reduction of the drain junction area and improved electrostatic control. The exponential enhancement of the leakage current with temperature is the main limitation of the circuit functionality in conventional CMOS operation at high temperature. SOI devices are fully functional up to over 300 °C and can be used in analog and digital circuits for automotive and aircraft applications.
5. In the memory field, the SOI structure also leads to improved performances for DRAM (dynamic random-access memory), SRAM (static random-access memory), and nonvolatile memories. For instance, a better soft error immunity and a reduction of the cell capacitor can be obtained. 1-T capacitorless DRAM is also promising for overcoming the challenge of the capacitor integration.
6. SOI MOSFETs are of great interest for niche applications, especially in the field of radiation hardness [advantage in transient ionizing and single-event upset (SEU)] for both space and nuclear radiation environments, owing to the small active volume. This was the first application of the SOI technology. Another example is low-temperature electronics for which the SOI structure prevents some harmful parasitic effects (e.g., the absence of the kink effect for thin-film SOI) observed in bulk transistors.
7. The dielectric isolation afforded by the buried insulator opens the way to integrate into state-of-the-art CMOS process complementary bipolar transistors, power switches, or sensors (flow sensor, magnetic sensor, pressure sensor, etc.). The SOI technology leads to new "system on chip" concepts.
8. The special SOI structure can be advantageously used for developing new devices, like the volume-inversion MOSFET with fully inverted channel (with two gates, three gates, or a gate all around the active Si layer), the voltage-controlled bipolar MOS device (with gate and body connected), and quantum devices (extremely thin Si film, quantum wires, single-electron transistors, etc.).
9. Some process simplification is induced by the SOI technology, mainly in the isolation and high-energy implantation.
10. Short-channel and hot-carrier effects, which are detrimental to device integration, power consumption, and reliability, can be reduced in SOI MOSFETs.
11. Three-dimensional ICs can also be realized on SOI materials, allowing a further potential enhancement of the level of circuit integration.

However, some parasitic effects can also be observed in SOI devices. In this respect, floating body effects and self-heating effects are the main drawbacks of the SOI technology.

The first large-volume commercial market for SOI technology will certainly be low-voltage low-power applications. In this respect, the capability of SOI has been demonstrated for a power supply reduction in the sub-1 V range, down to 0.5 V, with good performances. SOI devices will thus be part of the wireless electronic revolution.

A relevant question is the availability of SOI wafers in large number. Various approaches have been used for fabricating SOI materials. Among all the techniques proposed to perform the SOI structure, three have been strongly developed: SIMOX (separation by implanted oxygen), WB (wafer bonding, or BESOI—bond and etch back SOI), and the Smart Cut technology (material called UNIBOND) which combined both hydrogen implantation and bonding. The last technique has a much higher throughput potential and is today the leading technology.

1.2.2 SOI MOSFETs

MOSFET operation is governed by the electron or hole transport between the source and the drain of the devices which leads to the drain current. The inversion (N+PN+ or P+NP+ structures) or accumulation (N+NN+ or P+PP+ structures) channel of electrons or holes can be formed at the silicon/oxide interface by biasing the gate of the transistors. The threshold of the MOSFET is reached when a high density of free carriers (equivalent to the doping density) is created at the Si surface (for a surface potential equal to $2\Phi_F$, Φ_F being the Fermi potential depending on the Si doping). The gate voltage for which this situation appears is called the threshold voltage of the MOS transistor. Above the threshold voltage V_t, the drain current I_d presents a linear or sublinear variation with the gate bias V_g (strong inversion or accumulation region). Below V_t, the drain current varies exponentially with V_g (weak inversion or accumulation region). The slope of the transfer $I_d(V_g)$ characteristic in a logarithmic scale is called the subthreshold slope (the term "subthreshold swing" is also used and is proportional to the inverse of the slope). This is a key parameter because it determines the switching characteristics between the off (very small-current) and on (high-current) states of the device (see Fig. 1.3).

Fig. 1.3 Typical drain-current–gate-voltage transfer characteristic of a MOSFET

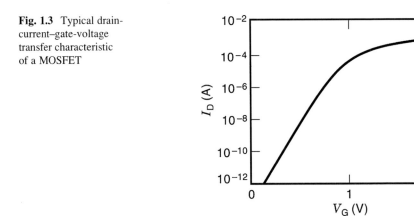

Fig. 1.4 Typical potential variation in the active silicon layer for (**a**) a bulk-like SOI MOSFET and (**b**) a fully depleted SOI MOSFET; a linearly varying potential is a good approximation for $V_g \leq V_t$ in thin-film fully depleted SOI devices

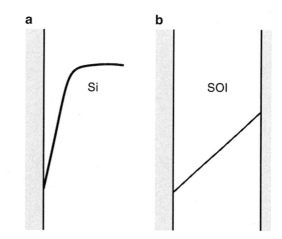

Two types of SOI MOSFETs can be obtained according to the thickness and the doping of the active silicon layer. When the depletion layer under the gate oxide, obtained by biasing the gate of the MOSFET, extends throughout the whole silicon film thickness, the device is called "fully depleted (FD)." If the maximum depletion width given by

$$x_{d\,max} = \sqrt{\frac{4\varepsilon_{Si}\Phi_F}{qN_a}} \qquad (1.1)$$

(which is observed for $V_g \geq V_t$) is smaller than the Si film thickness, the transistor is called "partially depleted (PD)" (Fig. 1.4) (ε_{Si} is the Si permittivity, Φ_F the Fermi potential, N_a the Si film doping, and q the electron charge). The first fully depleted SOI MOSFETs have been fabricated on SOS (silicon-on-sapphire) materials [8, 9], but a Si film of poor quality was obtained with this technology. The first fully depleted SOI MOSFET with an ideal swing of about 60 mV/dec was demonstrated by numerical simulation in 1985 [10]. The first experimental demonstration of a FD SOI MOSFET showing a swing of about 60 mV/dec was shown in 1986 [11]. SOI MOSFETs with full or partial depletion present very different electrical properties. A prominent advantage of fully depleted SOI MOSFETs formed on good-quality thin film is due to the possible improvement of the subthreshold swing as compared to partially depleted Si film or bulk silicon transistors. The equivalent circuits in weak inversion of a bulk-like MOSFET (bulk or partially depleted SOI transistor) and a fully depleted SOI MOSFET [12, 13] are shown in Figs. 1.5 and 1.6, respectively. For a fully depleted device (Fig. 1.6), the depletion capacitance C_d ($= dQ_d/d\Phi_S$, Q_d being the depletion charge and Φ_S the potential at the Si/SiO$_2$ interface) is suppressed because the depletion charge is limited by the thickness of the Si film and thereby does not vary with the gate voltage or the surface potential. C_d is replaced by a series of capacitances. If the FD SOI MOSFET is fabricated with a thick buried oxide

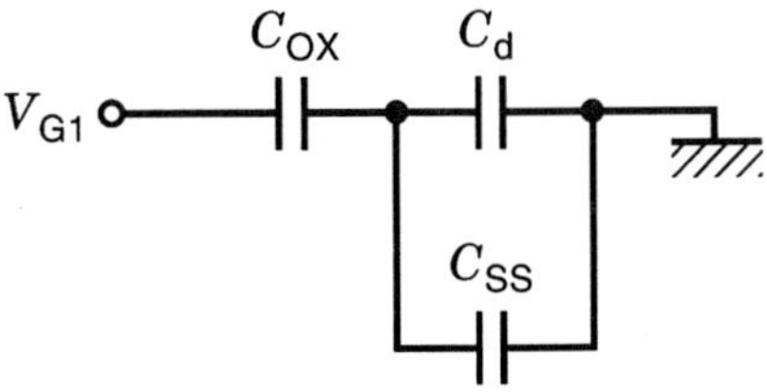

Fig. 1.5 Equivalent circuit
in weak inversion of a
bulk MOSFET

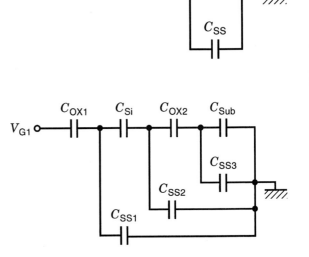

Fig. 1.6 Equivalent circuit
in weak inversion
of a fully depleted SOI
MOSFET [13]

and/or a low doping density in the silicon substrate (under the buried oxide) leading to
small buried oxide and substrate (depletion) capacitances, the subthreshold swing
S can be substantially lower than that observed in bulk devices. For small interface
state densities at the various Si/SiO$_2$ interfaces, which is usually the case for present
technologies, the swing can reach the minimum theoretical limit of about 60 mV/dec
at 300 K [10–14]. This offers the opportunity to achieve both a low-threshold voltage
and a small leakage current. These fully depleted SOI devices are very interesting for
high-performance, low-voltage, low-power integrated circuits.

The swing S for bulk-like [Eq. (1.2)] and FD SOI [Eq. (1.3)] MOSFETs [13] can
be described by

$$S_{\text{bulk}} = \frac{kT}{q}\ln(10)\left(1 + \frac{C_d + C_{ss}}{C_{ox}}\right) \tag{1.2}$$

$$S_{\text{soi}} = \frac{kT}{q}\ln(10)$$
$$\frac{(C_{ox1} + C_{ss1})[(C_{si} + C_{ss2})C + C_{ox2}(C_{sub} + C_{ss3})] + C_{si}[C_{ss2}C + C_{ox2}(C_{sub} + C_{ss3})]}{C_{ox1}[(C_{si} + C_{ss2})C + C_{ox2}(C_{sub} + C_{ss3})]} \tag{1.3}$$

where kT is the thermal energy; q is the electron charge; C_d ($= \varepsilon_{si}/x_d$, ε_{si} being the
silicon permittivity and x_d the width of the depletion charge under the gate) is the

Fig. 1.7 Subthreshold swing (proportional to the inverse of the subthreshold slope) of SOI MOSFETs showing the transition between partial and full depletion of the silicon layer; the Si film thickness for which this transition is obtained decreases with increasing the Si doping

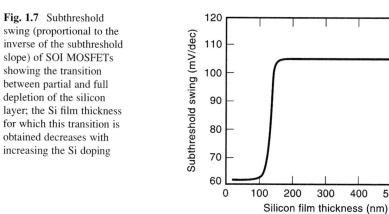

depletion capacitance; C_{ss} ($= qN_{ss}$, N_{ss} being the interface state density) and C_{ox} are the interface state and gate oxide capacitances for bulk structures, respectively; C_{si} ($=\varepsilon_{si}/t_{si}$, t_{si} being the Si film thickness) is the thin Si film capacitance for SOI devices; C_{ox1} and C_{ox2} are the gate oxide and buried oxide capacitances, respectively; C_{ss1}, C_{ss2}, and C_{ss3} are the interface state capacitances at the gate oxide/Si film, Si film/buried oxide, and buried oxide/Si substrate interfaces, respectively; C_{sub} is the substrate capacitance associated with the charge under the buried oxide; and $C = C_{ox2} + C_{ss3} + C_{sub}$.

Figure 1.7 presents the simulated variations of S as a function of Si film thickness t_{si}. A significant improvement of the swing is observed when the SOI device becomes fully depleted ($t_{si} \leq 100$ nm) [11].

1.2.3 Driving Current

Typical output drain-current–drain-voltage $I_d(V_d)$ characteristics are shown in Fig. 1.8. The curves present various operation regimes: the ohmic (or linear) operation for low V_d, the saturation for high V_d, and the beginning of the breakdown regime for the largest drain biases. The value of the driving current I_d in saturation (I_{dsat}) and the associated transconductance ($g_m = dI_{dsat}/dV_g$) are key electrical parameters for optimizing the performances of the MOSFETs.

In a fully depleted thin-film SOI MOSFET, the transverse electric field is reduced as compared to a partially depleted film (for a given surface potential). This decrease is accentuated for thinner layers, and the electric field has a quasi-linear variation with the Si film thickness. Therefore, the carrier profile extends deeper in the silicon film leading to higher carrier mobility μ owing to lower surface roughness and coulomb carrier scattering associated with the Si/SiO$_2$ or Si/high k interface. This interesting feature induces a substantial increase in drain current as compared to long-channel bulk-like devices operated at the same ($V_g - V_t$) bias [$I_{dsat} \propto W/L \, C_{ox} \, \mu \, (V_g - V_t)^2$;

Fig. 1.8 Typical drain-
current–drain-voltage
output characteristics
of a MOSFET

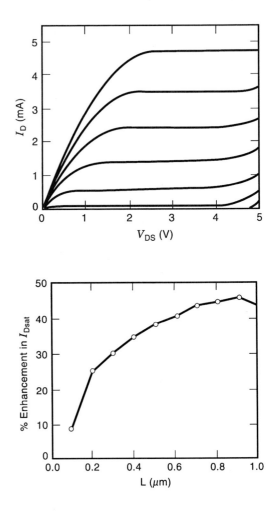

Fig. 1.9 Simulated relative
(to bulk counterparts)
enhancements of drain
current in saturation versus
channel length of fully
depleted n-channel SOI
MOSFETs at $V_d = 3.3$ V
and $V_g - V_t = 1.7$ V

W and L are the gate width and length, and C_{ox} is the gate oxide capacitance].
Therefore, this is also an advantage for low-voltage applications. However, carrier
velocity saturation driven by a high longitudinal electric field (the velocity maximum
being v_{sat}) limits the current enhancement when the devices are scaled down in the
deep submicron range [$I_{dsat} \propto W C_{ox} v_{sat} (V_g - V_t)$]. The simulations presented in
Fig. 1.9, fully supported by device measurements, show that the benefit of drain
current, relative to bulk Si or PD SOI, is partially lost as the channel length is reduced
and carrier velocity saturation occurs [15]. However, in an extremely thin Si layer
(<10 nm), a substantial increase of the driving current with scaling down the devices
has been pointed out in the deep submicron range. This effect is due to carrier
velocity overshoot occurring in sub-0.1 μm ultrathin SOI devices [16].

1.2.4 Short-Channel Effects

Short-channel effects can become severe in the deep sub-0.1 μm range. In this respect, two phenomena have to be optimized in order to obtain a reliable device and circuit operation. The charge-sharing effect is due to the increased influence of the depletion region at the source and drain junctions with scaling down the MOSFETs. This leads to a reduction of the depletion charge controlled by the gate and thereby a decrease of the threshold voltage of the transistor which can induce substantial leakage currents. The drain-induced barrier lowering (DIBL) is due to the electrostatic influence of the drain potential on the source/Si film barrier height at high V_d. This phenomenon has been shown to jeopardize deep submicron device operation with, in particular, a significant drain leakage current.

SOI structures offer unique options for the reduction of short-channel effects. However, a careful adjustment of the SOI parameters is necessary to improve the performance of deep submicron devices. Figure 1.10 presents the simulated variations of the threshold voltage in SOI MOSFETs due to charge-sharing effects [17]. The roll-off of the threshold voltage between long (1 μm) and short (0.1 μm) channel devices is a strong function of film thickness and doping. The worst-case condition (i.e., peak in the characteristics) is obtained for an intermediate silicon layer thickness, which roughly corresponds to the transition between full and partial depletion. In this region, short-channel effects are aggravated for fully depleted devices as compared to partially depleted SOI and bulk Si transistors. The silicon layer thickness must be far from the transition between partial and full depletion and substantially smaller than the junction depth of a comparable bulk MOSFET in order to improve short-channel effects in fully depleted SOI devices. A similar condition is obtained for attenuating the DIBL effect in deep submicron SOI MOSFETs [17].

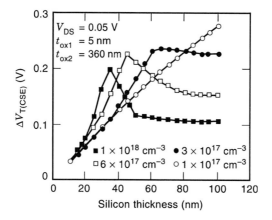

Fig. 1.10 Charge-sharing effect versus silicon layer thickness for various dopings. $\Delta V_t(\text{CSE}) = V_t(L_{\text{eff}} = 1\ \mu m) - V_t(L_{\text{eff}} = 0.1\ \mu m)$

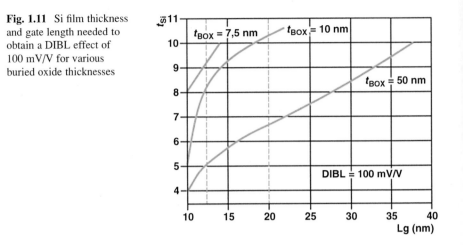

Fig. 1.11 Si film thickness and gate length needed to obtain a DIBL effect of 100 mV/V for various buried oxide thicknesses

The buried oxide thickness is another important parameter for alleviating short-channel effects. A significant attenuation of the threshold voltage shift is observed for ultrashort SOI devices when we reduce the buried oxide thickness down to sub-10 nm (Fig. 1.11) [18].

1.2.5 Alternative Device Architectures

1.2.5.1 Buried Oxide Engineering

Ground Plane

By using the SOI flexibility, a ground plane (highly doped or metallic) can be realized under the buried oxide, which leads to interesting behaviors in the sub-0.1 μm range. For instance, the DIBL is shown in Fig. 1.12a for traditional and ground plane (GP) SOI MOSFETs. This short-channel effect is substantially reduced in the case of a GP structure, in particular for thin buried oxides [19].

Low-k Buried Insulator

Figure 1.12b shows the variations of the DIBL as a function of the Si film thickness for various device architectures. The DIBL decreases for a low-k buried insulator and a ground plane as compared to the case of conventional SOI transistors whatever the silicon layer thickness is [19].

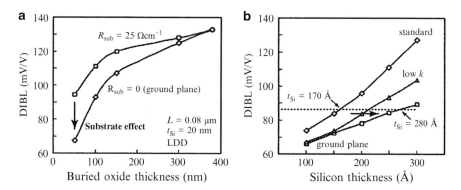

Fig. 1.12 (a) DIBL in traditional and GP SOI MOSFETs ($L_g = 0.08$ μm) as a function of buried oxide thickness; (b) DIBL in traditional, low-k buried insulator and GP SOI MOSFETs ($L_g = 0.08$ μm) versus Si layer thickness

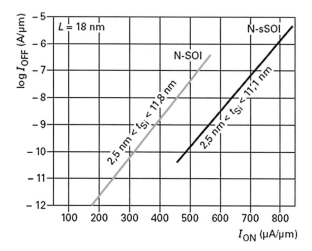

Fig. 1.13 Impact of strain Si on advanced n-channel SOI MOSFETs realized with various Si film thicknesses

1.2.5.2 Strained Channels

The strain engineering (uniaxial or biaxial, tensile or compressive) allow to change the band structures of Si or other channel materials (Ge, III–V) and to boost electron or hole carrier mobility and velocity.

Figure 1.13 shows an example of the impact of strain of 18 nm gate length n-channel SOI MOSFET fabricated on ultrathin SOI films ($t_{si} = 2.5–11.8$ nm). The ON current in strained devices (sSOI) is about 40 % higher in SOI for a given OFF current for similar SOI films, even for ultrathin Si layers down to 2.5 nm. This behavior has been attributed to a higher carrier injection velocity at the source [20].

Another booster for transport parameters is to use alternative channel or substrate orientations [21].

Fig. 1.14 Potential profile inside the silicon layer showing the volume inversion regime for thin-film ($\leq$100 nm) DG SOI MOSFETs; the occurrence of a volume inversion (with a quasi-constant potential $>2\Phi_F$ in the whole Si layer) depends on the Si film thickness and doping

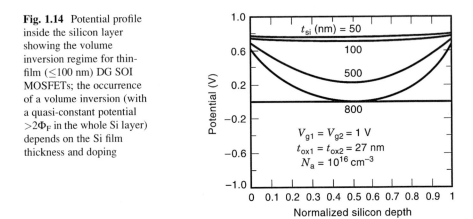

1.2.6 Ultimate Device Architectures

The inversion channel induced by the gate of a MOSFET is located at the Si/SiO$_2$ interface with a typical length of a few nanometers. The double-gate (DG) (or multi-gate—MG) control of an SOI MOSFET allows forcing the whole silicon film (interface layers and volume) in strong inversion and gives rise to the "volume inversion (VI)" concept for "fully inverted" devices (Fig. 1.14) [22]. The facts that the current drive of the VI-MOSFET is governed by two, three, or four gates and that carriers are no longer confined at one interface present remarkable advantages: enhancement of the number of minority carriers, increase in carrier mobility and velocity due to reduced influence of scattering associated with oxide charges and surface roughness, increase in drain current and transconductance, reduced short-channel effects, and ideal subthreshold slope [22, 23]. The subthreshold slope S_{vi} and threshold voltage V_t of the VI-MOSFET are calculated using a constant potential in the whole silicon layer [13, 24], which is a very good approximation for $V_g \leq V_t$:

$$S_{vi} = \frac{kT}{q}\ln(10)\left(\frac{C_{ox}+C_{ss}}{C_{ox}}\right) \tag{1.4}$$

$$V_t = V_{FB} + 2\Phi_F + \frac{qt_{si}N_a}{C_{ox}} - \frac{Q_{ss}(2\Phi_F)}{C_{ox}} \tag{1.5}$$

where C_{ss} ($= qN_{ss}$) is the interface state capacitance, C_{ox} is the gate oxide capacitance (similar interface state densities $N_{ss} = Q_{ss}/q$ and oxide thickness are assumed at the upper and lower Si film interface), V_{fb} is the flat-band voltage for both front

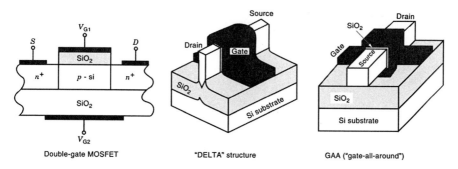

Fig. 1.15 Various SOI structures using the concept of volume inversion in order to obtain fully inverted devices: double, gate, delta or FinFET, and gate all around

and back gate, N_a is the silicon layer doping, Φ_F is the Fermi potential, and $Q_{ss}(2\Phi_F)$ is the interface state charge for a surface potential of $2\Phi_F$.

It is worthwhile noting that, in the threshold expression, the inversion charge Q_i is usually neglected with regard to the depletion charge Q_d. However, this is not possible in the case of volume inversion as, at threshold, $Q_i = Q_d$.

A dramatic reduction of hot-carrier effects (substrate current, photon emission, and hot-carrier-induced degradation) has also been reported in SOI MOSFETs with volume inversion operation [25]. Various SOI structures [double gate [26], DELTA or bulk/SOI FinFET [27], GAA (gate all around) [28]] have been proposed in order to take advantage of this original feature (Fig. 1.15). Figure 1.16 shows an example of the performance improvement during volume inversion operation as compared to a conventional single-gate SOI operation [28]: The GAA device exhibits a transconductance up to three times larger. The VI-MOSFET seems to be an ideal device for alleviating short-channel effects in ultimate ultrashort-channel MOS transistors [29]. A substantial reduction of the DIBL [Fig. 1.16b] and charge-sharing phenomena in volume inversion operation has been observed [30]. These advantages are obtained together with very large driving current and very small leakage currents for drain bias up to 1.5 V in a wide Si film thickness range. The short-channel effects (DIBL and charge sharing) are also reduced with decreasing the silicon film thickness (down to 5 nm) and doping whatever the architecture is [single and double gate; Fig. 1.16b]. Besides, a reduced sensitivity on the silicon film thickness and doping is observed for the double-gate devices that is very interesting for the optimization of their electrical properties.

This transistor seems to be the best candidate for the ultimate integration of silicon and could be proposed for devices down to sub-10 nm.

Indeed, fully inverted gate-all-around nanowires realized with strain Si or InGaAs channels show very good controls of the electrostatic properties and short-channel effects down to 5 nm gate length (Fig. 1.17), inducing almost ideal

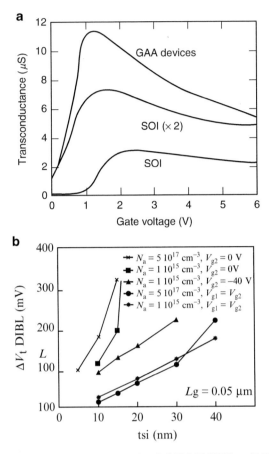

Fig. 1.16 (a) Transconductance in a conventional SOI MOSFET, a DG transistor without volume inversion, and a GAA device with volume inversion ($W/L = 3$ μm/3 μm, $V_d = 100$ mV); (b) DIBL effect versus silicon film thickness for 0.05 μm single-gate SOI MOSFETs ($t_{ox2} = 380$ nm $t_{ox1} = 3$ nm) with high doping ($N_a = 5 \times 10^{17}$ cm^{-3}, $V_{g2} = 0$ V), low doping ($N_a = 10^{15}$ cm^{-3}, $V_{g2} = 0$ V), and back-channel accumulation ($N_a = 10^{15}$ cm^{-3}, $V_{g2} = -40$ V) and for DG SOI MOSFETs ($t_{ox1} = t_{ox2} = 3$ nm, $N_a = 10^{15}$ cm^{-3}, and $N_a = 5 \times 10^{17}$ cm^{-3} with $V_{g1} = V_{g2}$)

subthreshold swing (75 mV/dec for 5 nm L_g), better that double-gate or carbon nanotube devices [31].

Multichannel MOSFETs fabricated with multi-gate devices, realized with the silicon-on-nothing technology, have also been proposed to boost the driving current using several channels. Very good transfer and output characteristics have been obtained for N- and P-MOS MCFET using high-k TiN/HfO$_2$ gate oxides (Fig. 1.18) [32].

Fig. 1.17 Subthreshold swing obtained by quantum simulations as a function of gate length for various MOSFET architectures: Ω-gate Si and InGaAs nanowires (3 nm diameter), gate-all-around carbon nanotubes (0.63 and 1 nm diameter), and double-gate ultrathin-body MOSFET (3 nm Si film thickness)

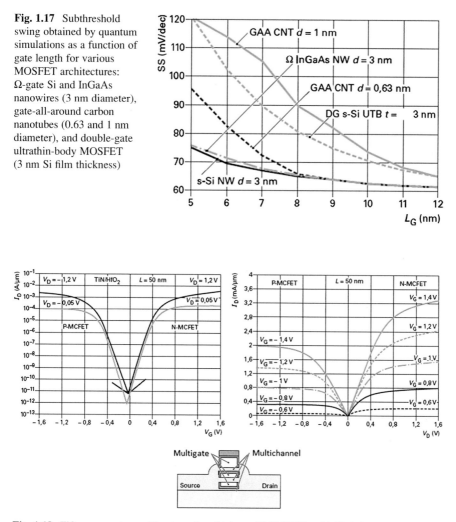

Fig. 1.18 Fifty nanometer multi-gate and multichannel MOSFETs with high-k gate oxide showing very good I_{on} and I_{off}

1.3 Beyond-CMOS Nanodevices for Ultralow-Power Operation

The slowdown of V_{dd} scaling and the substantial increase of the subthreshold leakage lead to a dramatic enhancement of the dynamic and static power consumption. This power challenge is due to the subthreshold slope limit S, which is 60 mV/dec at room temperature for MOSFETs.

Indeed, a lower limit in energy per operation E_{min} exists for an applied bias V_{ddmin}, both parameters depending on the subthreshold swing S (Fig. 1.19) [33]. The optimal

Fig. 1.19 Lower limit in energy per operation obtained for the optimal V_{dd}

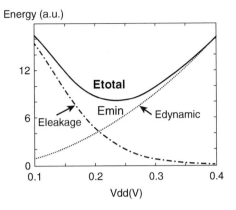

V_{dd} is given by $V_{ddmin} \sim S$ and the lower limit in energy per operation can be expressed as $E_{min} \sim C \times S^2$.

There are two ways for reducing the swing S:

$$S = \frac{\partial V_g}{\partial (\log I_d)} = \underbrace{\frac{\partial V_g}{\partial \psi_S}}_{m} \underbrace{\frac{\partial \psi_S}{\partial (\log I_D)}}_{n} = \left(1 + \frac{C_s}{C_{ins}}\right) \frac{kT}{q} \ln 10 \qquad (1.6)$$

1. Decrease of the transistor body factor m, using ultrathin body SOI; multi-gate or nanowire MOSFETs; carbon nanotube or graphene channels (leading to $m \sim 1$), as discussed above, and negative capacitance FETs; or MEMS/NEMS structures (leading to $m < 1$)
2. Reduction of n, using a low-temperature operation, which cannot be applied for traditional applications or using a modification of the carrier injection mechanisms with impact ionization or band-to-band tunneling.

The best MOSFET devices leading to S close to its minimum value are using fully depleted channels or fully inverted ones, with volume inversion, that is even better to optimize the control of the electrostatics of the structure (e.g., double gate, bulk or SOI tri-gate/FinFET, gate-all-around MOSFET, or nanowire FET), as discussed in Sect. 1.2.

However, other innovative devices are needed to overcome the sub-60 mV/dec barrier [34].

To outperform CMOS, these new devices, called the small swing/slope switches (Fig. 1.20), need an Ion in the range of hundreds of µA, an average subthreshold swing far below 60 mV/dec for at least four to five decades of I_d, and a ratio I_{on}/I_{off} larger than 10^5 and V_{dd} lower than 0.5 V, which is a very difficult challenge.

Fig. 1.20 Comparison
between the MOSFET
switch, the small swing/
slope switch, and the ideal
switch

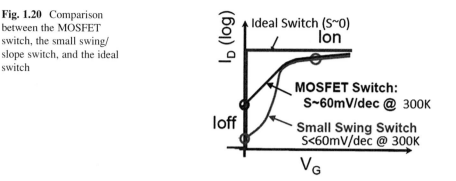

1.3.1   Tunnel FETs

The tunnel FETs use interband tunneling in heavily doped p+n+ junction with a
control of band bending with V_g and a reversed bias p-i-n junction. In these
transistors, the ambipolar effect has to be suppressed by an asymmetry in the
doping level or profile or the use of heterostructures in the channel between source
and drain.

In order to optimize the transmission probability for increasing the band-to-band
tunneling, we need to propose structures with the aim of reducing the effective
tunnel mass m^*, the source bandgap E_g and/or the tunneling length λ, and increasing
the band offset $\Delta\Phi$ between the conduction and valence band of the source/channel,
for enhancing I_{on} and reducing S for several decades of current.

For the optimization of E_g, m^*, $\Delta\Phi$, a change of materials is needed, and λ can be
decreased by a change of device dimension, doping, gate capacitance, gate overlap
on tunnel region, and bandgap using a reduced T_{ox} and t_{si}, a high-k gate, an abrupt
doping profile, a high source doping, a multi-gate structure, or other materials than
silicon.

It is also worth noticing that L_g has a reduced effect on I_d compared with
MOSFET, a TFET (tunnel field-effect transistor) with a heterostructure is prefer-
able (small bandgap at S and large bandgap at D), and complementary TFETs are
needed for logic circuits.

The same boosters that have been developed for ultimate CMOS can be proposed
to boost the performance of tunnel FETs, for which opposite source and drain doping
are used. Indeed, these devices have been shown to overcome the 60 mV/dec limit
for the subthreshold swing, which is the minimum of MOS transistors at 300 K.
Multi-gate, high-k dielectrics, ultrathin films, alternative materials (Ge, III–V), and
strains are leading to improved TFET properties (Fig. 1.21). Very good swing down
to sub-20 mV/dec has been obtained by numerical simulation, inducing possible
very low-voltage operation in the sub-0.5 V together with very small leakage
currents. The main challenge is to obtain at the same time high drain current,
which seems difficult experimentally. Therefore, these tunneling devices could be
very interesting for ultralow-power operation.

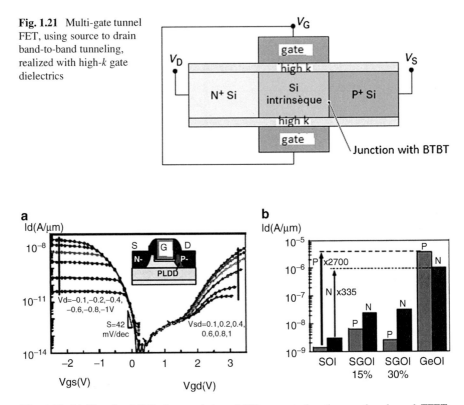

Fig. 1.21 Multi-gate tunnel FET, using source to drain band-to-band tunneling, realized with high-*k* gate dielectrics

Fig. 1.22 (**a**) Transfer $I_d(V_g)$ characteristics of 100 nm gate length n- and p-channel TFETs ($t_{si} = 20$ nm) and (**b**) I_{on} of 400 nm L_g SOI, SGOI, and GOI n- and p-TFETs ($t_{SiGe} = 20$ nm, $t_{Ge} = 60$ nm, $V_d = \pm 0.8$ V, $V_g = \pm 2$ V)

Figure 1.22a shows the experimental characteristics of SOI TFETs. A very good subthreshold swing down to 42 mV/dec is obtained for p-channel operation [35]. Using SGOI and GOI TFETs, strong improvements of the driving current, between several hundreds and thousands, are obtained compared with SOI transistors (Fig. 1.22b). I_{off} is very low but I_{on} is only of the order of several μA/μm for the best GOI TFETs.

As it is the case of MOSFETs, technology boosters can also be applied for TFETs, in particular: high k, abrupt doping profile at tunnel junction, thinner body, high S doping, multi-gate, gate oxide aligned with *i*-region, and shorter L_g/*i*-region. The results obtained by numerical simulation are illustrated in Fig. 1.23, where we can see that short-channel thin Si film double-gate TFET with high-k gate dielectrics and stress at the source junction leads to the best performance with a substantial increase of I_{on} and reduction of S [36].

GeSn TFETs have also shown better performance compared with Ge TFETs (Fig. 1.24). A strong increase of the driving current together with a substantial reduction of the swing is obtained [37].

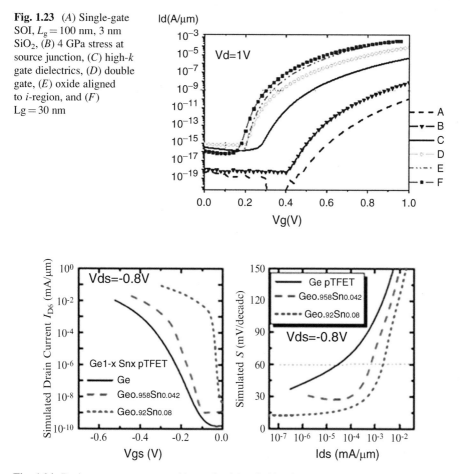

Fig. 1.23 (A) Single-gate SOI, $L_g = 100$ nm, 3 nm SiO$_2$, (B) 4 GPa stress at source junction, (C) high-k gate dielectrics, (D) double gate, (E) oxide aligned to i-region, and (F) Lg = 30 nm

Fig. 1.24 Drain current versus gate bias and subthreshold swing versus drain current for GeSn TFETs as compared to Ge TFETs obtained by numerical simulation

Feedback TFETs [38] and Z2-FET [39], with a forward biased PIN diode, have shown experimentally very small S (a few mV/dec) together with good I_{on}, however for biases larger than 0.5 V.

Interesting performance has been presented for strained Ge double-gate TFETs with asymmetric S/D, especially with Si drain, with a swing down to 50 mV/dec and I_{on} up to 300 μA/μm for high V_d [40]. However, these experimental results have been obtained with different biases applied to the front and back gates due to the different front- and back-gate oxides, and we have shown that an application of V_{g2}/V_{g1} proportional to t_{ox2}/t_{ox1} gives an overestimation of the performance of the devices [41].

Single-gate, double-gate, and GAA InAs TFETs lead to very good simulated results (for $V_g = V_d = 0.2$ V), with S lower than 20 mV/dec for small wire diameter (down to 2 nm), due to the small effective masses and bandgap of these III–V materials.

For very short gate lengths, small wire diameters are needed, down to 5 nm or below, in order to obtain a good subthreshold swing (Fig. 1.25) [42].

High performances, with I_{on} up to 1 mA/μm at low V_d, have been demonstrated by quantum transport simulation on strained InAs NW TFET (Fig. 1.26), the best result being shown for a biaxial strain [42].

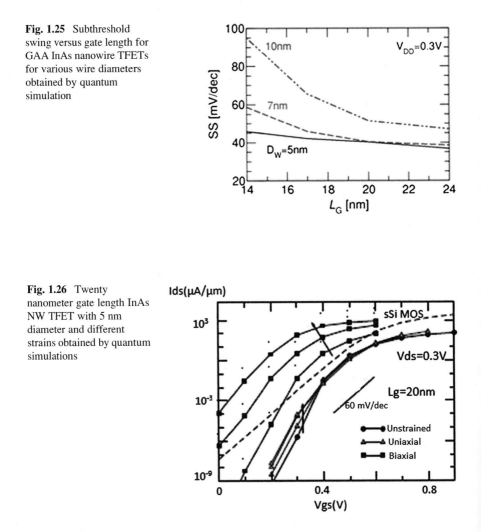

Fig. 1.25 Subthreshold swing versus gate length for GAA InAs nanowire TFETs for various wire diameters obtained by quantum simulation

Fig. 1.26 Twenty nanometer gate length InAs NW TFET with 5 nm diameter and different strains obtained by quantum simulations

However, the on-current improvements can be frustrated by the degradation of the swing in the presence of traps. Both traps and surface roughness can also be a relevant source of device variability for tunnel FETs [43].

The first experimental demonstration of S lower than 60 mV/dec (on less than one decade of I_d) using III–V materials has been shown with InGaAs heterojunction 150 nm gate length TFET using thin EOT and high source doping, with I_{on} of a few $\mu A/\mu m$ [44].

Carbon-based tunnel FETs were the first device showing experimentally $S < 60$ mV/dec, down to 40 mV/dec using CNT structures, but with very low I_{on} (<1 nA) [45]. Carbon materials have many advantages, in particular small effective masses, small and direct bandgap, and excellent electrostatic control (UTB), and could therefore be the best material choice. Graphene has similar properties as CNT but with planar processing compatibility. However, only theoretical studies have been performed so far. The high potential for graphene nanoribbons or bilayers has been shown by simulation [46] ($I_{on} = 100$ s of $\mu A/\mu m$, $I_{off} =$ few pA/μm, $S = 20$ mV/dec, $E_{gap} = 200$–300 meV, $I_{on}/I_{off} > 10^3$ at $V_d = 0.1$ V) but has not been confirmed experimentally up to now.

The best trade-off obtained by simulation for complementary TFET performance has been shown with Ge/InAs TFETs, leading to an increase if I_{on} of several hundreds compared to all-Si TFETs (Fig. 1.9) [47].

However, it is worth noting that to date some of the best TFET performance has been obtained with Si-based TFET and the experimental results are very poor compared with numerical simulation. No experimental demonstration of TFET with $I_{on} > 100$ $\mu A/\mu m$ and $S < 60$ mV/dec has been shown so far. Therefore, substantial improvements of the TFET technology (defects, traps, surface roughness, etc.) are needed, together with advanced numerical simulation taking into more realistic nanodevices with various types of defects.

Tunnel FETs seem up to now better in energy efficiency for applications with frequency lower than 1 GHz (Fig. 1.27) [48].

Fig. 1.27 Comparison of energy frequency between CMOS and TFET

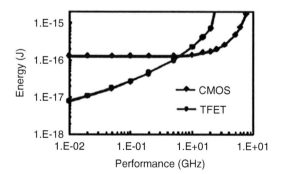

1.3.2 NEMS

NEM (nano-electromechanical) switch has potentially superior characteristics, especially quasi-zero leakage behavior and excellent density capability and operation in harsh environments. However, NEMS faces many challenges, in particular:

- High operating voltage (4–20 V)
- Fabrication difficulties
- Irreversible switching failure caused by surface adhesion

Recently sub-1-V NEM switch (~0.5 V) with nanogaps has been demonstrated [49], which could be interesting for several applications, e.g., NEM switch for ULP ICs, resonators, and sensors; however, a low on-current is obtained (Fig. 1.28).

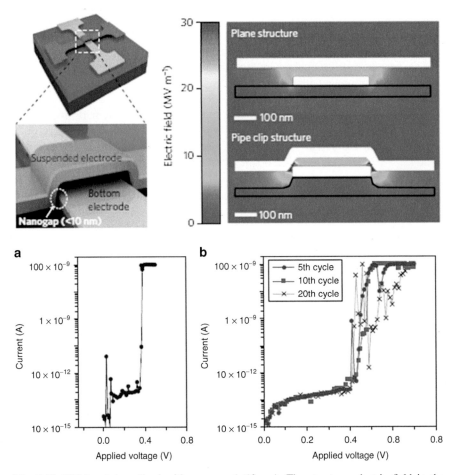

Fig. 1.28 NEM switch realized with nanogap (<10 nm). The structure, electric field in the NEMS, and the $I(V)$ characteristics with many cycles are shown, highlighting the very good commutation between off and on states at low voltage (<0.5 V)

1.3.3 FeFETs

Exploiting the ferroelectric polarization in the gate stack to control the inversion charge of a FET channel provides a compact transistor with an active gate stack for which many studies have been reported to date concerning memory applications. Sallahudin et al. [50] have proposed to use ferroelectrics in order to exploit their unique nonlinear energy dependence on polarization to provide a negative capacitance effect. First attempts to experimentally demonstrate such a ferroelectric abrupt switch (Fe-FET) were reported by Salvatore et al. [51] and Rusu et al. [52]. A key advantage of the Fe-FET is that, in contrast with tunnel FETs, its on-current is as high as the one of a conventional MOSFET. However, the challenge is to get an abrupt switch for several decades of drain current. An interesting recent study is the proposal of the combination of TFET and Fe-FET on the same device, using ferroelectric materials in the TFET gate (Fig. 1.29a).

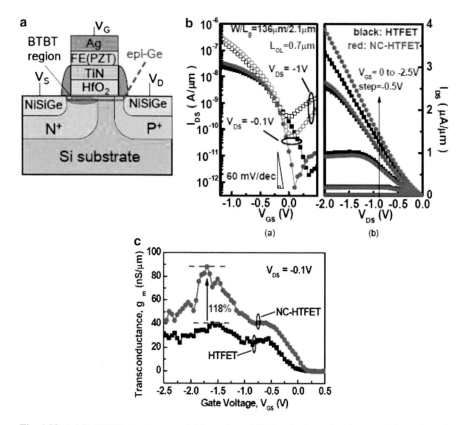

Fig. 1.29 (**a**) Fe-TFET structure, combining a tunnel FET and a ferroelectric gate, (**b**)transfer and output $I(V)$ characteristics showing an improvement of the drain current for negative capacitance heterostructure TFET (NC-HTFET) compared with HTFET only, and (**c**) transconductance of NC-HTFET with more than a factor 2 improvement compared with HTFET

A substantial improvement of the swing, drain current, and transconductance has been shown compared with TFET (Fig. 1.29b, c), however with a subthreshold swing which has to be reduced (about 60 mV/dec in this study) [53].

1.4   Conclusion

The reduction of energy consumption is the main challenge for future electronic systems due to the very big amount of exchanged and stored data, which is exponentially increasing. A number of innovations are needed in the following fields: transistor, memory, devices and interconnects technologies, circuit design techniques, systems architectures, and embedded software.

In the device domain, new physics, materials, and device structures are required. This will enable to continue scaling and performance improvement.

Multi-gate nanowire MOSFETs with volume inversion (especially with Si, sSi) lead to the best short-channel effects, S, I_{off}, V_{dd}, P, E_{min} for MOSFET architectures.

Tunnel FETs are among the best small slope switches up to now, with BTBT allowing for sub-60 mV/dec subthreshold swing, obtained by simulation and experimental results. TFET simulations show promise for very good S, substantial V_{dd} reduction, and high I_{on}, but additional process improvements are needed to improve real device performance.

Alternative materials or heterostructures TFET (with Ge, III–V) could be viable solutions for $I_{on} > 100$ µA/µm, $I_{on}/I_{off} > 10^5$, and $V_{dd} < 0.5$ V with improving the technology.

All Si-TFETs have shown theoretically very good I_{off} and poor I_{on}, but some of the best experimental performances have been reported to date with these devices.

Performance boosters require a good design, the best choice of materials, and integration on Si platforms.

It is also worth noticing that the variability of TFET compared with CMOS is reduced for doping or gate length fluctuations, but is increased as a function of the high-k gate process, the abruptness of doping at tunnel junction, and the film thickness in ultrathin body.

TFETs have better energy efficiency at low- or moderate-performance level. They are ideally suited so far for low power and low standby power at moderate f on the order of 100 s of MHz.

The biggest challenge is the improvement of I_{on} without degrading I_{off}, with $S < 60$ mV/dec for more than four decades of drain current. For that purpose, many technology boosters are needed together with a substantial improvement of the technology. The heterostructure TFET offers the best performance for complementary logic in ultrathin body or nanowire and can be considered as an add-on ultralow-power device option on advanced CMOS platforms.

Recent advances in the field of NEMS and the combination of several small-slope switches with a ferroelectic gate in a tunnel FET (Fe-TFET) are also promising.

Acknowledgments The author would like to thank the SiNANO Institute members and the FP7 Nanosil and Nanofunction European Networks of Excellence Partners.

References

1. D. Kahng, M.M. Atalla, Silicon–silicon dioxide field induced surface devices. IRE Solid-State Device Res. Conf., Carnegie Institute of Technology, Pittsburgh, PA, 1960
2. D. Kahng, A historical perspective on the development of MOS transistors and related devices. IEEE Trans. Electron. Devices **ED-23**, 655 (1976)
3. R. Landauer, IBM J. Res. Dev. **5**(3), 183–191 (1961)
4. C.H. Benett, IBM J. Res. Dev. **17**(6), 525–532 (1973)
5. G.P. Boechler, J.M. Whitney, C.S. Lent, A.O. Orlov, G.L. Snider, Appl. Phys. Lett. **97**, 103502 (2010)
6. S. Cristoloveanu, F. Balestra, Introduction to SOI technology and transistors, in *Physics and operation of Silicon devices and Integrated circuits*, ed. by J. Gautier (ISTE-Wiley, London, 2009)
7. F. Balestra (ed.), *Nanoscale CMOS: innovative materials, modeling and characterization* (ISTE-Wiley, London, 2010)
8. F. Balestra, J. Brini, P. Gentil, Simulation of deep depleted SOI MOSFETs with back potential control, Proc. ESSDERC '84, Lille, France, 1984, Physica, **129B**, 296 (1985)
9. F. Balestra, J. Brini, P. Gentil, Comparison between experiment, analytical models and numerical simulation for threshold voltages of deep depleted SOI MOSFETs, Proc. ESSDERC '85, Aachen, Germany (1985), p. 232
10. F. Balestra, Characterisation and simulation of SOI MOS transistors with back potential control, PhD Thesis, Grenoble Institute of Technology, Apr 1985
11. J.P. Colinge, Subthreshold slope of thin-film SOI MOSFETs. IEEE Electron. Devices Lett. **EDL-7**, 244 (1986)
12. F. Balestra et al., Analytical modelling of single and double gate thin film SOI MOSFETs, Proc. ESSDERC '89, Berlin, Germany, Sept 1989 (Springer, Berlin, 1989), p. 889
13. F. Balestra et al., Analytical models of subthreshold swing and threshold voltage for thin- and ultra-thin-film SOI MOSFETs. IEEE Trans. Electron. Devices **ED-37**, 2303 (1990)
14. D.J. Wouters, J.P. Colinge, H.E. Maes, Subthreshold slope in thin-film SOI MOSFETs. IEEE Trans. Electron. Devices **ED-37**, 2022 (1990)
15. J.G. Fossum, S. Krishnan, P.C. Yeh, Performance limitations of deep-submicron fully depleted SOI MOSFETs, Proc. IEEE Int. SOI Conf. (1992), p. 132
16. Y. Omura et al., Quantum mechanical transport characteristics in ultimately miniaturized MOSFETs/SIMOX. Electrochem. Soc. Proc. **96–3**, 199 (1996)
17. L.T. Su et al., Short-channel effects in deep submicrometer SOI MOSFETs, Proc. IEEE Int. SOI Conf. (1993), p. 112
18. F. Andrieu, O. Weber, J. Mazurier, O. Thomas, J.-P. Noel, C. Fenouillet-Béranger, Low leakage and low variability ultra-thin body and buried oxide (UT2B) SOI Technology for 20 nm low power CMOS and beyond, Proc. Symposium of VLSI Technology (2010)
19. T. Ernst, S. Cristoloveanu, The ground-plane concept for the reduction of short-channel effects in fully depleted SOI devices, Proc. ECS 9th Symp., SOI Technology and Device, Seattle, WA, May 1999
20. V. Barral, Proceedings of IEDM'2007, Dec 2007

21. I. Ben Akkez, C. Fenouillet-Beranger, A. Cros, P. Perreau, S. Haendler, O. Weber, F. Andrieu, D. Pellissier-Tanon, F. Abbate, C. Richard, R. Beneyton, P. Gouraud, A. Margain, C. Borowiak, E. Gourvest, K.K. Bourdelle, B.Y. Nguyen, T. Poiroux, T. Skotnicki, O. Faynot, F. Balestra, G. Ghibaudo, F. Boeuf, Impact of 45° rotated substrate on UTBOX FDSOI high-k metal gate technologies, Proc. VLSI-TSA, Taiwan, Apr 2012
22. F. Balestra et al., Double-gate silicon-on-insulator transistor with volume inversion: a new device with greatly enhanced performance. IEEE Electron. Devices Lett. **EDL-8**, 410 (1987)
23. F. Balestra et al., Optimum parameters for high performance volume-inversion MOSFETs in ohmic and saturation regions, Proc. Eur. SOI Workshop (1988), p. F-05
24. J. Brini et al., Threshold voltage and subthreshold slope of the volume-inversion MOS transistor. IEE Proc. G **138**, 133 (1991)
25. J. Jomaah, F. Balestra, G. Ghibaudo, Hot-carrier effects in single- and double-gate thin film SOI MOSFETs, Proc. ESSDERC '95 (1995), p. 809
26. K. Suzuki et al., Scaling theory for double-gate SOI MOSFETs. IEEE Trans. Electron. Devices **ED-40**, 2326 (1993)
27. D. Hisamoto et al., A fully depleted lean-channel transistor (DELTA)—a novel vertical ultrathin SOI MOSFET. IEEE Electron. Devices Lett. **11**, 36 (1990)
28. J. P. Colinge et al., Silicon-on-insulator gate-all-around device, IEDM Tech. Dig. (1990), p. 595
29. D.J. Frank, S.E. Laux, M.V. Fischetti, Monte Carlo simulation of a 30 nm dual-gate MOSFET: how short can Si go? IEDM Tech. Dig. (1992), p. 553
30. E. Rauly, O. Potavin, F. Balestra, C. Raynaud, On the subthreshold swing and short channel effects in single and double gate deep submicron SOI MOSFETs. Solid State Electron. **43**, 2033 (1999)
31. M. Luisier, M. Lundstrom, D.A. Antoniadis, J. Bokor, Proceedings IEDM (2011), p. 251
32. V. Barral, Proceedings VLSI'2008
33. S. Hanson, M. Seok, D. Sylvester, D. Blaauw, Nanometer device scaling in subthreshold logic and SRAM. IEEE Trans. Electron. Devices **55**, 175–185 (2008)
34. F. Balestra, *Beyond CMOS nanodevices 1 & 2* (ISTE-Wiley, London, 2014)
35. F. Mayer, C. Le Royer, J.-F. Damlencourt, K. Romanjek, F. Andrieu, C. Tabone, B. Previtali, S. Deleonibus, Impact of SOI, Si1-xGexOI and GeOI substrates on CMOS compatible Tunnel FET performance, Proc. IEDM (2008), pp. 1–5
36. K. Boucart, W. Riess, A.M. Ionescu, Proceedings ESSDERC (2009)
37. Y. Yang, S. Su, P. Guo, W. Wang, X. Gong, L. Wang, K.L. Low, G. Zhang, C. Xue, B. Cheng, G. Han, Y.-C. Yeo, Towards direct band-to-band tunneling in P-channel tunneling field effect transistor (TFET): technology enablement by germanium-tin (GeSn), Proc. IEDM (2012), p. 379
38. A. Padilla, C.W. Yeung, C. Shin, C. Hu, T.-J. King Liu, Feedback FET: a novel transistor exhibiting steep switching behavior at low bias voltages, Proc. IEDM (2008), p. 171
39. J. Wan, C. Le Royer, A. Zaslavsky, S. Cristoloveanu, A compact capacitor-less high-speed DRAM using field effect-controlled charge regeneration. IEEE Electron. Devices Lett. **33**, 179–181 (2012)
40. T. Krishnamohan, D. Kim, S. Raghunathan, K. Saraswat, Double-gate strained-Ge heterostructure tunneling FET (TFET) with record high drive currents and ≪ 60 mV/dec subthreshold slope, Proc. IEDM (2008), p. 947
41. J. Brini, M. Benachir, G. Ghibaudo, F. Balestra, Subthreshold slope and threshold voltage of the Volume Inversion MOS transistor. IEE Proc. G Circuits Devices Syst. **138**, 133 (1991)
42. F. Conzatti, M.G. Pala, D. Esseni, E. Bano, L. Selmi, A simulation study of strain induced performance enhancements in InAs nanowire Tunnel-FETs, Proc. IEDM (2011), p. 95
43. F. Conzatti, M.G. Pala, D. Esseni, Surface-roughness-induced variability in nanowire InAs tunnel FETs. IEEE Electron. Device Lett. **33**, 806–808 (2012)
44. G. Dewey, B. Chu-Kung, J. Boardman, J.M. Fastenau, Fabrication, characterization and physics of III–V heterojunction tunneling FET for steep subthreshold swing, Proc. IEDM (2011), p. 785

45. J. Appenzeller, Y.-M. Lin, J. Knoch, P. Avouris, Band-to-band tunneling in carbon nanotube field-effect transistors. Phys. Rev. Lett. **93**, 193805 (2004)
46. G. Fiori, G. Iannaccone, Ultralow-voltage bilayer graphene tunnel FET. IEEE Electron. Devices Lett. **30**, 1096–1989 (2009)
47. A. Ionescu, L. De Michielis, N. Dagtekin, Ultra low power: emerging devices and their benefits for integrated circuits, Proc. IEDM (2011), p. 378
48. A. Ionescu, H. Riel, Tunnel field-effect transistors as energy-efficient electronic switches. Nature **479**, 329–337 (2011)
49. J.O. Lee, Y.-H. Song, M.-W. Kim, M.-H. Kang, O. Jae-Sub, H.-H. Yang, J.-B. Yoon, A sub-1-volt nanoelectromechanical switching device. Nat. Nanotechnol. **8**, 36 (2013)
50. S. Salahuddin, S. Datta, Use of negative capacitance to provide voltage amplification for low power nanoscale devices. Nanoletters **8**(2), 405–410 (2008)
51. G.A. Salvatore, D. Bouvet, A.M. Ionescu, Demonstration of subthreshold swing smaller than 60 mV/decade in Fe-FET with P(VDF-TrFE)/SiO$_2$ gate stack, 2008, IEDM Tech. Dig., Electron Devices Meeting (2008), pp. 1–4
52. A. Rusu, G.A. Salvatore, D. Jimenez, A.M. Ionescu, Metal-ferroelectric-metal-oxide-semiconductor field effect transistor with sub-60 mV/decade subthreshold swing and internal voltage amplification, IEDM Tech. Dig. (2010), pp. 16.3.1–16.3.4
53. M.H. Lee, J.-C. Lin, Y.-T. Wei, C.-W. Chen, W.-H. Tu, H.-K. Zhuang, M. Tang, Ferroelectric negative capacitance hetero-tunnel field-effect-transistors with internal voltage amplification. Proc. IEDM **2013**, 104 (2013)

Chapter 2
Ultralow-Power Pseudospintronic Devices via Exciton Condensation in Coupled Two-Dimensional Material Systems

Xuehao Mou, Leonard Franklin Register, and Sanjay Kumar Banerjee

Abstract "Pseudospintronic" device concepts, novel "beyond-CMOS" device proposals targeted toward revolutionizing the current semiconductor technology based on MOSFETs and CMOS logic, are addressed in detail. These pseudospin devices include the voltage-controlled *Bi*layer pseudo*S*pin *F*ield-*E*ffect *T*ransistor (BiSFET) and the current-controlled *Bi*layer pseudo*S*pin *J*unction *T*ransistor (BiSJT). MOSFETs are confronted by the intractable physics of thermionic emission and resulting source-to-drain leakage that limits voltage scaling. As a result, CMOS faces an "energy crisis" much as the one faced by bipolar junction transistor-based logic that led to CMOS. As for many other beyond-CMOS concepts, these pseudospintronic devices are based on a completely different physics of switching, potentially allowing much lower voltage and power operation. These pseudospintronic device concepts employ possible room-temperature interlayer electron–hole exciton condensates between two dielectrically separated layers of two-dimensional (2D) materials for subthermal voltage (sub-k_BT/q) switching, specifically from a nearly shorted interlayer conductance state to a highly resistive interlayer conductance state with increasing interlayer voltage. These collective exciton states with their "which-layer" degree of freedom are somewhat analogous to collective spin states in magnets, which is the origin of the "pseudospintronics" moniker. Device performance *in the presence* of such condensates is the primary focus of this work; the possibility of room-temperature condensates, itself, is addressed by other research still in progress. We begin with a discussion of the underlying physics. Graphene-based pseudospintronic systems then are analyzed using quantum transport simulations incorporating the nonlocal exchange interaction. However, the essential transport physics should be much the same for other 2D material systems, including transition metal dichalcogenides for which the realization of the condensate may be easier. The BiSFET and BiSJT device concepts are presented in detail, and basic logic gate designs are illustrated for each. Compact device models are developed and SPICE-level circuit simulations are performed to

X. Mou (✉) • L.F. Register • S.K. Banerjee
The University of Texas at Austin, Austin, TX, USA
e-mail: xmou@utexas.edu; register@austin.utexas.edu; banerjee@ece.utexas.edu

© Springer International Publishing Switzerland 2015
A. Korkin et al. (eds.), *Nanoscale Materials and Devices for Electronics, Photonics and Solar Energy*, Nanostructure Science and Technology, DOI 10.1007/978-3-319-18633-7_2

demonstrate possible switching energies on the scale of or below a tenth of an attojoule, well below even end-of-the-road-map CMOS. However, like many other beyond-CMOS concepts, these devices remain concepts without solid experimental embodiments. The fabrication concerns of such novel devices are also discussed along with recent experimental progress.

2.1 Introduction

The "pseudospintronic" devices considered in this chapter are novel "beyond-CMOS" device proposals targeted toward revolutionizing the current semiconductor technology, which is currently based on MOSFETs and CMOS logic. These pseudospin devices include the *Bi*layer pseudo*S*pin *F*ield-*E*ffect *T*ransistor (BiSFET), the *Bi*layer pseudo*S*pin *J*unction *T*ransistor (BiSJT), and potential applications of near-perfect Coulomb drag. MOSFETs are confronted by the intractable difficulty of thermionic emission and resulting source-to-drain leakage that limits voltage scaling. As a result, CMOS faces an "energy crisis" much as the one faced by bipolar junction transistor-based logic that leads to CMOS. As for many other beyond-CMOS concepts, these pseudospintronic devices are based on a completely different physics of switching. These pseudospintronic device concepts employ possible room-temperature interlayer electron–hole exciton condensates in bilayer two-dimensional (2D) material systems for enhanced interlayer conduction and sub-k_BT/q switching voltages. In this chapter, device performance *in the presence* of such condensates is the main focus, but not the possibility of room-temperature condensates, itself, which is addressed by other research still in progress. Graphene-based pseudospintronic devices are simulated by quantum transport simulations for the essential physical properties, and SPICE-level circuit simulations are performed to demonstrate possible super low-power switching that could greatly exceed even so-called "end of the road map" CMOS. The fabrication concerns of such novel devices are also discussed along with recent experimental progress. However, also like many other beyond-CMOS concepts, these devices remain concepts without solid experimental embodiments. Though this device research has been carried out based on graphene systems until now, the essential physics should be translated to other 2D materials that may yet be more favorable than graphene for the realization of such devices.

2.2 Pseudospintronics

In this section, we will address the physics that underlies the proposed pseudospin devices. These devices rely on intrinsically quantum mechanical behaviors, in contrast to the CMOS device whose basic function can be explained classically. They also depend on many-body quantum effects that contrast them even to some other beyond-CMOS device concepts, such as tunnel FETs, which depend on single-particle tunneling.

2.2.1 Quantum Mechanics Prequel/Refresher

This subsection is specifically prepared for novices to quantum mechanics or those who have not dealt with the subject for some time, to introduce the reader to or remind them of necessary concepts going forward in this work; those well versed in the subject may skip this part. (Indeed, the material constitutes common knowledge and one may turn to one of many textbooks on quantum mechanics, solid-state physics, etc., for this information, so we will refrain from using citations here for the most part.) Although it is not necessary to understand the circuit applications of the pseudospintronic devices—knowledge of the simple I–V behavior of this pseudospintronic device should be enough for readers to browse through Sect. 2.5 addressing circuit applications—quantum mechanics is essential for understanding the why's and how's. The exotic physics of pseudospintronic devices is the foundation for their exotic device characteristics as compared to those of conventional electronic devices.

2.2.1.1 Description of the World with Quantum Mechanics

The development of quantum mechanics perhaps began in the late nineteenth century. Ludwig Boltzmann suggested that the energy states of a physical system could be discrete, and soon after Max Planck proposed energy quanta to successfully interpret blackbody radiation. However, quantum mechanics was not designed on the basis of "quanta," but rather "quanta" were the products of quantum mechanics in which, as De Broglie postulated, there is particle-wave duality of, strictly speaking, everything. The particle-wave duality, which was first used to describe light (the "photon" as the corresponding particle), was experimentally justified in the early twentieth century as in the electron interference pattern exhibited using single-crystal nickel [1]. Nevertheless, this duality is dominated by Planck's constant h, a constant so minute (6.63×10^{-34} J s) that only microscopic particles, such as electrons, protons, neutrons, and atoms/molecules composed thereof, manifest significant particle-wave duality. In this regard, quantum mechanics is mainly applied in the research of microscopic systems. In the macroscopic world, classical mechanics is sufficiently accurate.

In Schrödinger's wave mechanics, which is widely known, the wave property of particles is represented by a "probability wave" or "wavefunction" in real-space $\varphi(\mathbf{r})$, which may have real and imaginary components. Analogous to the sound wave for which how "loud" the sound—how much energy is contained in the sound wave—is proportional to the magnitude squared of the amplitude of the wave as a function of location, in quantum mechanics, the *probability* to find the particle in a certain location $\mathbf{r}$, another type of strength, is proportional to $|\varphi(\mathbf{r})|^2$, which remains an intrinsically real quantity.

Just as sound waves are delocalized in space, quantum mechanical wavefunctions generally are also delocalized, which is equivalent to saying that the location of the particle is *uncertain*. Similarly, there can be uncertainty in other physical observables, such as momentum, energy, angular momentum, etc., associated with the quantum mechanical wavefunction. However, there is a set of wavefunctions for each physical observable, the "eigenfunctions," $\alpha_n(\mathbf{r})$ for which there is no uncertainty in that observable. The associated precisely known value of the observable is the "eigenvalue," γ_n. However, for many, if not most, pairs of observables, being in an eigenstate of one observable will be mutually exclusive of being in the eigenstate of the other, as is the case for real-space position and momentum. This general principle is known as the *Heisenberg uncertainty principle* (which comes with a simple prescription for determining whether such uncertainty exists, which need not be addressed here).

Critically, the set of eigenfunctions of any observable $\{\alpha_n(\mathbf{r})\}$ provides a mathematically complete and orthogonal (or at least can be made orthogonal in the case of different eigenfunctions with the same eigenvalue) basis set in which any quantum mechanical wavefunction $\varphi(\mathbf{r})$ can be expanded, such that one may write $\varphi(\mathbf{r}) = \sum_n c_n \alpha_n(\mathbf{r})$. Thus, we now also may talk about the expansion coefficient c_n or the (complex) amplitude for finding the particle at any "location" in "momentum space," "energy space" or "angular momentum space." In other words, the c_n represent the momentum, energy or angular momentum space wavefunction, respectively. The use of such different basis sets is somewhat analogous to describing real-space position in Cartesian coordinates (x, y, z), with cylindrical coordinates (ρ, θ, z) or with spherical coordinates (r, θ, φ). However, the number of axes/dimensions can be infinite—and the sum over n above becomes an integral—in these spaces (which are examples of Hilbert spaces). A closer analogy would be transforming a real-space function to Fourier space or vice versa. Indeed, the transformation of a quantum mechanical wavefunction from real-space $\mathbf{r}$ to momentum space $\mathbf{p}$, or vice versa, is precisely equivalent to a Fourier transform.

In contrast to the wavefunction, the *state* of a particle, however, must be independent of its coordinate system/representation. In the coordinate-system-independent Dirac "bra-ket" notation, the state can be represented by the "ket" $|\varphi\rangle$. The also necessary coordinate-system-independent integral, $\int \varphi_b^*(\mathbf{r})\varphi_a(\mathbf{r})d^3r = \int \varphi_b^*(\mathbf{p})\varphi_a(\mathbf{p})d^3p = \ldots$, is written as $\langle\varphi_b|\varphi_a\rangle$, where the "bra" $\langle\varphi|$ is the "Hermitian adjoint" of the ket $|\varphi\rangle$. We emphasize that expression does not represent just the product of two wavefunctions; the integration is implicit in the notation. For example, the orthogonality condition between two basis states $|\alpha_n\rangle$ and $|\alpha_{n'}\rangle$ takes the form $\langle\alpha_{n'}|\alpha_n\rangle = \delta_{n,n'}$.

To convert from the coordinate-system-independent bra-ket notation to any desired coordinate system, one may simply use the "projection operator" $\sum_n |\alpha_n\rangle\langle\alpha_n| \equiv 1$, where $\mathbf{1}$ is the identity matrix. The equivalence to the identity matrix means that applying the projection operator to a state does not change the state in any way; however, it does "project" it onto the basis of the eigen-kets, $|\alpha_n\rangle$. (We note that there is a common convention to suppress the summation notation such that the expression $|\alpha_n\rangle\langle\alpha_n|$ implies the summation, but we shall refrain from doing so here for clarity.) That is, $|\varphi\rangle \equiv 1|\varphi\rangle \equiv \sum_n |\alpha_n\rangle\langle\alpha_n|\varphi\rangle$, where $\langle\alpha_n|\varphi\rangle \equiv c_n$ are the expansion/ amplitude coefficients in the basis $|\alpha_n\rangle$ of the chosen physical observable. For example, to find the coordinate in real space, we write, now with a continuum of eigenstates, $|\varphi\rangle = \int d^3r |\mathbf{r}\rangle\langle\mathbf{r}|\varphi\rangle$, where the $|\mathbf{r}\rangle$ are eigenstate of position $\mathbf{r}'$. The expansion coefficients $c(\mathbf{r}) = \langle\mathbf{r}|\varphi\rangle$ can be evaluated in any coordinate system, but we shall choose real space where the eigenstates of position are just delta functions $\delta(\mathbf{r} - \mathbf{r}')$. Therefore, $c(\mathbf{r}) \equiv \langle\mathbf{r}|\varphi\rangle \equiv \int d^3r' \, \delta(\mathbf{r} - \mathbf{r}')\varphi(\mathbf{r}') \equiv \varphi(\mathbf{r})$. That is, as a specific case of what was noted above, the expansion coefficients in the basis of the real-space eigen-kets are the real-space wavefunction.

The time evolution of the state is described by the *time-dependent Schrödinger's equation*,

$$i\hbar \frac{\partial}{\partial t}|\varphi\rangle = H|\varphi\rangle, \tag{2.1}$$

where H is the *Hamiltonian*, in analogy to the classical picture of *total energy* but in *operator* form. In the real-space representation, this equation becomes

$$i\hbar \frac{\partial}{\partial t}\varphi(\mathbf{r}, t) = \left(-\frac{\hbar^2}{2m}\nabla^2 + V(\mathbf{r}, t)\right)\varphi(\mathbf{r}, t), \tag{2.2}$$

which is a more well-known form. If the Hamiltonian does not explicitly depend on time, i.e., $V(\mathbf{r}, t) = V(\mathbf{r})$, Eq. (2.1) can be decoupled as

$$|\varphi\rangle = \sum_n c_n |\alpha_n\rangle e^{-iE_n t/\hbar} \tag{2.3a}$$

where

$$H|\alpha_n\rangle = E_n|\alpha_n\rangle \tag{2.3b}$$

is the total energy eigenvalue equation for the set of the energy eigenstates. This latter equation, whether written in real space or bra-ket notation as in Eq. (2.3b), is better known as the *time-independent* or *stationary Schrödinger's equation*. With the time-dependent Schrödinger's equation, the evolution in time of the state is uniquely described and, therefore, so is with the "movement" or "motion" of particles, at least in terms of their probability distributions for whatever observable.

In quantum mechanics, using Eq. (2.2), one may readily obtain (in real-space coordinates) the quantum mechanical version of the (in this case probability) current continuity equation for any given state $\varphi(\mathbf{r}, t)$ as

$$\frac{\partial}{\partial t}|\varphi(\mathbf{r}, t)|^2 = \frac{i\hbar}{2m}\nabla \cdot [\varphi^*(\mathbf{r}, t)\nabla\varphi(\mathbf{r}, t) - \varphi(\mathbf{r}, t)\nabla\varphi^*(\mathbf{r}, t)] = -\nabla \cdot \mathbf{j}. \tag{2.4}$$

Since $|\varphi(\mathbf{r}, t)|^2$ is the probability density, $\mathbf{j}$ is the probability flux/current which determines, with the addition of a factor of the electron charge $-q$, the charge current for electrons. Even for "stationary states," a nonzero current will flow if the product $\varphi(\mathbf{r})\nabla\varphi^*(\mathbf{r})$ has a nonzero imaginary part, such as the case for the (unnormalized) real-space wavefunction of a free electron $\varphi(\mathbf{r}) = e^{i\mathbf{k}\cdot\mathbf{r}}$. The term "stationary" indicates only that the observable properties of the energy eigenstates do not evolve in time, but not necessarily that the particle is not moving. Think of the example of steady-state current flow in electrical circuits; the charge distribution and the current flow remain static, but electrons are moving.

2.2.1.2 Many Particle Systems, Indistinguishability, Fermions and Bosons, and the Density Matrix

Spin is another concept in quantum mechanics that distinguishes different states. Although an accurate description of spin requires a lot more mathematics, in this chapter, *spin* can be viewed as an *independent coordinate* describing the particle's intrinsic angular momentum in units of $\hbar$. There are only two values of spin for electrons, $\pm\hbar/2$, which are commonly referred to as "spin up" and "spin down." For other "particles," there can be half-integer or integer spin values in terms of $\hbar$, and the number of allowed values can range from 1 upward. "Spinning" charge along with the orbital angular motion generates magnetic fields, the accumulation of which is the origin of macroscopic magnetic fields, with up spins producing fields in the opposite direction as down spins.

One essential property of fundamental particles such as electrons is their indistinguishability. For example, if you exchange the locations of any two electrons, there can be no observable difference. Mathematically, if $\varphi(\mu_1,\mu_2)$ describes a two-particle system with one particle at location μ_1 and the other at μ_2, within any coordinate system μ, there are only two allowed forms for this wavefunction to account for the indistinguishability under exchange, $\varphi(\mu_1,\mu_2) = \pm\varphi(\mu_2,\mu_1)$, where the possible sign change has no effect on the observable properties of the state. Whether or not the sign changes under exchange, however, leads to two fundamental classes of particles. With the plus sign, the particles are considered to have "even parity" and are called *bosons*, for example, photons or hydrogen$_1$ atoms. With the negative sign on the other hand, the particles are considered to have "odd parity" and called *fermions*. As it turns out, half-integer spin particles such as electrons and protons are fermions, and integer spin particles such as hydrogen$_1$ atoms, not coincidentally composed of two fermions, as well as photons are bosons.

A critical property of fermions is that for $\mu_1 = \mu_2 = \mu$ the two-particle wavefunction, which must satisfy $\varphi(\mu,\mu) = -\varphi(\mu,\mu)$ under exchange, must therefore vanish. This means that no identical fermions can occupy the same quantum mechanical state at the same time, in whatever coordinate system μ one may consider, which is the well-known *Pauli exclusion principle*. This results lead to, e.g., Fermi statistics for which the maximum occupation probability of quasi-single-particle states is unity.

Moreover, to the extent that the many-body fermion wavefunction is a continuous function of the coordinates $\mathbf{\mu}$, the probability of being found close together also will be less than otherwise expected. In other words, identical fermions such as electrons tend to avoid one another.

On the other hand, bosons are subject to no such restriction. Indeed, they actually tend to be found closer together than one would otherwise expect classically due to exchange considerations. That bosons, with their much different take on being close to each other, can be created from paired fermions is the basis for superconductivity.

The *density matrix* is a particularly useful tool to help describe the state of a statistical or probabilistic ensemble of particles. For many particles in quasi-single-particle states $\alpha_n(\mathbf{r})$ but with no coherence among the single-particle states, the density matrix takes the form $\rho(\mathbf{r}, \mathbf{r}') = \sum_n f_n \alpha_n(\mathbf{r}) \alpha_n^*(\mathbf{r}')$ in real-space coordinates, where f_n is the occupation probability of the state. For example, f_n would be obtained from Fermi statistics for electron energy eigenstates $\alpha_n(\mathbf{r})$ in equilibrium. The diagonal elements $\rho(\mathbf{r},\mathbf{r})$ are just the total probability density as a function of $\mathbf{r}$. The off-diagonal elements represent the degree of overall coherence in the many-body quantum mechanical state between the points $\mathbf{r}$ and $\mathbf{r}'$ and will generally decrease with increasing magnitude of $\mathbf{r} - \mathbf{r}'$. In the classical limit, the off-diagonal elements vanish and the density matrix is diagonal. In bra-ket notation, the density matrix *operator* takes the form, $\rho = \sum_n f_n |\alpha_n\rangle\langle\alpha_n|$, which is the form of the projection operator except for the occupation probability weightings. (Consistent with this latter relation, if the occupation probabilities are identical, $f_n = f$, the density operator reduces to f times the identity matrix).

2.2.1.3   Crystals, Bloch's Theorem, and Band Structure

Most of the solids we are interested in are crystalline, in which the arrangement of atoms is highly ordered. Mathematically, they exhibit *translation symmetry* such that one can always assign three vectors $\mathbf{R}_1$, $\mathbf{R}_2$, and $\mathbf{R}_3$ for a real three-dimensional crystal such that the atom arrangement is identical if observed at arbitrary positions $\mathbf{r}$ and $\mathbf{r} + n_1\mathbf{R}_1 + n_2\mathbf{R}_2 + n_3\mathbf{R}_3$, where the n are integers and the $\mathbf{R}$ are called the translation vectors. The set of points $\{\mathbf{R} = n_1\mathbf{R}_1 + n_2\mathbf{R}_2 + n_3\mathbf{R}_3\}$ defines a "Bravais lattice." The parallelepiped constructed by $\mathbf{R}_1$, $\mathbf{R}_2$, and $\mathbf{R}_3$ is one representation of a lattice "unit cell," which can be translated by the vectors $\mathbf{R}_1$, $\mathbf{R}_2$, and $\mathbf{R}_3$ to fill the entire space. The atoms contained within each unit cell and their positions are the "atomic basis." The smallest possible atomic basis (e.g., containing two silicon atoms in silicon crystals), the associated lattices and translation vectors are called "primitive." However, even the primitive unit cell of a given crystal is not unique. The unit cell that best exhibits the *point group symmetry* (symmetry under rotation, reflection, and inversion) about a lattice point is the "Wigner-Seitz" unit cell. For every real-space lattice, there is a corresponding "reciprocal space lattice" defined by the reciprocal space vectors $\mathbf{G}_1$, $\mathbf{G}_2$, and $\mathbf{G}_3$ determined by the requirement that $\mathbf{G}_i \cdot \mathbf{R}_j \equiv 2\pi\delta_{ij}$. The Wigner–Seitz unit cell of the reciprocal lattice is known as the "(first) Brillouin zone."

These symmetries are reflected in the quantum mechanical Hamiltonian H. As a result, after some work, it can be shown that the electron single-particle energy eigenfunctions of this system must be of the form, $\alpha_{\mathbf{k},\sigma}(\mathbf{r}) = e^{i\mathbf{k}\cdot\mathbf{r}} u_{\mathbf{k},\sigma}(\mathbf{r})$, where $\mathbf{k}$ is a wavevector within the first Brillouin zone, σ labels the "energy band" as there will be multiple energy eigenstates with the same value of $\mathbf{k}$ but with different energies, and $u_{\mathbf{k},\sigma}(\mathbf{r})$ has the periodicity of the real-space primitive unit cell. This result is Bloch's theorem, and the functions $e^{i\mathbf{k}\cdot\mathbf{r}} u_{\mathbf{k},\sigma}(\mathbf{r})$ are called "Bloch functions." The eigen-energy $E_{\mathbf{k},\sigma}$ vs. $\mathbf{k}$ relationship, which is continuous within each energy band, is known as the "band structure." *Various properties of electrons in crystals can be obtained entirely from the band structure, or at least depend on the band structure*, including carrier velocity and rate of acceleration due to applied fields (and associated "effective mass") as a function of "crystal momentum" $\hbar\mathbf{k}$, and "densities of states" per unit energy per unit volume for the Bloch states.

In terms of the types of crystals, an insulator results when there is a substantial energy "band gap" between an essentially fully occupied energy band—a "valence band"—where the electrons, loosely speaking, cannot move due to Pauli exclusion, and an essentially empty band—a "conduction band." A metal results when there is one or more partially full (or equivalently, partially empty) bands with many electrons free to move. A semiconductor is a type of insulator, such that it remains a poor conductor when intrinsic, but with a relatively small band gap such that it is perhaps no longer a particularly good insulator either. However, the materials we commonly refer to as semiconductors can be converted to good conductors by adding certain types of impurity atoms called "donors" and "acceptors," by attracting charge carriers to the surface by application of electric fields, or by absorption of light. Of importance, instead of keeping track of the occupied electron states in the valence band, it is generally more productive to keep track of any empty ones. In a formally defined way, one can associate positively charged quasi-particle "holes" with these empty states. Thus, we can consider two types of charge carriers, electrons and holes. Often electrons and holes are represented simply as "e" and "h," respectively. For example, electron–hole pair generation due to absorption of light becomes "e–h pair generation." Similarly, negatively and positively charged regions can be referred to as "n type" and "p type," respectively.

2.2.1.4 Tight-Binding Approximation and Current Flow

In practice, the Schrödinger's equation can be somewhat daunting to solve and even more so when considering many-body exchange effects. Therefore, there are many approximations developed to simplify the mathematics while still yielding reliable physics. The tight-binding approximation is one of them, which typically takes an outer-shell electron energy eigenstate of freestanding atoms (their "valence" "orbitals"), or a reasonable facsimile thereof, as a basis set for describing the actual emergent electronic states of the crystal. Using this basis, Schrödinger's equation can be transformed from a differential equation in real space, Eq. (2.2), to a matrix equation in orbital space with interorbital "hopping potentials." Although intrinsically incomplete with a finite number of orbitals, this approximation can be sufficient to yield accurate results, while also providing an efficient discretization of Schrödinger's equation that is compatible with numerical analysis and retaining intuitive clarity.

For example, assume that the atoms are positioned at $\{\mathbf{R}\}$, each with $n_{\mathbf{R}}$ orbitals $|\varphi_{\mathbf{R},i}\rangle$ where $i = 1, 2, \ldots, n_{\mathbf{R}}$, which (for convenience although not a necessity) have been modified enough to be orthogonal among the atoms. Any state spanned from this basis can be written as

$$|\Phi\rangle = \sum_{\mathbf{R}} \sum_{i=1\ldots n_{\mathbf{R}}} C_{\mathbf{R}i} |\varphi_{\mathbf{R}i}\rangle \tag{2.5}$$

Substituting Eq. (2.5) into Eq. (2.3b) and operating on both sides with $\langle \varphi_{\mathbf{R}',i'}|$ produce

$$\sum_{\mathbf{R}} \sum_{i=1\ldots n_{\mathbf{R}}} C_{\mathbf{R}i} \langle \varphi_{\mathbf{R}'i'} | H | \varphi_{\mathbf{R}i} \rangle = E C_{\mathbf{R}'i'}, \tag{2.6}$$

which defines an eigenvalue problem of *the Hamiltonian matrix* defined by the matrix elements or hopping potentials, $H_{\mathbf{R}'i';\mathbf{R}i} = \langle \varphi_{\mathbf{R}'i'} | H | \varphi_{\mathbf{R}i} \rangle$. The coefficients $C_{\mathbf{R}i}$ are the eigenvectors/eigenfunctions or "tight-binding wavefunctions" in this tight-binding space. These matrix elements can be obtained, e.g., from the first principles using the density functional theory (DFT) or by adjusting the hopping potentials $H_{\mathbf{R}'i';\mathbf{R}i}$ to match the results of other band structure calculation methods and/or experimental band structure characteristics. Once the hopping potentials are known, complex structures can be created and real-space variations in potentials can be modeled via additions to the onsite ($\mathbf{R}'i = \mathbf{R}i$) and, in the case of exchange interactions (see below), interatom and interorbital hopping potentials.

Following the same approach used in deriving Eq. (2.4), the tight-binding version of the probability current continuity equation

$$\sum_{\mathbf{R}} \sum_{i=1\ldots n_{\mathbf{R}}} \frac{2}{\hbar} \text{Im}\left(C^*_{\mathbf{R}'i'} H_{\mathbf{R}'i',\mathbf{R}i} C_{\mathbf{R}i}\right) = \frac{\partial}{\partial t} |C_{\mathbf{R}'i'}|^2 \tag{2.7}$$

can be derived from Eq. (2.6). This result implies that the current flowing between orbitals $\mathbf{R}i$ and $\mathbf{R}'i'$ is proportional to the imaginary part of the product $C_{\mathbf{R}'i'}^* H_{\mathbf{R}'i',\mathbf{R}i} C_{\mathbf{R}i}$.

2.2.2 Interlayer Electron–Hole Exciton Condensation

Excitons are simply an electron and a hole bound together by their Coulomb interaction [2, 3]. However, the exciton, composed of two half-integer spin particles, has integer spin and, thus, is a boson, much like that Cooper pairs of electrons in superconductors are bosons [4]. This bosonic nature allows for the possibility of "condensation" where the bosons occupy the same quantum mechanical state, which is not possible for the individual fermions. Such condensation is sometimes considered as a state of matter in parallel with the commonly known solid, liquid and gas states [5, 6]. Condensation, in turn, can lead to novel macroscopic physical effects such as superfluidity and superconductivity [7, 8]. However, the formation of condensates is something generally reserved for cryogenic temperatures [9–11]. Such pairing is known as Bose–Einstein condensation in the limit of tightly bound pairs and Bardeen–Cooper–Schrieffer (BCS) pairing in the weak pairing limit. In the exciton system, perhaps the strongest effects may be found near the BEC–BCS crossover.

Commonly, excitons are composed of electrons and holes in the same material (although different energy bands, of course), such as created by the absorption of light (photons) in bulk semiconductors. These excitons can be described as "spatially direct." The excitons considered in this work are "spatially indirect" interlayer excitons, where the electrons and the holes *nominally* are spatially separated in different material layers. Generally the electrons and holes of spatially indirect excitons reside in quasi-two-dimensional systems, such as double quantum wells or, of particular interest here, various 2D materials [12–16]. Condensation of indirect excitons has been observed experimentally in double quantum wells of III–V materials at cryogenic temperatures (~mK) and high magnetic fields (~T) [17–20]. The condensation was explained theoretically via a coherent interlayer many-body exchange interaction, which is a purely quantum effect arising from the indistinguishability of electrons as noted above, with the holes modeled as just the absence of electrons, as is the case in reality [13]. Multiple expected novel interlayer transport behaviors for such condensation were observed experimentally, including greatly enhanced interlayer transmission and near-perfect Coulomb drag between layers [20].

Graphene was the first truly 2D material to be considered, the unique gapless and linear dispersion relation of which allowed physicists to carry on experimental research which had only been theoretically predicted [21, 22]. Applying the interlayer many-body exchange interaction to the bilayer graphene system, the group of MacDonald et al. estimated that the interlayer e–h exciton condensation may be possible at room temperature [14], due to not only the unique physical properties of graphene including symmetric electron and hole band structures [23, 24] (which otherwise are obtained only under high magnetic fields with Landau levels in the conduction band of III–V materials where a partially empty Landau level serves as a hole band [18]) but also the possibility of placing the physically 2D layers much closer together and within a low-dielectric constant environment that is primarily extrinsic to the graphene and, thus, separately optimizable. Recent calculations by the same group suggest that room-temperature condensate formation in such graphene bilayers will be more difficult than initially expected, but may still be possible [25]. Because graphene was the first material system considered, it currently hosts most of the theoretical and experimental efforts toward "pseudospintronic devices."

However, room-temperature condensate formation may be more readily possible employing other 2D material systems such as monolayer transition metal dichalcogenides (TMDs) [16]. TMDs are compounds composed of a transition metal plus a column VI material such as MoS_2, WTe_2, and WSe_2.

If room-temperature condensates can be formed in these novel 2D materials, it is expected that the same novel transport properties observed in III–V materials at very low temperatures and high magnetic fields [20] could be reproduced in these bilayer 2D material systems absent magnetic fields.

2.2.3 Pseudospin and Pseudospintronics

Pseudospin can be defined for any parameter with only two options. In the bilayer systems of interest here, the "which-layer" degree of freedom for the electrons can be described as a "pseudospin" consisting of "top layer" (T) and "bottom layer" (B) analogous to "spin-up" and "spin-down" electrons. Single-electron states are therefore a mixture of "pseudospin-up" and "pseudospin-down" states coupled by any bare interlayer coupling and, of critical importance here, the interlayer many-body exchange interaction. More generally, in a collective state consisting of many single-particle electron states, the magnitude and phase of top-to-bottom interlayer density matrix $\rho(\mathbf{R_T}, \mathbf{R_B}) = \sum_{\varphi} f_{\varphi} \varphi(\mathbf{R_T}) \varphi^*(\mathbf{R_B})$ reflect an overall coherence strength and phase relationship between the top and bottom layers and are described as a collective "pseudospin strength" and "pseudospin phase," respectively [13, 14]. This coherent pseudospin state is analogous to an actual collective spin state composed of coherent spin up and spin down contributions. An essential property of exciton condensation is greatly enhanced—even entirely self-consistently supported in the conceptual "spontaneous" limit—interlayer coherence, i.e., density matrix/pseudospin strength.

To go further, we introduce the Hartree–Fock potential, which addresses the above-discussed many-body exchange interaction within a single-particle frame-work using the density matrix and the Coulomb interaction among the electrons. It is employed to explain and characterize the formation of the condensation, as well as associated novel transport effects. For graphene, the conduction and valence band properties can be reasonably represented considering only one spin degenerate $2p_z$ orbital per carbon atom. The tight-binding version of the Hartree–Fock model, explicitly considering only interlayer exchange, is for the top layer [26],

$$H_{TB}\varphi_\beta(\mathbf{R}_T) + V(\mathbf{R}_T)\varphi_\beta(\mathbf{R}_T) + \sum_{\mathbf{R}_B} V_{Fock}(\mathbf{R}_T, \mathbf{R}_B)\varphi_\beta(\mathbf{R}_B) = E_\beta\varphi_\beta(\mathbf{R}_T), \quad (2.8)$$

where

$$V_{Fock}(\mathbf{R}_T, \mathbf{R}_B) = -\frac{e^2}{4\pi\varepsilon|\mathbf{R}_T - \mathbf{R}_B|}\sum_{\beta'} f_{\beta'}\varphi_{\beta'}(\mathbf{R}_T)\varphi_{\beta'}^*(\mathbf{R}_B), \quad (2.9)$$

with a similar expression for the bottom layer. Here the $\varphi_\beta(\mathbf{R})$ are the quasi-single-particle tight-binding energy eigenstates, where β labels the energy band α, wavevector $\mathbf{k}$, and real spin s. H_{TB} is the nominally single-particle tight-binding Hamiltonian with only intralayer and (bare) interlayer hopping between nearest-neighbor atoms (although many-body exchange and correlation effects are implicitly included in the modeled band structure). The second term on the left-hand side is the "Hartree" term, where $V(\mathbf{R})$ is the electrostatic potential energy at $\mathbf{R}$ including potential energy contributed by both externally applied electric fields and the charge distribution within the layers (and even implicitly variations in the intralayer exchange interaction, which would need to be more carefully considered for self-consistent calculations of layer carrier densities as a function of applied gate voltages). The third term, or the "Fock" term, on the left-hand side of Eq. (2.8) is the exchange interaction, which is proportional to the interlayer density matrix as shown in Eq. (2.9), where we have approximated the Coulomb interaction using an effective dielectric constant that represents the interlayer dielectric and—even dominated by—the dielectrics above and below the 2D layers and self-consistent dynamic free-carrier screening [25]. Physically, the exchange interaction reduces the otherwise expected Coulomb interaction, consistent with the electrons being further away from each other than expected classically.

For this multistate β system, extending Eq. (2.7) for single states, the *total* interlayer current is calculated as in the presence of the exchange interaction as

$$j(\mathbf{R}_T, \mathbf{R}_B) = \frac{2e}{h}\mathrm{Im}\left\{\sum_\beta f_\beta\varphi_\beta^*(\mathbf{R}_T)[H_{bare}(\mathbf{R}_T, \mathbf{R}_B) + H_{Fock}(\mathbf{R}_T, \mathbf{R}_B)]\varphi_\beta(\mathbf{R}_B)\right\}$$

$$(2.10)$$

With a little work, it can be shown both that the total interlayer current due to the interlayer Fock interaction alone is identically zero and that the remaining contribution reduces to

$$j(\mathbf{R}_T, \mathbf{R}_B) = \frac{2e}{h}H_{bare}(\mathbf{R}_T, \mathbf{R}_B)\mathrm{Im}\left[\rho^\dagger(\mathbf{R}_T, \mathbf{R}_B)\right]. \quad (2.11)$$

Indeed, this is the expression for current flow in a multistate system in terms of the density matrix regardless of whether or not the Fock exchange interaction is considered. However, the interlayer exchange interaction can greatly increase the pseudospin/interlayer density matrix $\rho(\mathbf{R}_T,\ \mathbf{R}_B)$. Unlike the terminology *spintronics*, the pseudospintronic devices do not rely on manipulating any real spins. In the bilayer graphene system, different real spins can even be decoupled as just two identical channels for interlayer current. However, such a decomposition cannot be performed in TMD systems, where strong spin–orbit coupling breaks the degeneracy and mixes the spin states within individual energy eigenstates [27, 28].

2.3   Pseudospintronic Devices and Analysis

In this section, we will consider pseudospintronic devices: BiSFET, BiSJT, and Coulomb drag devices. The essential physics and device concepts should be independent of which bilayer system is considered. However, much of the analysis of this work has been performed for graphene systems for reasons already noted, so we provide detailed information on this system. In addition, we first consider condensate behavior in "bulk" (large area) graphene–dielectric–graphene systems, to provide a reference for device structures and behavior.

2.3.1   *Condensate Formation in Bulk Graphene Bilayers*

2.3.1.1   About Graphene

Graphene is a single layer of graphite. Pristine graphene has a honeycomb structure, where carbon–carbon bond length is $a_{cc} \approx 1.42$ Å and, within a tight-binding description, a nearest-neighbor nominally bare hopping potential of $t_0 \approx -2.7$ eV. The primitive cell includes two atoms labeled by A and B, as shown in Fig. 2.1a, and the set of A atoms and the set of B atoms constitute two sublattices. The linear band structure in graphene resembles the dispersion relationship for light and relativistic particles, as shown in Fig. 2.1b and, therefore, electrons in graphene are sometimes called the "massless Dirac fermions" [21]. However, the fixed carrier velocity magnitude in graphene is roughly 10^8 cm/s, only one three-hundredth the speed of light in vacuum. Specifically, the band structure of graphene has cone-shaped valence and conduction bands (the "Dirac cones") touching each other at the pinnacle (the "Dirac point"), located at the corners of the first Brillouin zone. There are two such cone-shaped pairs in total, centered at the K and K' points. (The "pseudospin" terminology also has been applied to the "which-atomic-sublattice" and to the "which-valley" degrees of freedom. In monolayer TMDs, while there are band gaps and the carriers are massive, the valleys are also located at the K and K' points. However, the valley pseudospin and real spin are coupled. The band edges are also split between real spin up and down, particularly in the valence band, but with the K and K' valleys having opposite spins at the valence band edge [29]).

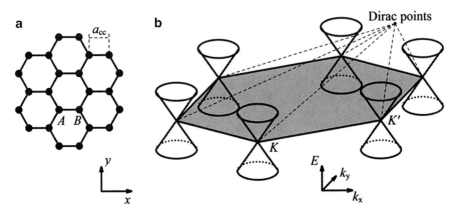

Fig. 2.1 (**a**) A graphene flake with a perfect hexagonal honeycomb structure. A and B label the two atoms within the primitive unit cells and the associated A and B sublattices. (**b**) The conical band structure of graphene around the K and K' points in the hexagonal Brillouin zone. Note that the orientations of the Wigner–Seitz unit cells in real space and in momentum space are offset by 30°

For our two-graphene-layer system, there are several stacking patterns that can be considered, as shown in Fig. 2.2. Even assuming rotationally aligned graphene for the moment, there are multiple translational alignments that can be considered. "Bernal stacking" as shown in Fig. 2.2b refers to the situation that A sublattice in one layer is aligned with B sublattice in the other, as naturally occurs in graphite. However, particularly with an intervening dielectric—as modeled for our pseudospin device analysis—and in terms of interatomic coupling, one may also consider hexagonal stacking (Fig. 2.2a), which consists of two perfectly aligned graphene layers with A sublattice upon A sublattice and B sublattice upon B sublattice. In terms of bare coupling through a dielectric, one may even consider just A sublattice-to-A sublattice *or* B sublattice-to-B sublattice coupling or linear combinations of all of these. The layers also may be rotated with respect to each other, particularly when one is physically placed upon another or upon an intervening dielectric. For commensurate rotation (Fig. 2.2c), one layer is rotated with respect to the other, but a global periodicity is still preserved for the bilayer system, which is convenient for numerical analysis, although on a larger geometric scale which is less convenient. The global periodicity observed is also named the Móire pattern. The possible commensurate rotation angles are infinite but still can be mathematically enumerated by two integer indices [30].

2.3.1.2　Simulation Technique

A full-quantum Hartree–Fock simulator can be developed from Eq. (2.8) to study the formation of exciton condensates in bulk and to simulate transport properties in the presence of the condensates in nanoscale bilayer systems as to be

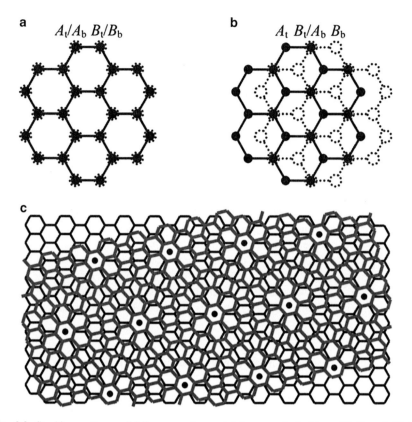

Fig. 2.2 Stacking patterns of bilayer graphene: (**a**) hexagonal stacking, (**b**) Bernal (A–B) stacking, and (**c**) commensurate rotational stacking. In (**c**), the two rotationally misaligned graphene honeycomb lattices are sketched with different colors and line widths, and the resulting honeycomb superlattice is represented by *dots*, which also shows the Móire pattern resulting from the angular rotation. The relative rotation between layers shown here is 13.17°, producing a total of 76 carbon atoms per unit cell counting both layers

considered later. For graphene layers with periodicity, either rotationally aligned or commensurately rotated, Bloch's theorem can be applied to Eq. (2.8) such that the tight-binding time-independent Schrödinger's equation takes the form,

$$H_{TB}\varphi_\beta(\mathbf{R}_T) + V_T\varphi_\beta(\mathbf{R}_T) + \sum_{\mathbf{r}_B} A_{Fock,\beta}(\mathbf{R}_T, \mathbf{R}_B)\varphi_\beta(\mathbf{R}_B) = E_\beta\varphi_\beta(\mathbf{R}_T), \quad (2.12)$$

$$A_{Fock,\beta}(\mathbf{R}_T, \mathbf{R}_B) = \sum_{\mathbf{R}} V_{Fock}(\mathbf{R}_T, \mathbf{R}_B + \boldsymbol{\xi})e^{i\mathbf{k}\cdot\boldsymbol{\xi}}$$

$$= \sum_{\mathbf{k}'} C(\mathbf{R}_T - \mathbf{R}_B, \mathbf{k} - \mathbf{k}')\sum_{\alpha', s'} f_{\beta'}\varphi_{\beta'}(\mathbf{R}_T)\varphi_{\beta'}^*(\mathbf{R}_B), \quad (2.13)$$

$$C(\mathbf{R_T} - \mathbf{R_B}, \mathbf{k} - \mathbf{k'}) = -\sum_{\mathbf{R}} V_{\mathrm{col}}(\mathbf{R_T}, \mathbf{R_B} + \boldsymbol{\xi}) e^{i(\mathbf{k} - \mathbf{k'}) \cdot \boldsymbol{\xi}}, \qquad (2.14)$$

where $\boldsymbol{\xi}$ are the lattice vectors and $\mathbf{R_T}$ and $\mathbf{R_B}$ are the locations of the top- and bottom-layer atoms within the primitive cell [31, 32]. $V_{\mathrm{col}}(\mathbf{R_T}, \mathbf{R_B} + \boldsymbol{\xi})$ describes the Coulomb interaction within (unlike what is shown in Eq. (2.9)) a possibly nonuniform dielectric environment. The time-independent Schrödinger's equation is consequently reduced to the primitive unit cell size. We solve this equation using an iterative method in which the exchange interaction and energy eigenstates β are obtained self-consistently. (If no actual bare coupling is considered, we "seed" the simulation with a small bare interlayer coupling for the first step only.) However, the calculation of density matrix ρ involves the summation over all states β, and the Hamiltonian contains the long-range (non-nearest-neighbor) interlayer Fock interaction. The resulting matrix equation is dense (has many nonzero elements) and different for each different energy eigenstate β. Equations (2.13) and (2.14) imply that the Fourier transform of the Coulomb interaction can be pre-calculated, making each single $\mathbf{k}$ state separable and assignable to different CPUs for parallel computing. However, even with supercomputing resources, this calculation can be computationally challenging for large unit cells under commensurate rotations.

2.3.1.3   Characterization and Properties of the Bulk Condensate

Consider first two rotationally aligned graphene layers with no bare interlayer coupling. In this case, the translational alignment is of no real consequence. Electrons in one layer and holes in the other are conceptually created via external gates. For simulation purposes, the required carrier concentration are adjusted by on-layer electrostatic potentials $V(\mathbf{R_T})$ and $V(\mathbf{R_B})$ as in Eq. (2.8) with respect to the equilibrium Fermi level E_F, which is chosen as the zero energy reference ($E_F \equiv 0$) for simplicity. Setting $V(\mathbf{R_T}) = -V(\mathbf{R_B}) = V_{\mathrm{diff}}/2$ produces equal electron and hole concentrations, which is ideal for condensate formation. The resulting band structure of such a bilayer graphene system absent significant interlayer coupling is shown in Fig. 2.3a for the interpenetrating cones of the conduction band of the n-type layer and valence band of the p-type layer. The intersection of the cones is at the Fermi level.

The formation of an exciton condensate is accompanied by a band-gap opening/anti-crossing where the bands otherwise would have crossed, as shown in the simulation results of Fig. 2.3b as a function of interlayer separation [14, 26]. For large separation, the band structure is unaffected and therefore remains two essentially decoupled and overlapping Dirac cones representing the top and bottom layers. As the separation decreases and the Coulomb interaction increases, a band gap opens up which can be understood in this way: There can be no interlayer coherence in the absence of interlayer coupling of any type. That is, the interlayer pseudospin is zero. However, if there is any exchange interaction, there will be

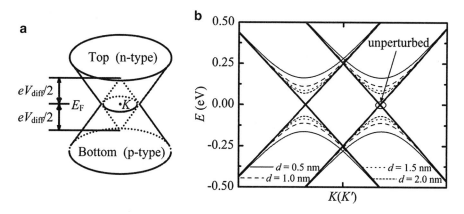

Fig. 2.3 Bilayer graphene band structures. (**a**) Without any interlayer interaction, with the n-type layer valence band and p-type layer conduction band suppressed for clarity. (**b**) As calculated with only the many-body exchange interaction at 0 K as a function of interlayer separation in units of nm, with an *effective* dielectric constant ε of 2.2. The latter represents the actual dielectric constants of the interlayer dielectric plus—even dominated by—those above and below the graphene layers, plus free-carrier screening which should be calculated self-consistently, but which is beyond the scope of this work [25]. The band structure in (**b**) is plotted along a line of **k** vectors that crosses the K point. The full band structure can be envisioned as that of (**b**) pivoting around a z-axis that passes through the K point. While interlayer bare coupling can lead to a band gap opening [33], the opening of such a "condensate gap" does not rely on interlayer bare coupling [14, 26]

interlayer coherence/nonzero pseudospin. In the end, we can find a self-consistent solution for interlayer exchange coupling and interlayer coherence even in the absence of bare interlayer coupling. (Bare coupling still would be required to allow any interlayer current flow, per Eq. (2.11)). This latter solution represents the condensed state, a "spontaneous" one in this case, with coherent interlayer coupling between the n-type and p-type layers, mediated by the exchange interaction. Thus, the excitonic condensate does not result simply from the interlayer Coulomb interaction, but rather from the many-body exchange correction to the Coulomb interactions when, and only when, there is interlayer quantum coherence. The band splitting/anti-crossing about the Fermi level and associated lowering of the occupied states near the anti-crossing reduce the overall energy of the system, making the condensed state the energetically favored one [26].

The establishment and strength of the condensate/pseudospin is subject to a positive feedback loop: the stronger the exchange interaction, the stronger the interlayer pseudospin; the stronger the exchange interaction, the stronger the pseudospin; and so forth and so on and vice versa. As a result, the condensate is a sensitive function of dielectric environment, layer separation, and, as shown in Fig. 2.4, temperature [26]. The result shows a clear "critical temperature" T_c for the formation of these condensates. Moreover, the critical temperature is strongly correlated to the zero-temperature band gap, $4k_B T_c \approx E_{g0}(E_g$ at $T = 0)$, where k_B is Boltzmann's constant and T_c is the critical temperature in Kelvin. However, the

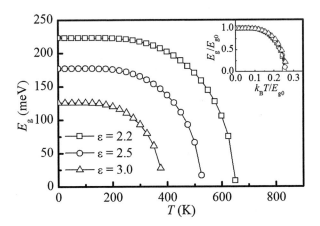

Fig. 2.4 Temperature dependence of the induced band gap/anti-crossing for spontaneous condensates. The layer separation about the interlayer dielectric is $d = 1$ nm. Various effective dielectric constants were considered as shown. Balanced electron and hole concentrations are created by $V_T = -V_B = 0.25$ V. The *inset* is plotted in energies normalized by induced zero-temperature band gap $E_g(T=0) \equiv E_{g0}$

shape of these band gaps vs. temperature curves is not unique to excitonic condensates but rather is much as exhibited in conventional superconductors.

There is also a Kosterlitz–Thouless transition temperature for 2D systems to consider, which is beyond the scope of this work to explain but which is expected to be approximately $8k_B T_{K-T} = |E_F - E_D|$ for the graphene system, where E_D is the Dirac point energy at the pinnacle of the Dirac cones; the condensate can only exist if both conditions are met [14]. However, the zero band gap and low graphene density of states make it relatively easy to move the Fermi level well into the bands with limited carrier concentrations and, thus, manageable gating fields. This condition can be satisfied for electron and hole concentrations of somewhat less than $5 \times 10^{12}/\text{cm}^2$ for condensation at $T = 300$ K [34]. The carrier density in graphene layers has been electrostatically modulated to as high as $10^{13}/\text{cm}^2$ using independent gates [35].

We also have found that the strength of the spontaneous condensate depends little on the pattern of interlayer alignment, including rotational misalignment even when the size of the resulting unit cell is large compared to the in-plane coherence length of the condensate [32], as seen in Fig. 2.5. The latter insensitivity results because if we consider exchange coupling between top and bottom layers whose K valleys are separated by $\mathbf{K_B} - \mathbf{K_T} = \Delta\mathbf{K}$, the resulting uniform phase shift of $e^{+i\Delta\mathbf{K}\cdot\mathbf{R_B}}$ introduced within the bottom layer to each wavefunction β is canceled by a corresponding uniform phase shift $e^{-i\Delta\mathbf{K}\cdot\mathbf{R_B}}$ introduced into the exchange interaction operating on the bottom layer as required to determine the effect on the top layer and vice versa for the exchange interaction acting on the bottom layer. However, the rotation of one layer with respect to the other can greatly alter the interlayer bare hopping.

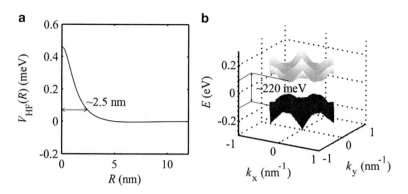

Fig. 2.5 (**a**) Magnitude of the Fock exchange interaction $V_{HF}(\mathbf{R_T},\mathbf{R_B})$ as a function of R, the magnitude of the in-plane component $|\mathbf{R_T} - \mathbf{R_B}|_{\text{in-plane}}$. The interlayer separation is $d = 1$ nm, the effective dielectric constant is $\varepsilon = 2.2$, and $T = 0$ K. The Fock interaction, limited both by in-plane coherence in the density matrix and the Coulomb interaction, is only significant over a radius in R of approximately 3 nm or less. (**b**) Band structure of the conduction and valence bands in the first Brillouin zone of commensurately rotated bilayers with a 2.65° rotation angle. The shown condensate-induced band gap is essentially unchanged as a result of the rotation. The real-space unit cell is of approximately 6 nm on an edge, containing 1,876 atoms, and there is a correspondingly small Brillouin zone. All other parameters are the same as in (**a**)

Balanced charge distributions are optimal for exciton condensate formation as already noted. Conversely, charge imbalance can impact condensate formation negatively. The result of Fig. 2.6 shows that small charge imbalances are tolerable, but larger charge imbalances can completely prevent condensation [26]. Thus, strong "charge puddling," charge imbalance created by electrostatic potential variations resulted from nonideal substrates, impurities, etc. [36], is to be avoided. We note, however, that variations in the electrostatic potentials on scales comparable to or smaller than the coherence length have a more limited effect, as shown in Fig. 2.7 [32].

2.3.2 Pseudospin Transport Properties

2.3.2.1 Simulation Technique

The simulated structure for modeling transport properties is shown in Fig. 2.8. This four-terminal structure includes two coupled graphene sheets with both bare coupling and Fock exchange interaction within, and only within, a channel of length L, through an interlayer dielectric with interlayer separation d within an effective dielectric environment ε. The channel is connected to four semi-infinite, noninteracting leads. The pattern of interlayer bare coupling, as discussed above, does not have a great effect on the condensate formation but does influence the interlayer current profoundly according to Eq. (2.11). However, this pattern

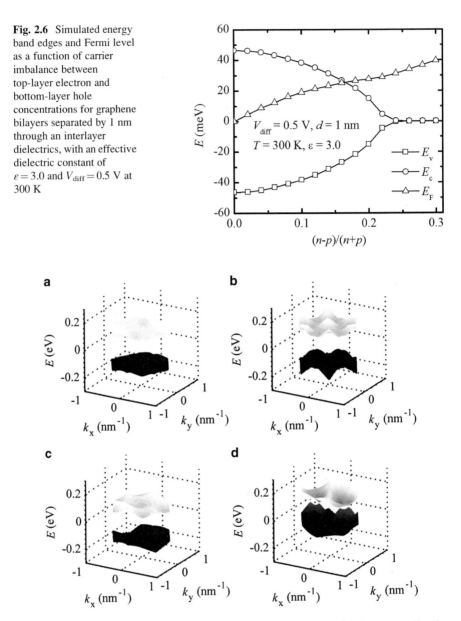

Fig. 2.6 Simulated energy band edges and Fermi level as a function of carrier imbalance between top-layer electron and bottom-layer hole concentrations for graphene bilayers separated by 1 nm through an interlayer dielectrics, with an effective dielectric constant of $\varepsilon = 3.0$ and $V_{\text{diff}} = 0.5$ V at 300 K

Fig. 2.7 Typical band structures for condensates in graphene bilayers with short-range disorder with a commensurate rotation of $\theta = 2.65°$ making the unit cell larger than the coherence length of the exchange interaction. $d = 1$ nm, $\varepsilon = 2.2$, and $T = 0$ K. The disorder potentials are characterized by standard deviation σ_v and correlation length l, both extending less than one unit cell. (**a**) $l \approx 0.6$ nm and $\sigma_v \approx 0.4$ V resulting in $E_g \approx 225$ meV, (**b**) $l \approx 2.5$ nm and $\sigma_v \approx 0.1$ V resulting in $E_g \approx 214$ meV, (**c**) $l \approx 2.5$ nm and $\sigma_v \approx 0.4$ V resulting in $E_g \approx 178$ meV, and (**d**) $l \approx 2.5$ nm and $\sigma_v \approx 0.5$ V resulting in $E_g \approx 27$ meV

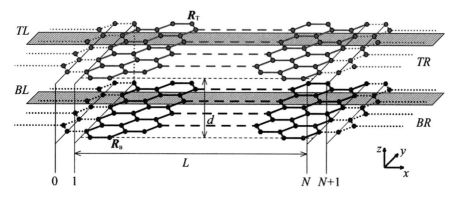

Fig. 2.8 Simulated structure. Two graphene layers (infinite in the y direction) are coupled in a channel of length L and with dielectric of thickness d. The effective interlayer dielectric constant is ε. Primitive cells are represented by the *longitudinal shaded stripes*, and atoms are located within slices with consecutive integer-indexed labels, starting from 1 and ending with N in the channel. There are four semi-infinite, perfectly absorbing/injecting leads connected to the graphene layers at the top left, bottom left, top right, and bottom right, labeled by TL, BL, TR, and BR, respectively

becomes at least conceptually tunable through interlayer dielectrics, although bringing about additional challenges to fabrication, a topic that will be discussed later. However, for mixed coupling patterns, either A–A hopping (coupling between the A-sublattice atoms) or, equivalently, B–B hopping is expected to dominate the contributions to interlayer current flow because of the way they align with the natural pattern of the "spontaneous" condensate [37]. In this work, we assume A–A coupling for modeling purposes. The transport is assumed along the direction of armchair chains of atoms.

In such a structure, the single-electron states β are now characterized by the lead of injection α, the predetermined/chosen energies of injection E, the transverse wavevector k_y (also referred to as a "mode" in transport problems), and the real spin s (which can be neglected here as in the bulk for the graphene system). A thorough tutorial of quantum transport simulations can be found in Supriyo Datta's well-known book on the subject [38]. Conceptually, an electron can be injected from any lead and then be reflected into the same lead or be transmitted to other leads. The wavefunction including the injected, reflected, and transmitted states in the leads, as well as the potentially complicated wave profile in the channel, is called the "scattering state."

To calculate scattering states, we start with a Bloch state that describes an incoming wave in one lead, for which the wavevector $\mathbf{k}$ is determined from the E–$\mathbf{k}$ dispersion relation in the leads and subject to the required direction of propagation being *into* the lead. In this open simulation region, the *incident* current (not counting reflection) for each of the scattering states β, as calculated from Eq. (2.7), is necessarily normalized consistent with the ballistic Landauer–Büttiker limit for conductance per transverse mode e^2/h. The nominal number of occupied transverse modes per unit width is then $4k_F/\pi$ including a factor of 4 for valley and spin degeneracy, where k_F is the Fermi wavevector, $k_F = E_F/(\hbar v)$, E_F is the Fermi energy, and v is the fixed carrier velocity in graphene [34]. With the device separated into slices in the longitudinal direction, the Schrödinger's equation, Eq. (2.12), can be then generalized into a slice-by-slice form, $\sum_j \mathbf{H}_{ij}\boldsymbol{\varphi}_j = E\boldsymbol{\varphi}_i$, where $\boldsymbol{\varphi}_i$ represents the wavefunction of slice i (a 2×1 vector representing the top and bottom atoms) and $\mathbf{H}_{ij}$ couples slices i and j via bare and any Fock exchange interaction.

The next step is to determine the boundary conditions between leads and channel, i.e., the variation of the wavefunction between slices 0 and 1 for the left leads and N and $(N+1)$ for the right leads. Consider a left lead as the lead of injection. On the boundaries between the semi-infinite leads and the channel, beyond the region of interlayer coupling within channel, the injected and reflected waves must satisfy [31]

$$\begin{aligned}
\boldsymbol{\varphi}_0 &= \boldsymbol{\varphi}_{0,i} + \boldsymbol{\varphi}_{0,r} = \boldsymbol{\varphi}_{0,i} + \mathbf{P}_{10,r}\boldsymbol{\varphi}_{1,r} = \boldsymbol{\varphi}_{0,i} + \mathbf{P}_{10,r}\left(\boldsymbol{\varphi}_1 - \boldsymbol{\varphi}_{1,i}\right) \\
&= \mathbf{P}_{10,r}\boldsymbol{\varphi}_1 + (\mathbf{I} - \mathbf{P}_{10,r}\mathbf{P}_{01,i})\boldsymbol{\varphi}_{0,i}
\end{aligned} \tag{2.15}$$

Here, $\boldsymbol{\varphi}_{k,i}$ and $\boldsymbol{\varphi}_{k,r}$ are incident (i) and reflected (r) components of the wavefunction on slice k, respectively, such that $\boldsymbol{\varphi}_k = \boldsymbol{\varphi}_{k,i} + \boldsymbol{\varphi}_{k,r}$. $\mathbf{P}_{01,i}$ and $\mathbf{P}_{10,r}$ are 2×2 diagonal matrices for the incident and reflected waves, respectively, the diagonal elements of which are the corresponding additional phase factors from slice 0 to 1 (incident) and 1 to 0 (reflected) determined by $\mathbf{k}$. If this lead is not the lead of injection and there is only an outgoing wave to consider, Eq. (2.15) reduces to simply $\boldsymbol{\varphi}_0 = \mathbf{P}_{10,r}\boldsymbol{\varphi}_1$. In the lead of injection, the wavevector is necessarily real. However, it is possible that the energy of injection E will fall within the band gap/forbidden region in one or more of the outgoing leads such that the corresponding wavevector of propagation $\mathbf{k}$ will be complex. Physically meaningful solutions require that the probability density *decay* into those leads.

With Eq. (2.15), the Schrödinger's equation for transport simulation is then completely described within the channel on the left boundary as

$$(E\mathbf{I} - \mathbf{H}_{10}\mathbf{P}_{10,r} - \mathbf{H}_{11})\boldsymbol{\varphi}_1 - \mathbf{H}_{12}\boldsymbol{\varphi}_2 = \mathbf{H}_{10}(\mathbf{I} - \mathbf{P}_{10,r}\mathbf{P}_{01,i})\boldsymbol{\varphi}_{0,i} \tag{2.16}$$

A similar equation is obtained on the right boundary (slices N and $N+1$). This equation is equivalent to the nonequilibrium Green's function (NEGF) equation $(E\mathbf{I} - \mathbf{H} - \boldsymbol{\Sigma})\boldsymbol{\varphi} = \mathbf{S}$ with "self-energy" $\boldsymbol{\Sigma}$ corresponding to boundary absorption and the "source" $\mathbf{S}$ corresponding to the injected wavefunction. Specifically, $\boldsymbol{\Sigma}_{1,1} = \mathbf{H}_{10}\mathbf{P}_{10,r}$, $\boldsymbol{\Sigma}_{N,N} = \mathbf{H}_{N(N+1)}\mathbf{P}_{N(N+1),r}$, and all other $\boldsymbol{\Sigma}_{ij} = \mathbf{0}$, and $\mathbf{S}_1 = \mathbf{H}_{10}(\mathbf{I} - \mathbf{P}_{10,r}\mathbf{P}_{01,i})\ \boldsymbol{\varphi}_{0,i}$, $\mathbf{S}_N = \mathbf{H}_{N(N+1)}\ (\mathbf{I} - \mathbf{P}_{N(N+1),r}\mathbf{P}_{(N+1)N,i})\boldsymbol{\varphi}_{N+1,i}$, and all other $\mathbf{S}_i = \mathbf{0}$ [31]. However, we do not explicitly solve for the full Green's function matrix $(E\mathbf{I} - \mathbf{H} - \boldsymbol{\Sigma})^{-1}$, which would be required only if we were to consider "injection" from within the channel, such as when considering inelastic (and, thus, phase-breaking) phonon scattering within the simulation region. However, the scattering in a nanoscale graphene device can be expected to be very limited even at room temperature [39].

The occupation probabilities f_β for the states β, as required to calculate current flow and to obtain the exchange interaction, follow the Fermi distribution in the lead of injection α, $f = \left(1 + e^{E-E_{F,\alpha}/k_B T}\right) = \left(1 + e^{E+qV_\alpha/k_B T}\right)$, where V_α is the applied voltage on lead α. Aiming for room-temperature device operation, transport simulations are performed at a temperature of $T = 300$ K unless otherwise noted. However, beyond the existence and strength of the condensate, temperature has little effect on simulation results.

As for the bulk calculation, the carrier concentrations in top and bottom layers are conceptually created by control gates, omitted in Fig. 2.8, producing opposite and fixed electrostatic voltages on the top and bottom layers. For proposed pseudospintronic device applications, the variations in carrier concentration even with external voltages applied to the leads are negligible compared to the concentration itself and, hence, the associated self-consistent changes in electrostatic potential can be neglected. Only the Fock interaction, $V_{\text{Fock}}(\mathbf{R}_T, \mathbf{R}_B)$, is obtained self-consistently for now allowing the isolation of essential physics associated with the condensate formation and transport absent such perturbations.

2.3.2.2 Condensate Formation in the Channel

To model the condensate formation in nanoscale regions as required for device applications, the *local* density of states (LDOS), which is defined as the available states per unit (here) area per unit energy *as a function of the position*, has been calculated for different channel lengths with fixed carrier concentrations and interlayer separation. The interlayer bare coupling V_b is also very weak, on the scale of 1 meV, to guarantee the band gap/anti-crossing is primarily the result of condensation. The LDOS in the center of the channel is shown in Fig. 2.9, where the LDOS about the Fermi level approaches zero with increasing channel length. In other words, an anti-crossing band gap, although incomplete due to exponentially decaying tails of the states originating outside the condensate region, forms in the channel due to the exchange-enhanced interlayer coupling in the presence of the condensate. The size of the band gap saturates to approximately that of the bulk condensate band gap [26], implying that the condensate is nearly fully formed in the channel center. The channel length needed for the formation of the condensate also depends on the strength of the bulk condensate, as adjusted here via the effective dielectric constant; the stronger is the bulk condensate, the shorter is the channel needed to fully form the condensate. The spatial extent of the condensate within the channel also is important for device applications. In the example of Fig. 2.10, the LDOS exhibits ~5 nm transition regions on either side of a ~5 nm bulk-like condensate region (consistent with the coherence length of the exchange interaction in Fig. 2.5a). However, even an incompletely formed condensate will give rise to the novel transport phenomena, as will be shown.

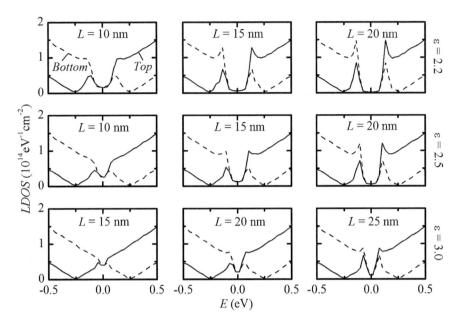

Fig. 2.9 The calculated local density of states (LDOS) in the top (*solid line*) and bottom (*dashed line*) layers in the center of the channel as a function of the effective dielectric constant ε. A band gap begins to form and saturates to its bulk value with increasing channel length. The other parameters are $n = p \approx 6 \times 10^{12}$ cm^{-2}, $V_b = 0.5$ meV, and $d = 1$ nm

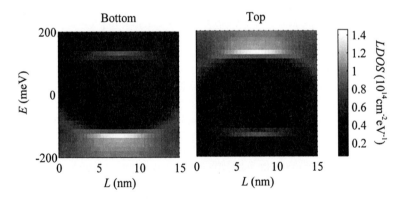

Fig. 2.10 Local density of states (LDOS) versus energy and position in the channel for $\varepsilon = 2.2$, $d = 1$ nm, $L = 15$ nm, $n = p \approx 6 \times 10^{12}$ cm^{-2}, and interlayer bare coupling energy $V_b = 0.5$ meV. LDOS for the bottom layer is plotted on the *left* and that for the top layer on the *right*. The region of the condensate within the channel in terms of both energy and position is exhibited by the *dark center* of each figure

2.3.2.3 Enhanced Interlayer Tunneling

As alluded to previously and consistent with Eq. (2.11), condensate formation with the associated enhanced interlayer density matrix/pseudospin can greatly enhance interlayer current for fixed (but nonzero) bare interlayer coupling.

First, we consider the interlayer transmission probabilities in the equilibrium limit. Calculated k_y-averaged transmission coefficients among the leads for injections from two leads on the same side of the channel, BL and TL, are plotted in Fig. 2.11. The simulated individual transmission and reflection coefficients for injection from either lead add up to unity, as required for current conservation. The individual transmission probabilities from the top to bottom layer also match the inverse transmission probabilities from the bottom to top layer, as required by "detailed balance," such that in equilibrium there is no current flow. The difference between the *average* transmission coefficients from TL to BL and from BL to TL is an artifact of the averaging process, where different numbers of states—the denominator—are available in the top and bottom leads as a function of energy except at the Fermi level. Absent the condensate, essentially 100 % of the injected carriers are transmitted to the other side of the same layer. With the condensate, about 75 % of the injected carriers are transmitted to the same side of the *other* layer, with another ~20 % reflected. In other words, the two layers would be nearly shorted together on the same end of the channel, reaching 75 % of the Landauer–Büttiker limit for ballistic current flow.

2.3.2.4 Detailed Current Flow Under Various Biasing Schemes

As already predicted theoretically [40] and demonstrated experimentally in cryogenic III–V systems, condensate formation can produce novel transport effects including but not limited to enhanced interlayer transport and near-perfect "Coulomb drag" in III–V double quantum wells [18–20]. In the case of Coulomb drag, the current in a biased layer produces a current of near-equal magnitude but opposite sign in the other layer. Again, the necessary interlayer electron–hole interactions—the interaction between electrons in one layer and their absences in the other layer—for these considered coherent transport effects are provided in simulation by the *many-body Fock exchange interaction within this otherwise single-particle formalism.*

The structure in Fig. 2.8 also allows for the consideration of each of these effects. To model Coulomb drag, we set $V_{BL} = -V_{BR} = V_{al}/2$ and $V_{TL} = V_{TR} = 0$. To model interlayer transport, as required for the BiSFET, we set $V_{TL} = -V_{BL} = V_{il}/2$ and $V_{TR} = V_{BR} = 0$. To model the BiSJT, we set $V_{TL} = -V_{BL} = V_{ctrl}/2$ and $V_{TR} = -V_{BR} = V_{drv}/2$. The intra- and interlayer current flows for the BiSFET-like biasing and Coulomb drag biasing (perhaps more accurately called "current counterflow" biasing [40]) are illustrated in Fig. 2.12. These intralayer and interlayer

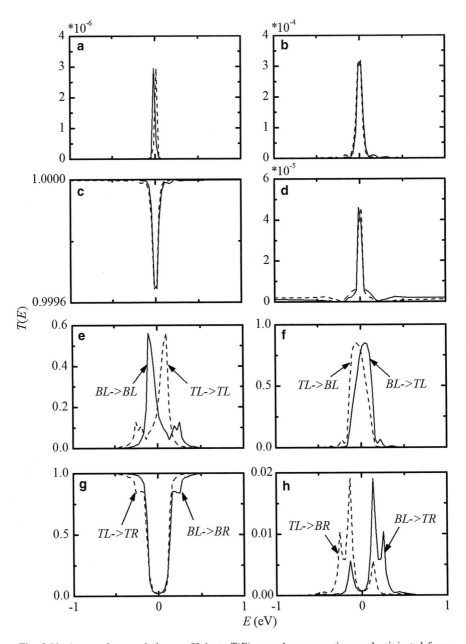

Fig. 2.11 Averaged transmission coefficients $T(E)$ over the propagating modes injected from leads BL and TL of the structure of Fig. 2.8. Transmission coefficients without the condensate are plotted in (**a**)–(**d**) and those with the condensate in (**e**)–(**h**). Device parameters are dielectric constant $\varepsilon = 2.2$, interlayer separation $d = 1$ nm, channel length $L = 15$ nm, interlayer bare coupling $V_b = 1$ meV, and carrier concentrations $n = p \approx 6 \times 10^{12}$ cm^{-2}. The *solid* and *dashed curves* in (**a**)–(**d**) correspond to the same transmission components as those in (**e**)–(**h**), respectively (but are difficult to label). Interlayer transmission on the *left side*, (**b**) and (**f**), is enhanced by more than a factor of 1,000 by condensate formation

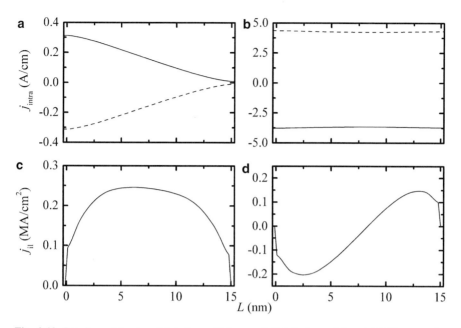

Fig. 2.12 Intralayer (j_{intra}) and interlayer (j_{il}) current distributions. (**a**) and (**c**) illustrate the BiSFET biasing condition with $V_{il} = 2$ mV, $V_b = 0.75$ meV; (**b**) and (**d**) exhibit the Coulomb drag biasing condition with $V_{al} = 50$ mV, $V_b = 0.5$ meV. Other parameters are $\varepsilon = 2.2$, $d = 1$ nm, $n = p \approx 6 \times 10^{12}$ cm^{-2}, and $L = 15$ nm. Intralayer current flowing to the right and interlayer current flowing to the bottom layer are considered positive. For the intralayer currents, those in the top layers are shown as *solid lines* and those in the bottom layer as *dashed lines*. A strong delocalization of both the intra- and interlayer current distribution is exhibited

currents as measured at the channel boundaries are entirely consistent with the above-noted prediction of novel behavior.

However, *how* the current flows is far from trivial. The injected current flows in the vicinity of and between the Fermi levels of the leads [38] and consequently into the anti-crossing band gap of the condensate. Under BiSFET biasing conditions, current flow *through* the condensate region is essentially eliminated. However, it penetrates further into the condensate region than one would expect based on very limited LDOS near the Fermi level in the channel. Moreover, interlayer current flow is actually maximized very near the center of the condensate region, not near the left edge. Under Coulomb drag biasing, the intralayer current is largely uniform across the channel despite the condensate-induced band gap. To help understand what is going on, a cutoff energy E_{co} is selected well below the Fermi levels and, thus, well below the region of current injection and also above the lower anti-crossing band-gap edge in the condensate region. As shown in Fig. 2.13, the currents above E_{co} are indeed largely restricted to the periphery of the condensate region. However, substantial currents flow below E_{co} and in a looping pattern within and between the layers. This closed current loop is just the "exciton flow"

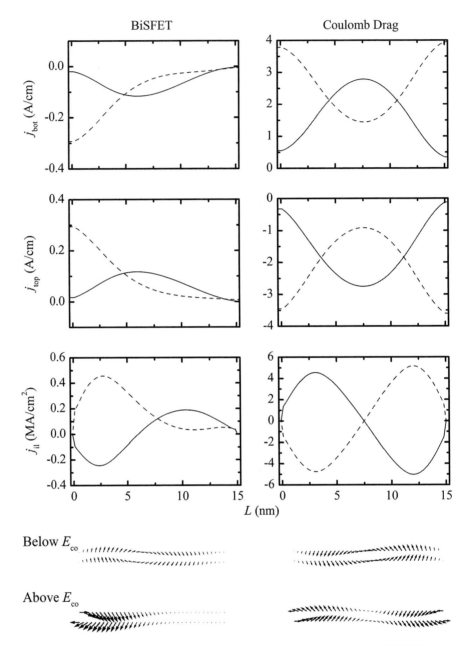

Fig. 2.13 Bottom-layer (j_{bot}), top-layer (j_{top}), and interlayer (j_{il}) distribution for BiSFET biasing condition in the *left column* and Coulomb drag in the *right column*, following the same device parameters, biasing conditions, and rules for signs of currents as those in Fig. 2.12. In the *top* six figures, current below the cutoff energy $E_{co} = -65$ meV is plotted in *solid curves* and that below E_{co} in *dashed curves*. In the remaining figures, these distributions are visualized with *arrows* that originate at the points they represent, where the vertical (horizontal) components are proportional to the interlayer (intralayer) current densities. With different units, however, the scaling between the vertical and horizontal components of the *arrows* is arbitrary and chosen for clarity here

within the channel. Thus, we find that on the same side of the condensate region, an injected (extracted) electron from one layer and hole from the other are annihilated (excited), creating (annihilating) a coherent electron–hole pair/interlayer exciton within the condensate region below the anti-crossing band gap. There is, however, *no* energy loss or gain in any of this process; annihilation (creation) of the incident (extracted) electron–hole pair and, separately, the creation (annihilation) of the coherent exciton within the condensate region are both entirely elastic processes. This entire process, including required exciton conservation within the closed current loop in the condensate region, is intimately coupled to pseudospin device behavior.

2.3.2.5 Interlayer Critical Current and Voltage Under Various Biasing Schemes

Consistent with Eq. (2.11), the interlayer current is determined by the interlayer bare coupling energy, the pseudospin magnitude, and the sine of pseudospin phase. For small applied interlayer biases, the pseudospin amplitude, as shown in Fig. 2.14a, is independent of the applied voltage but dependent of the position in the channel. The pseudospin phase, in contrast, is largely a position-independent property but dependent of the applied voltage as shown in Fig. 2.14b. Therefore, Eq. (2.11) can be approximated as simply $I \cong I_c \sin(\theta)$, which reaches its maximum amplitude I_c when $|\theta| = \pi/2$. Thus, only a maximum current of I_c can be carried in

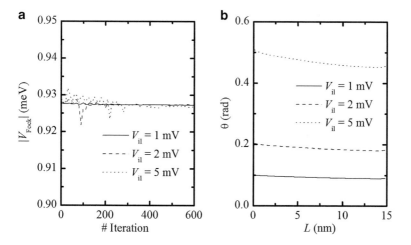

Fig. 2.14 (a) Interlayer Fock exchange interaction strength at the center of the channel vs. iteration and (b) converged pseudospin phase vs. position in channel, both for the nearest interlayer neighbor A–A coupling. The device parameters are $\varepsilon = 2.2$, $V_b = 0.5$ meV, $d = 1$ nm, $L = 15$ nm, and $n = p \approx 6 \times 10^{12}$ cm^{-2}. The transport problem is solved by starting from an initial equilibrium condition with grounded leads and ending with a self-consistent solution obtained iteratively with the BiSFET biasing condition

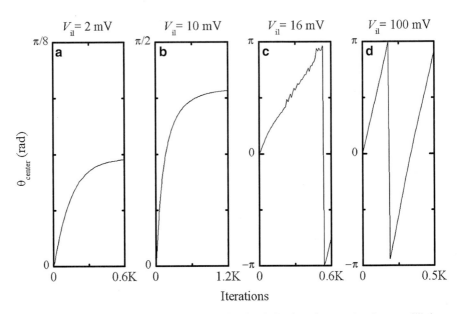

Fig. 2.15 Evolution of the pseudospin phase with simulation iteration, starting from equilibrium conditions. Device parameters and biasing configuration are the same as in Fig. 2.14. (**a**) $V_b = 0.5$ meV and $V_{il} = 2$ mV, (**b**) $V_b = 0.5$ meV and $V_{il} = 10$ mV, (**c**) $V_b = 0.5$ meV and $V_{il} = 16$ mV, and (**d**) $V_b = 0.5$ meV and $V_{hop} = 100$ mV. The converged solutions (**a**) and (**b**) are consistent with stable DC behavior below a critical voltage; (**c**) and (**d**) are consistent with the collapse of steady-state behavior and suggestive of the expected cyclic oscillatory behavior beyond a critical voltage

this way. I_c is called the "critical current" and the associated interlayer voltage V_c is herein called the "critical voltage" (although the same term is used for different purposes elsewhere). With further increases in the interlayer voltage, the steady-state current is expected to drop drastically with the onset of current oscillation on the scale of 1 THz (10^{12} Hz) per 2 mV of interlayer voltage. Such oscillations are qualitatively analogous to those in the Josephson junction [17, 19, 41, 42]. The simulation results in Fig. 2.15 are consistent with this expectation. Below a critical voltage (Fig. 2.15a, b), our iterative calculations readily converge to provide steady-state interlayer currents for $|\theta|$ below $\pi/2$. However, with higher voltages such that $|\theta|$ reaches and exceeds $\pi/2$ (Fig. 2.15c, d), no steady-state solution can be obtained and we observe cyclic oscillations (not just fluctuations) in θ with numerical iteration step in our calculations, which are suggestive of such time-dependent oscillations. Actual simulation of this beyond-critical voltage behavior, however, would require time-dependent NEGF capabilities not yet available.

The $\sin(\theta)$–V_{il} relation, or equivalently the I–V relation, is a linear function of interlayer voltage V_{il} both in principle and as shown in the simulation results of Fig. 2.16a (and Fig. 2.17a below). Consequently, the critical voltage can be obtained by linear extrapolation of V_{il} to $|\sin(\theta)| = 1$. (In principle, one could sample values of voltage to either side of, and increasingly close to, the transition

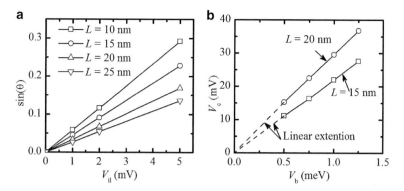

Fig. 2.16 (**a**) Linear $\sin(\theta)$–V_{il} relation (producing a linear I_{il}–V_{il} relation) for BiSFET biasing conditions, from which the critical voltage occurring at $|\sin(\theta)| = 1$ can be obtained. Here $\varepsilon = 2.2$, $d = 1$ nm, $V_b = 0.75$ meV, and $n = p \approx 6 \times 10^{12}$ cm^{-2}, with varying channel lengths. Although we have shown by direct simulation that the linear relationship extends at least very close to $|\sin(\theta)| = 1$ (and takes increasing long to converge in the process), we have shown only a limited region here. (**b**) Linear dependence of the thus obtained critical voltage V_c on interlayer bare coupling energy V_b under otherwise the same conditions as (**a**)

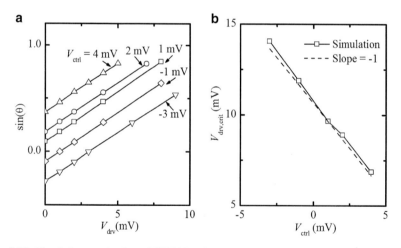

Fig. 2.17 Simulation results for a BiSJT-biased structure with $\varepsilon = 2.2$, $d = 1$ nm, $L = 15$ nm, $V_b = 0.5$ meV, and $n = p \approx 6 \times 10^{12}$ cm^{-2}. (**a**) The sine of pseudospin phase as a function of "drive-end" voltage drop from lead TR to BR, for different "control-end" voltage drops from lead TL to TR. (The slopes of the *straight lines* are proportional to the interlayer conductance.) (**b**) The effective drive-end critical voltage drop as a function of the control-end voltage drop. $V_{ctrl} + V_{drv,crit}$ is almost perfectly conserved, with perfect conservation represented by the reference *dashed line* of slope $= -1$

between convergence and oscillation in iterative numerical calculations. However, the time to converge or oscillate increases as the critical voltage is approached, as seen in Fig. 2.15, which makes this latter approach impractical outside of a few test cases in which we have got quite close.) Extracted critical voltages are presented in

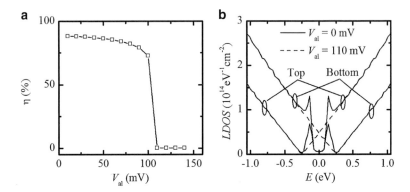

Fig. 2.18 (a) The Coulomb drag efficiency, defined as $\eta = (|I_{TL}| + |I_{TR}|)/(|I_{BL}| + |I_{BR}|)$ under Coulomb drag biasing/current counterflow biasing/intralayer biasing applied to the bottom layer, as a function of intralayer voltage V_{al}. Here, $\varepsilon = 2.2$, $d = 1$ nm, $L = 15$ nm, $V_b = 0.75$ meV, and $n = p \approx 6 \times 10^{12}$ cm^{-2}. Near-perfect Coulomb drag approaching 90 % is provided until the condensate, itself, collapses at relatively large intralayer voltages V_{al}, as confirmed by (b). (b) LDOS plotted in the center of the channel with the structure under low and high intralayer biasing with the same parameters as for (a), showing collapse of the anti-crossing gap, and thus the condensate, at high biases well beyond $k_B T/q$

Fig. 2.16b. This latter figure also shows the expected linear dependence of V_c on interlayer bare coupling energy and an increase in V_c with channel length. The latter is also to be expected due to increasing size of the condensate region.

With both ends of the condensate region biased as in the BiSJT biasing condition, we find that there is a conserved total critical current, $I_{ctrl} + I_{drv} \cong I_c$. In this symmetric structure, that critical current conservation translates to a conserved critical voltage, $V_{ctrl} + V_{drv} \cong V_c$, as shown in Fig. 2.17b. Indeed, it is expected that the critical current would be conserved for any number of leads and independent of input resistance, but consideration of additional leads is beyond our current simulation capabilities.

This latter result for BiSJT biasing suggests that for equal but opposite voltage drops across the lead pairs on opposite sides of the condensate, equal but opposite currents can be maintained that, individually, are much larger than the conserved critical current for the system as a whole. Although the voltages on either end of the condensate are not evenly split between the top- and bottom-layer leads as for the BiSJT biasing condition, the current counterflow biasing condition with equal but opposite voltages applied across one layer closely approaches this limit. Indeed, we find that much larger currents and voltages can be achieved for the same bare coupling [43]. Moreover, the current flow pattern is entirely consistent with the near-perfect Coulomb drag, where the current in the unbiased drag layer closely approaches the current in biased drive layer, as shown in Fig. 2.18a. We also note that the ultimate collapse of Coulomb drag seen in Fig. 2.18a is not due to eventually exceeding the critical current and inducing oscillatory behavior but due to achieving intralayer voltages large enough to collapse the condensate, as shown in Fig. 2.18b, via a process somewhat akin to raising the temperature.

Physically, independent of the biasing condition, the critical current limit results from the requirement for a stable condensate in the channel to enhance the interlayer transmission. To maintain stability, the population of excitons within the channel needs to be conserved. However, as discussed above, electron–hole pairs are injected (extracted) consistent with external voltages and excite (annihilate) excitons in the channel. Thus, for a stable condensate, the rate of injection (extraction) of electron–hole pairs at the left lead pair must be either balanced by extraction (injection) of electron–hole pairs at the right lead pair as in the case of Coulomb drag biasing or balanced by bare-tunneling-assisted recombination (generation) within the condensate region as in the case of BiSFET biasing or balanced by a combination of the two as for BiSJT biasing. When that balance cannot be maintained, the critical current is exceeded. This required balancing also provides a physical explanation for the dependence of the (conserved) critical current on the interlayer bare coupling strength.

As seen in Figs. 2.16b and 2.17b, the critical voltage can be less than the thermal voltage k_BT/q, perhaps well below k_BT/q, as determined by the interlayer coupling V_b, which is controllably via interlayer dielectric engineering in principle. Subthermal voltages offer the possibility of switching at much lower voltages than CMOS, the latter limited by single-particle thermionic emission. The transition from a high conductance state below V_c to a low conductance state beyond V_c is characterized by a region of negative differential resistance (NDR), and the ON/OFF ratio with regard to DC current may be quite high. However, a factor of 10 would probably due for logic applications considered later. The sharpness of the NDR transition is not known at this point but has been observed to be quite sharp in analogous III–V systems and in Josephson junctions [20]. However, it is the transition from ON to OFF, not the region of NDR per se that is important. Such a novel $I–V$ behavior is sketched in Fig. 2.19, which is the basis for device proposals such as the Bilayer pseudoSpin Field-Effect Transistor (BiSFET) and Bilayer pseudoSpin Junction Transistor (BiSJT) [34, 44]. In some ways, it is analogous to a resonant tunneling characteristic. However, the maximum interlayer conductance is always achieved at zero interlayer voltage, and the width of the "resonance" as limited by the critical voltage can be made very small.

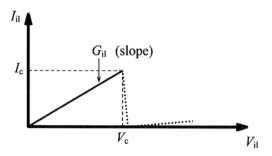

Fig. 2.19 Schematic representation of novel $I–V$ behavior of pseudospintronic devices. The enhanced interlayer conductance G_{il} (ON-state G_{il}) approaches 3/4 the Landauer–Büttiker ballistic limit. The critical voltage V_c can be lower than k_BT/q in principle. The approximate behavior beyond V_c is shown via the *dotted line*

2.3.2.6 Dependencies on Device Parameters

The above work on essential pseudospintronic device physics is sufficient to define device concepts. However, if such devices are ever to be practical, we must consider, beyond the ability to create a condensate, the ability to create it and control it in deeply scaled devices. Here we consider the dependencies of nanoscale condensates on device parameters. Modeling dependencies of critical current I_c and critical voltage V_c and interlayer *ON-state* conductance G_{il} on various parameters including channel length L, effective dielectric constant ε, interlayer separation d, interlayer bare coupling energy V_b, and carrier concentrations n and p are all within the capability of our simulator. For weak bare coupling that is of most interest here, the bare coupling has little effect on the condensate strength [37], but it remains crucial to determining critical currents, of course. Although only semi-quantitative, such studies can identify critical design considerations. The effects of these various parameters can perhaps most easily be interpreted in terms of *bulk* condensate strength as indicated by its critical temperature, $4k_BT_{c,\text{bulk}} \approx E_g(T=0)$, and the ratio $\eta = \lambda_F/\lambda_{\text{tran}}$ that characterizes the abruptness of the condensate region onset λ_{tran} relative to the Fermi wavelength $\lambda_F = 2\pi/k_F$ of injected carriers. λ_{tran} is quantitatively defined as the length over which the LDOS at the equilibrium Fermi level decays from 90 to 10 % of its channel-edge value, as pictured in Fig. 2.20, and k_F is the magnitude of the Fermi wavevector relative to the valley center. The stronger the bulk condensate (the larger $T_{c,\text{bulk}}$), the smaller the expected region required to

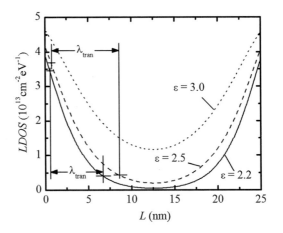

Fig. 2.20 Estimation of the transition region length λ_{tran}, quantitatively defined as the length over which the LDOS at the equilibrium Fermi level decays from 90 to 10 % of its channel-edge value, in a 25 nm channel via the LDOS at $E = 0$ eV (equilibrium Fermi level). The greater the effective dielectric constant ε, the greater the bulk condensate strength, the more abruptly the condensate forms in the channel. For $\varepsilon = 3.0$, the transition region as defined cannot be determined due to the incomplete band-gap formation. Other parameters for these simulations are $d = 1$ nm and $n = p \approx 6 \times 10^{12}$ cm^{-2}. The weak interlayer hopping energy V_b has little effect on the condensate formation, as aforementioned

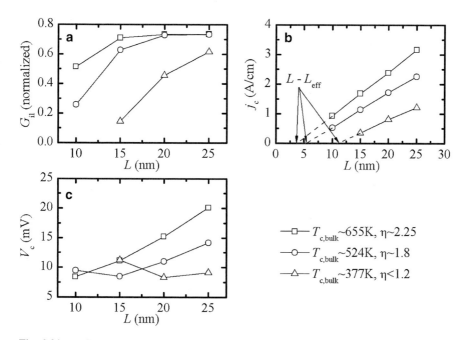

Fig. 2.21 (**a**) Interlayer conductance G_{il} normalized to the Landauer–Büttiker ballistic limit, (**b**) interlayer critical current density j_c, and (**c**) critical voltage V_c with varying dielectric constant ε. The dielectric constants sets are 2.2, 2.5, and 3.0. The other parameters are $n = p \approx 6 \times 10^{12}$ cm^{-2}, $V_b = 0.5$ meV, and $d = 1$ nm. The corresponding T_c are 655 K, 524 K, and 377 K, respectively. For $T_c = 377$ K ($\varepsilon = 3.0$), the band gap still does not fully form in the 25 nm channel, such that transition length λ_{tran} as defined cannot be determined

form the condensate. However, the more abrupt the onset of the condensate region (the larger η), the more reflection is expected and the smaller the interlayer conduction.

Typical trends for G_{il}, I_c, and V_c are shown in Fig. 2.21 [45, 46], where the devices are simulated with varying interlayer dielectric constants but fixed interlayer separation, interlayer bare coupling energy, and carrier concentrations. The interlayer conductance G_{il} in Fig. 2.21a increases and saturates to a constant value with increasing channel length, except for the smallest critical temperature (largest dielectric constant) for which the condensate is not fully formed; a longer channel length would be required to achieve saturation. The linear dependence of I_c on channel length L for a given $T_{c,bulk}$ in Fig. 2.21b implies that the average critical current per unit channel length is constant in terms of an *effective channel length* L_{eff} that is somewhat shorter than the actual channel length, even for the case in which the condensate is not fully formed. Note that the difference between actual channel lengths and effective channel lengths for the fully formed condensates is comparable to twice (two channel ends) the exchange potential coherence radius shown in Fig. 2.5a [26, 32]. Both L_{eff} and critical current per unit effective channel

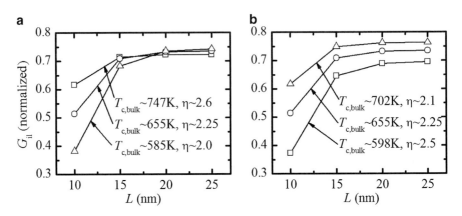

Fig. 2.22 Interlayer conductance G_{il} normalized to Landauer–Büttiker ballistic limit with (**a**) varying interlayer spacing d and (**b**) varying carrier concentrations $n = p$. In (**a**), the interlayer spacings d are 0.8, 1.0, and 1.2 nm for $T_{c,bulk} \sim 747$ K, 655 K, and 585 K, respectively. Other parameters are $\varepsilon = 2.2$, $V_b = 0.5$ meV, and $n = p \approx 6 \times 10^{12}$ cm^{-2}. In (**b**), the carrier concentrations $n = p$ are $\approx 3.8 \times 10^{12}$ cm^{-2}, 6.0×10^{12} cm^{-2}, and 8.6×10^{12} cm^{-2} for $T_{c,bulk} \sim 598$ K, 655 K, and 702 K, respectively. Other parameters are $\varepsilon = 2.2$, $V_b = 0.5$ meV, and $d = 1$ nm

length j_c/L_{eff} increase with condensate strength, consistent with stronger condensates forming more abruptly and with their larger interlayer density matrices allowing larger interlayer currents per unit area as per Eq. (2.11). The critical voltage, $V_c = I_c/G_{il}$, decreases with increasing L initially due to the faster increase in G_{il} than in I_c and increases with increasing L subsequently because of the saturation of G_{il} and continuous increase of I_c. G_{il} is normalized to the Landauer–Büttiker ballistic limit, about 210 S/cm for carrier concentrations of $n = p = 6.0 \times 10^{12}$ cm^{-2}, and appears largely independent of interlayer dielectric constant ε.

We also note that the interlayer conductance appears to saturate to about 75 % of the Landauer–Büttiker ballistic limit for conductance within the leads in Fig. 2.21. This value, however, is not fundamental. It is a result of reflection from the channel where the more abrupt the onset of the condensate as characterized by η, the more reflection. In Fig. 2.22a, where condensate strength is varied via interlayer spacing, the normalized G_{il} saturates more quickly with channel length for a stronger condensate with larger $T_{c,bulk}$; however, a weaker condensate provides a slightly larger value of G_{il} for larger channel lengths due to a smaller η (although ultimately a smaller I_c). In Fig. 2.22b, a stronger condensate is achieved by increasing the carrier concentration, which provides a shorter Fermi wavelength and, ultimately, a decreased value of η. As a result, the normalized value of G_{il} increases with increasing condensate strength (and the Landauer–Büttiker conductance to which it is normalized also increases). However, the differences in saturated normalized conductance are not large, particularly considering that G_{il} are likely larger than those of practically achievable contact conductances that likely will become the primary current limiters. Indeed, even the interlayer conductance of partially formed condensates, as seen in Figs. 2.21a and 2.22, which remain a significant fraction of the Landauer–Büttiker ballistic limit, may be comparable to or greater than lead conductances and remain orders of magnitude larger than the interlayer conductivity absent the condensate.

Collectively these results suggest that the novel interlayer $I–V$ characteristics diagramed in Fig. 2.19 for interlayer transport may well be obtainable at ultralow voltages in nanoscale devices.

2.3.3 Bilayer PseudoSpin Field-Effect Transistor (BiSFET)

2.3.3.1 BiS Junction/Diode

To realize device application of the exciton condensates, a well-localized region with exciton condensation is required. Theoretically, it can be defined by varying anything that affects the condensate strength, such as charge concentration (e.g., defined by the region of *fixed* voltage gates), dielectric environment, and interlayer spatial separation. Assuming that such a condensate can be formed in a bilayer graphene system with a pair of contacts established to the individual graphene layers, a structure much like Fig. 2.8 with floating right leads TR and BR, one already should have what can be called a "Bilayer pseudoSpin (BiS) junction" or "Bilayer pseudoSpin (BiS) diode" (BiS junction or BiS diode) with the novel $I–V$ behavior as depicted in Fig. 2.19. The BiS junction would be expected to behave much like a Josephson junction, producing a symmetric low-voltage-onset DC NDR characteristic with an AC/oscillatory current beyond. In this case, however, the oscillations would be on the multi-THz scale in this case, which would be expected to be filtered out via resistance–capacitance time-constant limits in logic circuits.

2.3.3.2 BiSFET

To create transistors, however, one must have one or more additional terminals to control the $I–V$ characteristic between the original two. For the logic circuits considered in this work, two or more devices are coupled in series. Once the device with the lowest critical current reaches that current, it will turn off, dropping most of the supply voltages across it, leaving the other devices ON. The resulting resistance divider network determines the inter-device node voltage(s) and, thus, the output voltage. Under these conditions, the control input need only vary the critical current and associated critical voltage of the devices, not turn them ON and OFF directly.

The original BiSFET proposal [34] (which can be labeled as BiSFET 1 in this work), as schematically diagramed in Fig. 2.23, incorporated field-effect gating to control charge balance within the condensate, which would change the condensate strength, consistent with the examples of Fig. 2.6. Varying the condensate

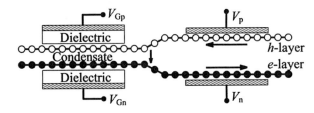

Fig. 2.23 Initial BiSFET design (BiSFET 1), where the electrons and holes to form the condensates are created by and between control gates voltages, V_{Gp} and V_{Gn}, and the interlayer current is driven through the device via separate contacts to each layer, V_p and V_n. Nominally the condensate would be established through work-function engineering, and gate voltages would be applied only to adjust the carrier balance and, thus, the condensate strength—consistent with Fig. 2.6—and, thus, the critical current and voltage, consistent with Fig. 2.21b and c and Eq. (2.11). The direction of current flow is marked with *arrows*. This schematic representation is intended to show basic elements, not physical layout. It is not to scale and vertical dimensions are exaggerated as compared to in-plane dimensions

strengths, in turn, would change the critical current and voltage, consistent with Eq. (2.11) and the examples of Fig. 2.21b, c. The intent was to supply the gate voltages required to create the charged regions between the gates in Fig. 2.23 effectively via work-function engineering. This approach, in turn, would leave the gates available for also adjusting the charge balance via small voltages on the scale of room-temperature $k_BT/q \approx 25$ mV or lower. As only small variations in the critical current should be sufficient to establish which BiSFET reaches its critical current first, even weakened input signals should be able to establish the logic state for serially connected devices as described above, providing signal restoration and/or fan-out.

However, with the improved understanding of the requirements of the electrostatic environment [25], namely, that the environmental window was much smaller than initial estimates, we realized that the gates would have to be moved further away to prevent them from partially screening the exchange interaction. As a result, work-function engineering would not be sufficient to create the required charge concentrations; actual and significant gate voltages would be required for this purpose, such that the same gates would no longer be available for application of low-voltage input signals.

To address this issue, we turned to a separate "control gate" in the design of *BiSFET 2*, as briefly illustrated in Fig. 2.24 and later discussed in this chapter and in more detail elsewhere [47]. This device is essentially two BiS diodes in parallel with gate control of the input resistance to at least one, although all incorporated into a single structure in this figure. In this way, for any given current through the device, the gate(s) can shunt more current to one or the other parallel BiS diodes/current paths, increasing the current in that path so that its critical current will be reached sooner for a given total current. However, once the critical current for that path is exceeded and that condensate region falls in a high resistance state, the current will be shunted back along the other path causing the critical current to be

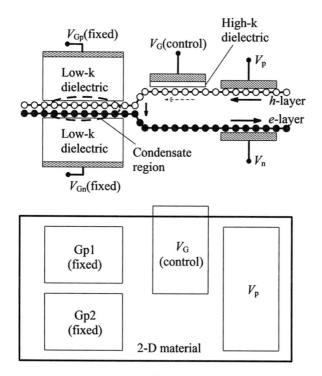

Fig. 2.24 Schematic representation of BiSFET 2, where the gate to control the critical current/voltage, labeled "V_G(control)," is distinct from the gates to create the condensate, labeled "V_{Gp}" and "V_{Gn}" in the side view, which are now separated further from the 2D material layers. Two separate condensate regions are required, and the control gate would cover only approximately half the channel width. The interlayer currents are still driven by separate contacts to each 2D layer of graphene or other 2D materials. This schematic representation is intended to show basic elements, not the physical layout. It is not to scale and vertical dimensions are exaggerated as compared to in-plane dimensions

exceeded along that path as well. As a result, the BiSFET as a whole is now turned OFF. (Simulating this process is beyond our NEGF simulations capabilities but has been demonstrated in circuit simulations with compact device models consistent with our NEGF simulation to date, as to be discussed.) Again, the gate control need only establish which series-connected BiSFET has the lowest critical voltage. Thus, the gate control need only overcome unintended variations in critical current within and among devices and noise, which appears possible even in gapless graphene with ±25 mV gates and perhaps easier in gapped TMDs. Note, however, that it is not sufficient simply to crowd the current into a small region of a larger condensate, as the critical current is expected to be a globally conserved property of each condensate region regardless of where it enters, as already discussed and consistent with the simulation results of Fig. 2.17. To minimize the screening in BiSFET 2, low-dielectric-constant dielectrics above and below the condensate, including perhaps air gaps, have been proposed, in addition to moving the gates used to create the layer charge carriers further away from the condensate [47]. With the gates used to create the electrons and holes required for condensation moved further away, the voltages on these "condensate gates" would need to be on the scale of perhaps 10 V. However, the voltages on these gates would be fixed, and the gates would be well insulated from the circuit and act as virtual grounds with only limited capacitive coupling to the circuit.

2.3.4 Bilayer PseudoSpin Junction Transistor (BiSJT)

Similar in this respect to a bipolar junction transistor (BJT), the proposed BiSJT, as schematically illustrated in Fig. 2.25, is a current-controlled pseudospintronic device. It is motivated by the search for an alternative method to field-effect gating—which is subject to device-to-device variations in the electrostatic potential under identical gate voltages—to allow even lower voltage operation and perhaps simpler device design. It is inspired by the critical current conservation for multiple leads as illustrated in Fig. 2.17. As already discussed in Sect. 2.3.2 for a symmetric device, the effective critical voltage across the drive-end leads, $V_{c,drv}$, is controlled by a smaller voltage applied across control-end leads, V_{ctrl}, such that $V_{c,drv} = V_c - V_{ctrl}$, where V_c is the conserved total critical voltage. The control end of the BiSJT, therefore, plays the same role as the "control gate" in BiSFET 2, differentiating the apparent critical current of the device by an amount sufficient to overcome noise and variations in critical current among devices. However, even at room temperature, applied voltage—vs. applied electrostatic potentials—can be extremely well controlled, so that further voltage reduction becomes plausible. The additional complexity in BiSFET 2 introduced by the control gate and the need for two separate condensate regions also is bypassed, although more contacts to the 2D layers are required. In simulations and the above discussion, the "control" and "drive" ends are interchangeable, but an asymmetry between "control" and "drive" ends may be desirable. The control end alone should not be able to turn OFF the BiSJT. The use of a larger input resistance (e.g., smaller contacts) on the control end can ensure this condition. However, fan-out could serve the same purpose in some cases.

Fig. 2.25 Schematic representation of BiSJT. Device elements are not to scale

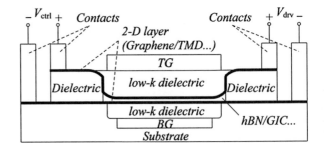

2.4 Fabrication Techniques

If the theory holds, then realization of the pseudospintronic devices may be defined by fabrication technology limitations, not theoretical ones. There are four basic elements required to create a pseudospintronic device: 2D material layers (graphene, TMD, or others) separated by a dielectric tunnel barrier, a region of condensation created and localized by some means, some means of controlling the critical current, and contacts to the individual 2D material layers. (For graphene devices, one can imagine all-graphene layouts where the contacts would just be graphene leads. However, other issues arise such as doping.) Among those, the greatest technology challenge is providing the required dielectric environment necessary for the creation of the condensate and control of the single-particle/bare interlayer tunneling via the interlayer dielectric, both required in BiSFET and BiSJT proposals. Creating good contacts to graphene, especially in BiSJT design, is still a nontrivial process, and the state of the art lags considerably for TMDs. This latter challenge is not specific for pseudospintronic devices, although independently contacting the 2D layers adds to the challenge.

The dielectric requirements for creating a condensate between dielectrically separated 2D layers are fundamentally different from those required for optimal gate control in a MOSFET. Low-k dielectrics are desired rather than high-k. The dielectric constants of the material above and below the dielectrically separated 2D layers generally will be more important than that of the dielectric in between. It also may be the case that the high-frequency dielectric constant is more important than the low-frequency one, which is usually smaller [25]. Despite the latter positive note, air gaps in the vicinity of the condensate may still be necessary. For this purpose, one may start with sacrificial oxides which are then etched away. While certainly technologically challenging, the use of air gaps between the gate and the source and the drain has already been proposed to reduce parasitic gate-to-source and gate-to-drain capacitances in conventional MOSFETs [48], and top- and bottom-gated suspended graphene bilayers have already been reported [49]. The dielectric requirements may be at least less stringent for TMDs as previously noted or even for the graphene-analogous materials of silicene or germanene (2D layers of silicon and germanium, respectively). However, achieving sufficient carrier concentrations in TMDs will require finding pairs of TMD materials with limited *interlayer* valence to conduction band gaps, and the challenges for producing these latter materials may be greater. In contrast, the dielectric requirements for the control gate region of BiSFET 2 scheme are much the same as for "conventional" MOSFETs made in these materials. Moreover, the amount of gate control required is perhaps significantly smaller than conventional CMOS technology. The challenge will be reproducibility.

The interlayer dielectric, while also preferably low-k, must also be thin and provide both limited and reproducible interlayer single-particle/bare coupling, since the critical current varies with the interlayer bare coupling. This requirement suggests naturally layered systems such as hexagonal boron nitride (hBN), TMDs or graphene intercalants (GIC), all of which are being explored currently. The interlayer coupling also must be on the appropriate scale.

Rotational alignment of the opposing 2D layers will also be important as all of the proposed materials have their energy valleys located at the periphery of the Brillouin zone. Both simple theory and computationally intensive numerical simulations indicate that the rotational alignment of the graphene layers, and likely other materials for the same reasons, will have little impact on the spontaneous condensate [32]. However, rotational alignment will certainly affect the single-particle tunneling and, thus, the critical current through the pattern of single-particle coupling. Indeed, simply rotating graphene layers may be enough to decouple them sufficiently to produce the condensate, and no interlayer dielectric may be required. However, with the corresponding increase in interlayer capacitance, gating the required charge densities to achieve room-temperature condensation would become more problematic.

Use of graphene intercalants as the interlayer dielectric would be advantageous in this regard, as one starts with crystallographically aligned bilayers into which the intercalant is then chemically inserted. As intercalants also heavily dope the graphene layers [50], their use could lead to the ability to produce the condensate with just one gate to create the condensate, which would be used to strongly invert one layer if while reducing the carrier concentration in the other in the process.

An illustrative rough layout proposal considering such factors is illustrated in Fig. 2.26 for BiSFET 2. While daunting, having technology as the primary roadblock to be chipped away over time, as for CMOS previously, would still represent a considerable improvement over the roadblock of intractable physics of thermionic emission and source-to-drain and channel-to-gate tunneling currently standing in the way of long-term progress in CMOS.

2.5 Logic Applications of Pseudospintronic Devices

We have found that it should be possible to create a full array of logical gates using BiSFETs or BiSJTs and novel clocked power supplies. In this section, we provide examples for each, beginning with compact device models and ending with circuit-level SPICE simulations.

2.5.1 BiSFET Logic

2.5.1.1 BiSFET Inverter

We symbolically represent the BiSFET 1 and BiSFET 2 as shown in Fig. 2.27. "N-type" and "p-type" versions of BiSFET 2 are differentiated by the use of a negative control voltage applied to an n-type device to decrease its critical current

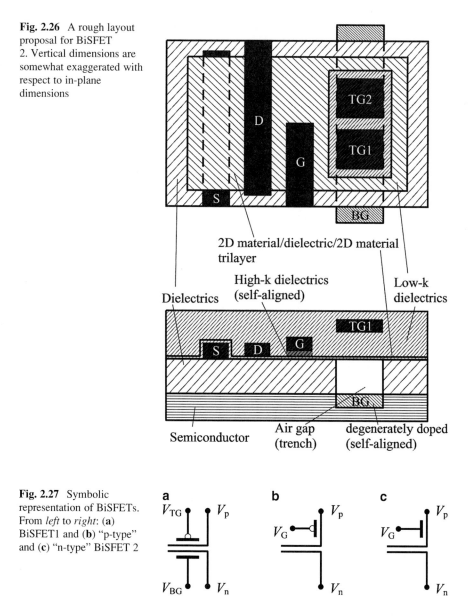

Fig. 2.26 A rough layout proposal for BiSFET 2. Vertical dimensions are somewhat exaggerated with respect to in-plane dimensions

Fig. 2.27 Symbolic representation of BiSFETs. From *left* to *right*: (**a**) BiSFET1 and (**b**) "p-type" and (**c**) "n-type" BiSFET 2

(which will ultimately lead to it being turned OFF) and the use of a positive control voltage applied to a "p-type" device to decrease its critical current, following the convention for CMOS devices. However, there is no need to invert the polarity of carriers in the channel for conduction, nor do "n" and "p" necessarily denote carrier type for conduction.

If BiSFETs are inserted into a conventional CMOS inverter circuit, a memory element results instead of a BiSFET inverter [51]. As schematically illustrated in Fig. 2.28a, there will be three separate operating points given fixed power supply voltages $\pm V_{sup}$. Two of the operating points will have one BiSFET or the other within its low conductance/OFF state beyond the NDR onset voltage V_c, while the remaining BiSFET will remain in its high conductance/ON state. The output voltage will be either high ("1") or low ("0"). For the remaining operating point, both BiSFETs will be in their low conductance/OFF state beyond their NDR onset voltages. The output will be in a perhaps not entirely stable intermediate state and with little available current to drive a load. Moreover, altering the input voltage to produce limited changes in the NDR onset voltages will have essentially no effect on the high- or low-output states and no apparent useful effect on the intermediate state. Such static operating points are common in NDR devices. Although such a static voltage supply scheme is not applicable for a BiSFET inverter, at least this indifference to input voltage with a fixed supply voltage is compatible with memory applications as noted [51].

Output control, however, can be achieved using a dynamic voltage supply scheme. If the inputs are set *first* to differentiate the critical currents of the two BiSFETs and the power supplies are then ramped up, the desired output logic state will be obtained, as schematically illustrated in Fig. 2.28b. In this way, initially there is only one possible operating point at $V_{out} = 0$ with both BiSFETs in their high conductance/ON state. As the balanced supply voltages are ramped, initially both BiSFETs remain in their high conductance/ON states and the current increases, while V_{out} remains approximately zero for matched BiSFETs with the same contact and ON-state interlayer resistances. However, at some point with the increasing supply voltages, the BiSFET with the lower critical current reaches its critical current and falls into its low conductance/OFF state beyond its critical voltage. With the same current funneled through both BiSFETs, the other BiSFET can never reach its critical current and, thus, remains in its high conductance/ON state. As a result, V_{out} follows the supply voltage connected to the ON device until $\pm V_{clk,peak}$ is reached and the desired inversion operation is achieved, although delayed by the power supply ramp time. Moreover, once the clock voltage is raised and the gate output is set, the input signal can be turned off with no effect. Each gate doubles as a latch since the output = HIGH and output = LOW states are stable with the now static supply as described above. The next operation cycle will begin when V_{clk} is ramped up again. Therefore, a BiSFET inverter needs a clocked power supply and is actively switches at the leading clock edge. Moreover, if multiple stages are cascaded to create complex logic gates, a delay in the clock signal between adjacent states is necessary to allow for the time lag between input and output of a single stage. Thus, BiSFET logic requires a sequential clocking scheme [34, 51]. Finally, we note that a BiSFET follower/buffer gate (BUF) can be created simply by flipping the n-type and p-type BiSFETs. Buffers can be used to synchronize inputs and/or outputs of elements with different numbers of stages.

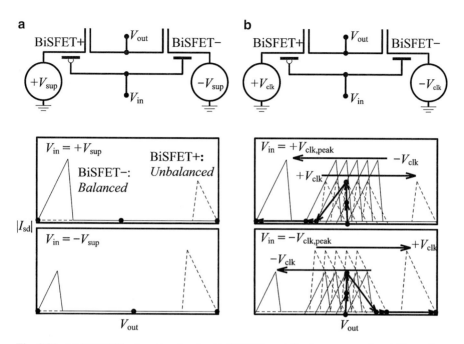

Fig. 2.28 (**a**) CMOS-like inverter circuit with BiSFETs and fixed power supplies $\pm V_{\text{sup}}$ and an associated output voltage V_{out} that is essentially independent of input voltage V_{in}. The *solid* and *dashed curves* illustrate the interlayer current magnitude vs. common source/output voltage V_{out}, for "BiSFET$-$" and "BiSFET$+$," respectively. The intersection points of the paired BiSFET *I–V* characteristics represent possible circuit operating points/values of V_{out}, of which there are three for each input voltage. These operating points remain largely fixed independent of V_{in}; there is no path between operating points with changes in V_{in}. This behavior offers opportunities for memory, but not logic. (**b**) CMOS-like inverter circuit with the BiSFETs and power supply ramped to $\pm V_{\text{clk},}$ $_{\text{peak}}$ *after the input voltage is set.* The *solid* and *dashed curves* move with respect to V_{out} with increasing $\pm V_{\text{clk}}$ and therefore so does the intersection/circuit operating point/value of V_{out} for logic application, as shown by the *arrows*

2.5.1.2   BiSFET Model

For circuit modeling purposes, simple compact device models for both BiSFET 1 and BiSFET 2 schemes were created to capture essential device elements and behavior, as shown in Fig. 2.29. These models were meant be qualitatively accurate and conservative when in doubt in terms of desired device operation, but not to be precise as would be required for conventional devices. Fractional changes in performance are of secondary importance at this stage of development.

For BiSFET 1, the models include capacitances C_{tg} and C_{bg} between each gate and the adjacent graphene layers, capacitances C_{il} between the two graphene layers, and intralayer quantum capacitances C_q associated with moving the Fermi level with respect to the band edge for each layer. Contact resistances R_c are included. Finally, a model of the interlayer current I_{il} as a function of the interlayer voltage (Fermi level difference) and, for gating in BiSFET 1, charge balance/imbalance is provided.

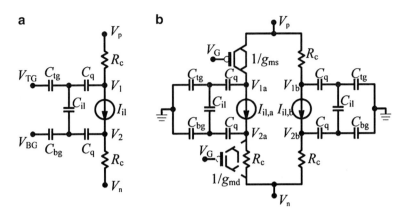

Fig. 2.29 BiSFET circuit models. (**a**) For BiSFET 1 and (**b**) for BiSFET 2

For BiSFET 2, gating has been modeled as an effective variation of the contact resistance of one (or more) leads, $1/g_{m,S}$ and perhaps $1/g_{m,D}$. In circuit simulations, to date, this gating has been modeled after a graphene FET with thermal and inhomogeneous smearing of the conductance minimum. The capacitive coupling C_{tg} and C_{bg} to the gates used to create the condensate in BiSFET 2 are substantially reduced due to their much greater separation from the layers. Moreover, while these gates have fixed large voltages applied, they should act as virtual grounds in the circuit and, thus, have little effect on power consumption. However, charging and discharging the control gate, which is inherently strongly coupled, becomes a new source of power consumption.

The current flow including the transition over the NDR region has been modeled via smooth functions as [34, 51]

$$I_{sd} = G_0 V_{sd} \left[1 + \frac{|V_{sd}|/V_c}{\exp(1 - |V_{sd}|/V_c)} \right]^{-1} \tag{2.17}$$

where I_{sd} is the source-to-drain interlayer current, G_0 is the Landauer–Büttiker ballistic conductance for injection into these highly charged regions, V_{sd} is the source-to-drain voltage drop across the condensate $-(E_{F,s} - E_{F,d})/q$ (distinguished from the source-to-drain voltage drops across the contacts), and V_c is the critical voltage at which NDR onset occurs.

For BiSFET 1, where charge imbalance induced through C_{tg} and C_{tg} is employed to tune the critical voltage, V_c is modeled as [34]

$$V_c = V_{c,n=p} \exp\left(-10 \frac{|p-n|}{p+n} \right) \tag{2.18}$$

For the BiSFET 2 compact model, the resistance of the gated lead as a function of the control gate voltage is assumed to be limited by thermal and inhomogeneous smearing about the Dirac point as [51]

$$R^{-1} = \sqrt{\sigma_g^2 + \sigma_{min}^2} \qquad (2.19)$$

where $\sigma_g = 8q^2 E_F W/(v_F h^2)$ is the quantum conductance due to the number of modes in a graphene layer of width W and Fermi level E_F, h is the Planck's constant, q is the electron charge, v_F is the fixed Fermi velocity magnitude of electrons in graphene, and W is the channel width [34]. σ_{min} represents the minimum conductivity due to thermal and inhomogeneous smearing about Dirac point. The Fermi level in graphene under the gate is calculated using an equivalent series capacitance network of gate oxide and graphene quantum capacitance. In practice, only a ± 50 mV range ($2V_{clock}$ in simulations to follow) is considered about the maximum resistance.

2.5.1.3 BiSFET Logic and Circuit Simulation

The BiSFET inverter has much the same structure as the CMOS inverter but a different clocking scheme. BiSFET NAND/NOR gates also can be similarly designed following the CMOS NAND/NOR gates. An example BiSFET 2 NAND gate is shown in Fig. 2.30, where the critical current of the "$2W$" device is twice of that of the "W" device, nominally by doubling the channel width. The maximum current allowed through the two $2W$ BiSFETs in series will be reduced by either negative $V_{in,A}$ or negative $V_{in,B}$ or both, while the total current of the two W devices in parallel is approximately the same when $V_{in,A}$ and $V_{in,B}$ are different or increased when both $V_{in,A}$ and $V_{in,B}$ are negative. Therefore, the $2W$ BiSFET(s) with the reduced critical current will be turned OFF first and the output will follow the $+V_{clk}$ going high. On the other hand, only two positive inputs will effectively increase the maximum current allowed through the two $2W$ BiSFETs in series, while the total current of the two W BiSFETs in parallel is reduced. Therefore, the output will follow the $-V_{clk}$ going low after the two W BiSFETs are turned OFF. Thus, the logical operation of a NAND gate is achieved.

The NOR gate can be designed similarly with two W n-type BiSFETs in parallel and connected to $-V_{clk}$ and two $2W$ p-type BiSFETs in series and connected to $+V_{clk}$. AND and OR gates can be achieved by switching the polarity but not sizes of the n-type and p-type BiSFETs of NOR and NAND gates, respectively. In general, arbitrary logic can be achieved by a NOT gate plus only one out of AND, OR, NAND, and NOR gates. However, unlike CMOS logic, BiSFET logic does not in principle have a preferred gate selection because of the ability to make "n-type" and "p-type" devices with leads to either carrier type. In CMOS logic, NAND gates are preferred because it takes much larger area for a PMOS to carry double the current of the NMOS devices due to the lower mobility of holes.

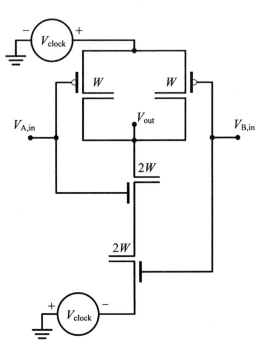

To evaluate the circuit performance, Verilog-A model-based circuit-level SPICE simulations were performed using the device models described in Fig. 2.29. Except for that of the gated lead in the compact model of BiSFET 2, the contact resistances were taken to be 320 Ω μm at zero gate voltage, corresponding to a Fermi level of $|E_F| = 50$ mV and a carrier concentration of about $2 \times 10^{11}/\text{cm}^2$ in the leads. The interlayer conductance is approximately 200 S/cm, or equivalently the interlayer resistance is approximately 50 Ω μm. The contact resistance of the gate lead was modeled consistent with Eq. (2.19) for the simulation results shown here. Even with a large $\sigma_{\min}$ producing an accompanying small ON/OFF ratio of 1.06, successful circuit operation has been achieved in simulation absent device-to-device variation. The interlayer capacitance was taken to be ≈ 0.35 fF/μm corresponding to a SiO$_2$-*equivalent* oxide thickness (EOT) of 1 nm and a channel length of 10 nm, given that the injected carriers penetrate only peripherally into the condensate, i.e., only the charge in the transitional region varies with current flow. The capacitance to control gate was taken to be ≈ 0.41 fF/μm corresponding to an EOT of 0.85 nm. The capacitance to each of the gates used to create the condensate was taken to be 0.03 fF/μm conservatively, corresponding to an EOT of 10 nm and representing a low-k gate dielectric environment shown in Fig. 2.26 with somewhat physically closer gates, which may not be the case. The interlayer current–voltage relation through the condensate is that of Eq. (2.17). The width W of the condensate regions, by which all of the above was scaled, was taken to be 20 nm unless otherwise noted. The peak value of the clocked power supply voltage $V_{\text{clk,peak}}$ was taken to be 25 mV in all cases for illustrative purposes, approximately $k_\text{B}T/q$ at $T = 300$ K.

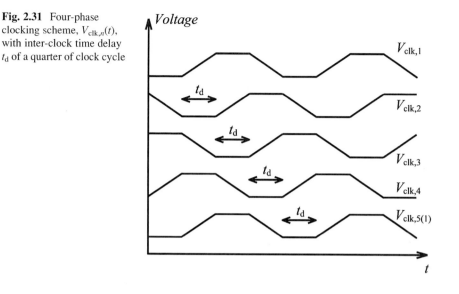

Fig. 2.31 Four-phase clocking scheme, $V_{clk,n}(t)$, with inter-clock time delay t_d of a quarter of clock cycle

The clocks are taken to have a quarter-cycle delay between stages as illustrated in Fig. 2.31. Simulation results of a BiSFET 2-based inverter and NAND gate are provided in Fig. 2.32, where dual 10 GHz, ± 25 mV clock signals are applied. The energy per BiSFET per clock cycle is ~10 zJ (1 zJ = 10^{-21} J) in the inverter. The total energy per clock cycle is ~140 zJ for the NAND gate (which has six times the device area of a single BiSFET). Therefore, the average energy per BiSFET per clock cycle in these simulations is on the order of tens of zeptojoules, orders of magnitude lower than end-of-the-road-map CMOS [52].

A 1-bit ripple-carry adder, with sum and carry, incorporating smaller logic elements of XOR, NAND, and buffer gates, was also simulated based on BiSFET 1 in earlier work. However, BiSFET 1 and BiSFET 2 behave essentially the same in logic circuits. The results are shown in Fig. 2.33, illustrating the ability to create complex, multistage BiSFET logic circuits.

2.5.2   BiSJT Logic

2.5.2.1   BiSJT Inverter and Model

A BiSJT inverter is shown in Fig. 2.34. Like the BiSFET inverter, the clock is ramped up *after* the input signal is set, and adjacent gates are clocked sequentially with a nominally quarter-cycle delay, with each gate serving as its own latch once set until re-clocked. If a positive input voltage is provided, the effective I_c seen at the drive end of BiSJT 1 is reduced, and that of BiSJT 2 is increased (BiSJT1 and BiSJT2 are just different devices in the same circuit, not to be confused with the different varieties of BiSJTs). When the two opposite-polarity clocks are ramped

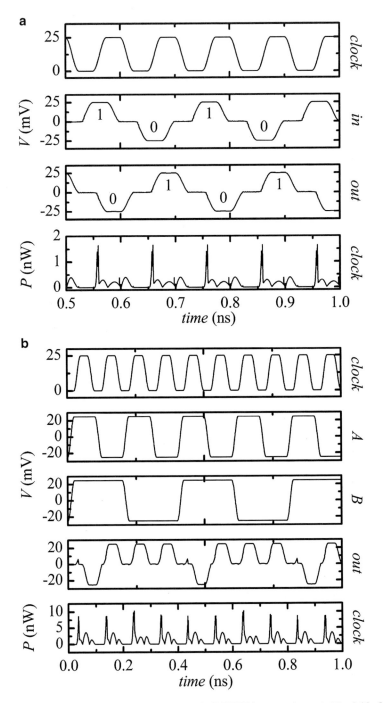

Fig. 2.32 SPICE-level simulated response of (**a**) the BiSFET inverter shown in Fig. 2.28. (**b**) The NAND gate shown in Fig. 2.30, both of which employ BiSFET 2 with dual 10 GHz, ±25 mV peak clocked power supplies ±$V_{clock}(t)$ driving a four-identical-inverter load (fan-out of four). The input signal was also taken from a preceding inverter in a similar fashion. The energy per operation/clock cycle per BiSFET is about 10 zJ in for the inverter, and the total energy consumption per operation cycle is about 140 zJ in for the NAND gate. The instantaneous power was conservatively calculated as the *magnitude* of the product of the current times voltage summed over both supplies. There actually is some power return during part of the clock cycle in simulations

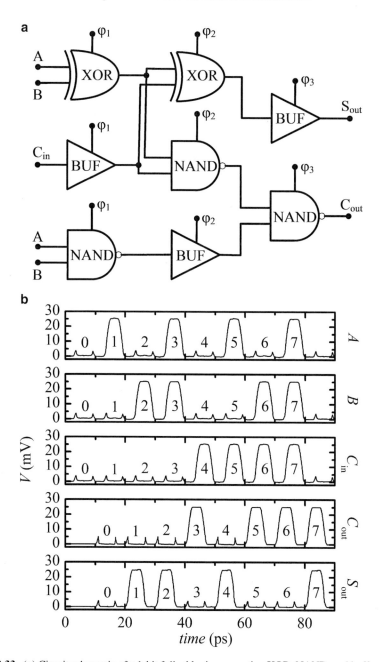

Fig. 2.33 (a) Circuit schematic of a 1-bit full adder incorporating XOR, NAND, and buffer gates, with clocked power supplies φ_1, φ_2, and φ_3. The buffer is employed to synchronize signals. (b) Output of the 1-bit adder, including sum S_{out} and carry C_{out}. These results were obtained with BiSFET-1-based logic elements and with 100 GHz clocked power supplies separated by a 2.5 ps delay. The higher speeds were allowed because the parasitic contact resistances were not considered in these early proof-of-concept simulations

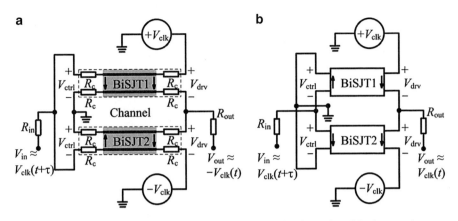

Fig. 2.34 (**a**) A low-output BiSJT inverter. Sequential clocking is employed for input and output signals with a delay $\tau = T/4$ namely. The *arrows* show the direction of the interlayer current flow. Contact resistances R_c are explicitly shown here, while capacitances, which affect operation speed and charging energy but not basic logic design, are implicit (but explicit in compact model of Fig. 2.35 below). The gates to create the condensate are also omitted here (and also included in Fig. 2.35). (**b**) A BiSJT buffer gate is achieved by switching the input and grounded nodes on the control side of the BiSJTs. Here the contact resistances are also implicit

Fig. 2.35 The BiSJT compact model used in circuit simulations, which corresponds to the boxed "BiSJT1" or "BiSJT2" in Fig. 2.34b

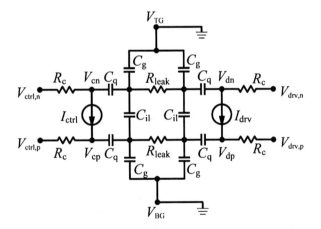

up, BiSJT 1 reaches its now-smaller critical current and turns OFF for both the control and drive ends. As a result, the output is coupled to the negative clock through BiSJT 2 which remains ON having never reached its now larger critical current. Consequently, the output is pulled negative. Similarly, if a negative input is provided, a positive output is achieved.

In contrast to BiSFETs, however, inputs to both the control and drive ends of the BiSJT are coupled to the condensate. The maximum input current to the control end must be limited to below the critical current. Nevertheless, if the output state is still set when the input signal is applied, the input signal may be able to hold output state even absent the clock signal to the output side given hysteresis in these NDR devices.

Thus, in addition to the input being set before the clocks begin their ramp up, the clocks also should be fully ramped down before the next input begins its ramp up. Although not required, such was already archived in BiSFET circuits in principle, but the timing was very close and subject to slight variances in time delays. However, more care should be taken for BiSJT circuits.

Another difference from the BiSFET logic is that there is always one low-resistance path on the input end (also analogous to the BJT logic), making BiSJT less suitable for static memory applications as compared to the BiSFET. However, logic calculations remain ultralow power, as shown in the SPICE simulation results to follow. Indeed, absent concerns over electrostatic variations in the gate region, which are much more difficult to control than voltage variations, it may be possible to operate the proposed BiSJT at substantially lower voltages than the proposed BiSFET for still lower energy operation.

For the inverter element of Fig. 2.34a, the shown input side external resistance represented by R_{in} combined with the input-end contact resistances of the BiSJT (both perhaps defined by contact dimensions) must be large enough to prevent the I_{ctrl} alone from reaching the critical current of either BiSJT. Similarly, the shown load resistance represented by R_{out} combined with the output side contact resistances must be large enough to, after ramping the clocks, prevent I_{drv} of the nominally ON BiSJT from reaching its critical current. Of course, the contact resistances themselves must also be small enough to allow the critical current to be reached in one device during the clock ramping. A signal follower/buffer can be created similarly by exchanging the signal and ground terminals on the input end, as shown in Fig. 2.34b.

The compact device model used to describe the BiSJT is shown in Fig. 2.35. The charge density in the center of the condensate region, even with exciton current flow, is not a function of the control and drive voltages since the injected carriers only penetrate a short distance into the channel. Otherwise, the Fermi level within the channel lies at the center of the associated anti-crossing band gap, such that the (differential) quantum capacitance, the interlayer capacitance and the capacitive coupling to the gates used to create the condensate are all essentially zero on the scales of interest. Consequently, the C_{il} represent interlayer capacitances in the leads plus a short extension of a few nanometers (taken as 5 nm here for simulation purposes) into the channel. The C_q are quantum (density of states) capacitances for each graphene layer over the same region as C_{il}. The R_c represent the contact resistances between metal contacts and the 2D layers. The R_{leak} account for leakage currents mediated by tunneling of normal currents through the condensate-induced band-gap region, which would be quite limited for a fully formed condensate, and normal current flow around the edges of the condensate region, which could be more significant. The C_g represent (weak) capacitive coupling from the relatively remote gates of fixed voltage, which would be used to electrostatically dope the 2D layers, again subject to penetration of the charge carriers only a short distance into the channel. The near shorting of the 2D layers at each end before the total critical current is reached and the collapse thereafter are modeled by qualitatively abrupt, but quantitatively smooth, functions $I_{\text{ctrl/drv}}(V_{\text{cn}},$ $V_{\text{cp}}; V_{\text{dn}}, V_{\text{dp}})$ as

$$I_{\text{ctrl}} = G_0\left(V_{\text{cn}} - V_{\text{cp}}\right)K\left(\left|V_{\text{cn}} - V_{\text{cp}} + V_{\text{dn}} - V_{\text{dp}}\right|\right) \tag{2.20}$$

$$I_{\text{drv}} = G_0\left(V_{\text{dn}} - V_{\text{dp}}\right)K\left(\left|V_{\text{cn}} - V_{\text{cp}} + V_{\text{dn}} - V_{\text{dp}}\right|\right) \tag{2.21}$$

$$K(V) = e^{-\left[\max\left(V, V_{\text{c,ttl}}\right) - V_{\text{c,ttl}}\right]^2 / V_0^2} \tag{2.22}$$

where G_0 is the ON-state/enhanced interlayer conductance, approaching the Landauer–Büttiker limit as discussed previously. In these simulations where the device continues to be modeled as symmetric, $V_{\text{c,ttl}} = I_{\text{c,ttl}}/G_0$ is the *total* critical voltage, which is a function of the adjustable bare interlayer coupling, as well as condensate region size. V_0 controls the sharpness of the NDR region, which is perhaps more important to the reliability of the numerical simulations than to the physical results.

2.5.2.2   Complex BiSJT Logic and Circuit Simulation

As discussed previously, to the extent that device-to-device variation can be neglected and current noise is limited, the magnitude of the input signal does not matter toward determining which BiSJT or BiSJTs turn(s) off and which remain (s) on, only the sign of the input. This behavior allows for fan-out, where the more gates are driven by a single gate, the less current flows to each single second-stage gate. In addition, if combined with multiple inputs, this behavior allows for simple realization of majority gates (MAJ) and AND/OR gate by essentially averaging three inputs together at the control-end input to a buffer, as shown in Fig. 2.36 along with fan-out. The complementary NAND/NOR gates can be created similarly by averaging the inputs to an inverter. Consider the MAJ gate of Fig. 2.36b as an example. If two or three of the three *balanced* inputs are HIGH, the effective input is HIGH, even if with two possible strengths. The output is therefore HIGH. Similarly, the output is LOW if two or more of the balanced inputs are LOW. The AND and OR logic can be obtained from the MAJ logic as AND(A,B) = MAJ (A,B,0) and OR(A,B) = MAJ(A,B,1), where the 0 or 1 reference state is provided directly from the clock of the previous stage, which is synchronized to the other input signals, if through some limiting resistances. Generally speaking, the level of the three input signals, resulting/achieved through fan-out or input resistances, must be of roughly the same scale. At least no one input signal can be larger than the other two combined. Input buffer gates could be used to balance signals if not otherwise possible. One also can imagine more than two distinct physical contacts to the condensate itself but that is, again, beyond the current capabilities of quantum transport simulation. Importantly, in addition to being qualitatively simple, this version of MAJ/AND/OR/NAND/NOR has the same device area as a simple inverter or buffer. This input arrangement does mean that the outputs of previous

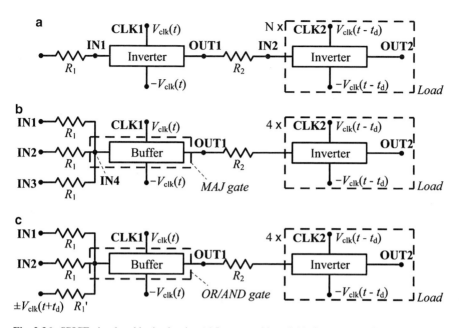

Fig. 2.36 SPICE-simulated logic circuits. (**a**) Inverter with variable fan-out $=N$, (**b**) majority gate with four-inverter fan-out loads, and (**c**) OR and AND gates with four-inverter fan-out loads. The majority gate consists simply of a buffer with three different input signals tied to its single input node. One also can imagine more than two distinct physical contacts to the condensate itself, but that is beyond this work as well as our current capabilities of quantum transport simulation. The OR and AND gates are formed by using the positive or negative clocks, respectively, of the previous stage as one of the three inputs, with a slightly increased input resistance R_1' representing the value of R_1 plus that of a series ON-state BiSJT to match the current level of the other two inputs. Simulation results for these systems are provided in Fig. 2.37. An inverted majority gate and NOR and NAND gates can be obtained simply by replacing the buffer in (**b**) and (**c**), respectively, with an inverter. $t_d = \tau/4$ where τ is the clock cycle

logic gates are coupled together, which, beyond requiring some circuit design care, provides additional leakage paths. However, particularly with the possibility of further reduced voltage, low power can be maintained. Note that the sequential clocking scheme provides input–output state isolation essentially independent of any leakage current between the inputs and outputs.

For circuit simulations, with about 20 nm long and 45 nm wide ($3 \times$ condensate channel length L) leads and an interlayer effective oxide thickness (EOT) of 1.77 nm (if closer but of lower dielectric constant), $C_{il} = 21.9$ aF for each side of the BiSJT. With $n = p \approx 6 \times 10^{12}$ cm^{-2} over the same region, $C_q = 86.6$ aF each. Assuming a (at least) 5 nm EOT for the gate dielectrics used to create the condensate in a conservative estimate and again an approximately 5 nm overlap under the gates into the condensate region for each lead, $C_g = 1.5$ aF each. Consistent with simulations, G_0 is taken to be 75 % of the Landauer–Büttiker ballistic limit for the graphene doping levels $n = p \approx 6 \times 10^{12}$ cm^{-2} or $G_0 = 0.70$ mS [46].

Both the R_c and R_{leak} are taken to be 6 kΩ (~270 Ω μm, approximately four times the ON-state interlayer resistance $1/G_0$). $V_{c,ttl}$ is taken to be 1.2 mV. (The abruptness factor V_0 is then taken as 0.4 % of $V_{c,ttl}$.) For the fan-out circuit of Fig. 2.36a and the majority gate circuit of Fig. 2.36b, $R_1 = 30$ kΩ and $R_2 = 15$ kΩ. For the OR/AND gates of Fig. 2.36c, R_1 and R_2 remain the same as the MAJ gate, but R_1' is raised to 43 kΩ, which is the series combination of resistance of R_1 and an ON-state BiSJT in these simulations, to match the current level of the other two inputs.

Figure 2.37 includes simulation results for the BiSJT circuits of Fig. 2.36. In Fig. 2.37a, the fan-out varies from 1 to 4; in Fig. 2.37b, c, the fan-out is fixed at 4. In Fig. 2.37a, the step-like shape of the input end voltage of the BiSJTs is due to the sudden change of input resistance due to the turning off of one of the two BiSJTs. In Fig. 2.37b, despite variations in the input signal strength, the output signal OUT1 depends only on the sign of the input signal IN4, not the magnitude. The clock signals have a peak magnitude of 15 mV, a 50 GHz frequency, and a one-quarter-cycle delay between gates. IN1-3 in Fig. 2.37b and IN1-2 in Fig. 2.37c are driven by buffer gates not shown. The average energy consumption per BiSJT per clock cycle is ~60 zJ for the MAJ based logic and ~30 zJ for the inverter, comparable with the BiSFET logic circuits.

2.5.3 Comment on Energy and Noise (BiSJT)

Despite the low voltages and energies considered here, BiSFET/BiSJT logic still appears to remain in a realm of reliable operation in terms of fundamental noise and energy benchmarks. For Johnson–Nyquist (thermal) noise associated with the random motions of the carriers due to scattering [53, 54], for example, taking a 10 kΩ resistance R and a 10 GHz clock representing the bandwidth Δf gives a value of voltage fluctuations of $\Delta V = \sqrt{4 k_B T R \Delta f} \cong 0.4$ mV, which remains small compared to the BiSFET/BiSJT critical voltage considered here for the devices as a whole. These particular input parameters are perhaps more representative of TMD systems that we have been looking at recently than of graphene systems. For graphene systems, however, the frequency would be perhaps larger, but the resistance would be much less (and only the resistance associated with the phonon system is relevant, not the Landauer–Büttiker contact resistance). We also note that cross-talk noise should scale with the overall operating voltage of the circuit. Therefore, as a percentage, it should remain constant and, thus, no more problematic for low-voltage circuits than high-voltage ones. In terms of energy, the few tens to 100-plus zJ consumed per logic gate in simulations to establish a binary logical output, although small, nevertheless remain large compared to Landauer's limit for irreversible computation, $k_B T \ln(2) \approx 2.9$ zJ at room temperature [55].

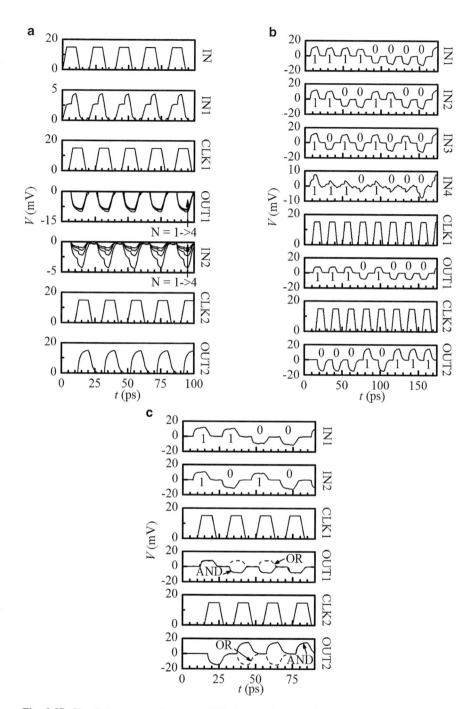

Fig. 2.37 Simulation results for the BiSJT circuits of Fig. 2.36, (**a**)–(**c**) here for (**a**)–(**c**) of Fig. 2.36, respectively. Note that in (**b**) the output signal OUT1 depends only on the sign of the input signal IN4, not the magnitude, showing signal restoration. In (**a**), the fan-out N is from 1 to 4. In (**c**), the *dotted* and *solid lines* are for OR and AND gates, respectively. The clock signals have a peak magnitude of 15 mV, a 50 GHz frequency, and a one-quarter-cycle delay between gates. IN1-3 in (**b**) and IN1-2 in (**c**) are driven by prior buffer gates not shown

2.6   Outlook and Conclusion

The goal of the pseudospintronic devices is to restart the "road map" beyond-CMOS. Of course, that is the goal of all "beyond-CMOS" work. However, with zeptojoule-scale switching energies and novel functionality, pseudospintronic devices could add a substantial amount to the road map.

Moreover, only applications to logic circuits have been considered; there should be more opportunities if pseudospintronic devices including but not limited to those discussed here can be realized. Some of these opportunities are suggested by the similarities of the I–V characteristics of the proposed BiSFETs/BiSJTs to those of gated resonant tunneling diodes but with lower voltages and larger ON/OFF ratios and to those of Josephson junctions but with gating and potentially room-temperature operation. The possibility of multiple contacts to the same coherent region of condensation or gate-controlled coupling and decoupling of coherent regions of condensation, even if at higher voltages, also suggests the possibility of quantum computing applications.

Theory continues to suggest that room-temperature superfluidity is possible in dielectrically separated graphene layers, although challenging. If not successful in graphene-based systems or even if successful, analogous systems of 2D materials such as TMDs [16], silicene, and germanene may prove better suited to the task of creating room-temperature condensates, even if for now these materials are more technologically challenging. Work on TMDs is underway by the authors and colleagues. In part, the experimental efforts of Ref. [20] in III–V systems were informed by theory of [40], and that theoretical effort was supported as part of the BiSFET/BiSJT effort. Qualitatively, almost all of the essential physics of interest has been seen experimentally in analogous physical systems including III–V double quantum wells and Josephson junctions, except, of course, room-temperature exciton condensation at the time of writing.

However, pseudospintronic devices are based on exotic physics in exotic material systems based on imperfect theory. Ultimately, the issue must be resolved experimentally. The challenges to observing the condensate are substantial. If we can observe the condensate in the laboratory, much more work will be required to extend those results to commercial production. Moreover, even if successful, initial efforts may not run at voltages comparable to $k_B T/q$ or lower due to technological limitations. However, technology would be expected to improve over time, much as it did for CMOS.

The latter scenario, however, is precisely the above-stated goal of this work, to restart the "road map," i.e., to turn making substantial long-term progress into a technological challenge rather than the physical impossibility that it is for CMOS. Moreover, room-temperature superfluidity in nanoscale devices could offer new application avenues, not just ongoing progress along existing avenues.

Acknowledgment This work was supported by the South West Academy of Nanoelectronics (SWAN), which in turn is supported by the Semiconductor Research Corporation (SRC) and the National Institute of Standards and Technology (NIST) through the Nanoelectronics Research Initiative (NRI).

References

1. C.J. Davisson, The diffraction of electrons by a crystal of nickel. Bell Syst. Tech. J. **7**, 90 (1928)
2. J. Frenkel, On the transformation of light into heat in solids. I. Phys. Rev. **37**, 17 (1931)
3. J. Frenkel, On the transformation of light into heat in solids. II. Phys. Rev. **37**, 1276 (1931)
4. L.N. Cooper, Bound electron pairs in a degenerate Fermi gas. Phys. Rev. **104**, 1189 (1956)
5. S.N. Bose, Plancks Gesetz und Lichtquantenhypothese. Z. Phys. **26**, 178 (1924)
6. A. Einstein, Quantentheorie des einatomigen idealen Gases. Sitzungsberichte der Preussischen Akademie der Wissenschaften **1**, 3 (1925)
7. J. Bardeen, L.N. Cooper, J.R. Schrieffer, Microscopic theory of superconductivity. Phys. Rev. **106**, 162 (1957)
8. D.R. Tilley, J. Tilley, *Superfluidity and superconductivity* (Institute of Physics Publishing, Bristol, 1990)
9. P. Kapitza, Viscosity of liquid helium below the lambda-point. Nature **141**, 74 (1938)
10. J.F. Allen, A.D. Misener, Flow phenomena in liquid helium II. Nature **142**, 643 (1938)
11. D.D. Osheroff, R.C. Richardson, D.M. Lee, Evidence for a new phase of solid He^3. Phys. Rev. Lett. **28**, 885 (1972)
12. I.O. Kulik, S.I. Shevchenko, Excitonic 'superfluidity' in low-dimensional crystals. Solid State Commun. **21**, 409 (1977)
13. J.P. Eisenstein, A.H. MacDonald, Bose-Einstein condensation of excitons in bilayer electron systems. Nature **432**, 691 (2004)
14. H. Min, R. Bistritzer, J.-J. Su, A.H. MacDonald, Room-temperature superfluidity in graphene bilayers. Phys. Rev. B **78**, 121401(R) (2008)
15. D.S.L. Abergel, M. Rodrigues-Vega, E. Rossi, S. Das Sarma, Interlayer excitonic superfluidity in graphene. Phys. Rev. B **88**, 235402 (2013)
16. Private communication with A.H. MacDonald: Exciton condensation in bilayer MoS$_2$ system
17. I.B. Spielman, J.P. Eisenstein, L.N. Pfeiffer, K.W. West, Resonantly enhanced tunneling in a double layer quantum Hall ferromagnet. Phys. Rev. Lett. **84**, 5808 (2000)
18. E. Tutuc, M. Shayegan, D.A. Huse, Counterflow measurements in strongly correlated GaAs hole bilayers: evidence for electron-hole pairing. Phys. Rev. Lett. **93**, 036802 (2004)
19. L. Tiemann, W. Dietsche, M. Hauser, K. von Klitzing, Critical tunneling currents in the regime of bilayer excitons. New J. Phys. **10**, 045018 (2008)
20. D. Nandi, A.D.K. Finck, J.P. Eisenstein, L.N. Pfeiffer, K.W. West, Exciton condensation and perfect Coulomb drag. Nature **488**, 481 (2012)
21. K.S. Novoselov, A.K. Geim, S.V. Morozov, D. Jiang, M.I. Katnelson, I.V. Grigorieva, S.V. Dubonos, A.A. Firsov, Two-dimensional gas of massless Dirac Fermions in graphene. Nature **438**, 197 (2005)
22. Y. Zhang, Y.-W. Tan, H.L. Stormer, P. Kim, Experimental observation of the quantum Hall effect and Berry's phase in graphene. Nature **438**, 201 (2005)
23. S. Reich, J. Maultzsch, C. Thomsen, P. Ordejón, Tight-binding description of graphene. Phys. Rev. B **66**, 035412 (2002)
24. A.K. Geim, K.S. Novoselov, The rise of graphene. Nat. Mater. **6**, 183 (2007)
25. I. Sodemann, D.A. Pesin, A.H. MacDonald, Interaction-enhanced coherence between two-dimensional Dirac layers. Phys. Rev. B **85**, 195136 (2012)

26. D. Basu, L.F. Register, D. Reddy, A.H. MacDonald, S.K. Banerjee, Tight-binding study of electron-hole pair condensation in graphene bilayers: gate control and system-parameter dependence. Phys. Rev. B **82**, 075409 (2010)
27. G.-B. Liu, W.-Y. Shan, Y. Yao, W. Yao, D. Xiao, Three-band tight-binding model for monolayers of group-VIB transition metal dichalcogenides. Phys. Rev. B **88**, 085433 (2014)
28. Feng-Cheng Wu, Fei Xue, A.H. MacDonald, Theory of spatially indirect equilibrium exciton condensates. arXiv:1506.01947. Submitted for publication (2015)
29. X. Xu, W. Yao, D. Xiao, T.F. Heinz, Spin and pseudospins in layered transition metal dichalcogenides. Nat. Phys. **10**, 343 (2014)
30. S. Shallcross, S. Sharma, E. Kandelaki, O.A. Pankratov, Electronic structure of turbostratic graphene. Phys. Rev. B **81**, 165105 (2010)
31. X. Mou, L.F. Register, S.K. Banerjee, Quantum transport simulation of Bilayer pseudoSpin Field-Effect Transistor (BiSFET) with tight-binding hartree-fock model. IEEE SISPAD 2013, 420, Glasgow, United Kingdom (2013)
32. X. Mou, D. Basu, L.F. Register, S.K. Banerjee, manuscript in preparation
33. M. Cheli, G. Fiori, G. Iannaccone, A semianalytical model of bilayer-graphene field-effect transistor. Trans. Electron Devices IEEE **56**, 2979 (2009)
34. S.K. Banerjee, L.F. Register, E. Tutuc, D. Reddy, A.H. MacDonald, Bilayer pseudospin field-effect transistor (BiSFET): a proposed new logic device. Electron. Device Lett. IEEE **30**, 158 (2009)
35. P. Avouris, Z. Chen, V. Perebeinos, Carbon-based electronics. Nat. Nanotechnol. **2**, 605 (2007)
36. J. Martin, N. Akerman, G. Ulbricht, T. Lohmann, J.H. Smet, K. von Klitzing, A. Yacoby, Observation of electron-hole puddles in graphene using a scan single-electron transistor. Nat. Phys. **4**, 144 (2008)
37. D. Basu, L.F. Regisetr, A.H. MacDonald, S.K. Banerjee, Effect of interlayer bare tunneling on electron-hole coherence in graphene bilayers. Phys. Rev. B **84**, 035449 (2011)
38. S. Datta, *Electronic transport in mesoscopic systems*. Cambridge studies in semiconductor physics and microelectronic engineering (Cambridge University Press, Cambridge, 1997)
39. E.H. Hwang, S. Das Sarma, Acoustic phonon scattering limited carrier mobility in two-dimensional extrinsic graphene. Phys. Rev. B **77**, 115449 (2008)
40. J.-J. Su, A.H. MacDonald, How to make a bilayer exciton condensate flow. Nat. Phys. **4**, 799 (2008)
41. K.K. Ng, *Complete guide to semiconductor devices* (Wiley, New York, 2002), pp. 569–574
42. Y. Yu, S. Han, X. Chu, S.-I. Chu, Z. Wang, Coherent temporal oscillations of macroscopic quantum states in a Josephson Junction. Science **296**, 889 (2002)
43. X. Mou, L.F. Register, S.K. Banerjee, Quantum transport simulations on the Feasibility of the Bilayer pseudoSpin Field-Effect Transistor (BiSFET). IEDM 2013, 4.7.1–4.7.4, Washington DC, United States (2013)
44. X. Mou, L.F. Register, S.K. Banerjee, Bilayer pseudospin junction transistor (BiSJT): a new "beyond-CMOS" logic device. Submitted for publication (2015)
45. X. Mou, L.F. Register, S.K. Banerjee, Quantum transport simulation of exciton condensate transport physics within a bilayer graphene system. Submitted for publication (2015)
46. X. Mou, L.F. Register, S.K. Banerjee, Interplay among Bilayer PseudoSpin Field-Effect Transistor (BiSFET) performance, BiSFET scale and condensate strength. Accepted by IEEE SISPAD 2014, Yokohama, Japan (2014)
47. L.F. Register, X. Mou, D. Reddy, W. Jung, I. Sodemann, D. Pesin, A. Hassibi, A.H. MacDonald, S.K. Banerjee, Bilayer pseudo-spin field effect transistor (BiSFET): concepts and critical issues for realization. ECS Trans. **45**, 3 (2012)
48. K. Wu, A. Sachid, F.-L. Yang, C. Hu, Toward 44% switching energy reduction for FinFETs with vacuum gate spacer. IEEE SISPAD 2012, 253, Denver, Colorado (2012)
49. R.T. Weitz, M.T. Allen, B.E. Feldman, J. Martin, A. Yacoby, Broken-symmetry states in doubly gated suspended bilayer graphene. Science **330**, 812 (2010)

50. P. Jadaun, H.C.P. Movva, L.F. Register, S.K. Banerjee, Theory and synthesis of bilayer graphene intercalated with ICl and IBr for low power device applications. J. Appl. Phys. **114**, 063702 (2013)
51. D. Reddy, L.F. Register, E. Tutuc, S.K. Banerjee, Bilayer pseudospin field-effect transistor: applications to Boolean logic. IEEE Trans. Electron. Devices **57**, 755 (2010)
52. http://www.itrs.net
53. J.B. Johnson, Thermal agitation of electricity in conductors. Phys. Rev. **32**, 97 (1928)
54. H. Nyquist, Thermal agitation of electric charge in conductors. Phys. Rev. **32**, 110 (1928)
55. R. Landauer, Irreversibility and heat generation in the computing process. IBM J. Res. Dev. **5**, 183 (1961)

Chapter 3
Graphene-Based Photonics and Plasmonics

Oleg L. Berman, Roman Ya. Kezerashvili, and Yurii E. Lozovik

Abstract The optical properties of graphene-based structures are discussed. The universal optical absorption in graphene is reviewed. The photonic band structure and transmission of graphene-based photonic crystals are considered. The spectra of plasmon and magnetoplasmon excitations in graphene layers and graphene nanoribbons (GNRs) are analyzed. The localization of the electromagnetic waves in the photonic crystals with defects, which play a role of a waveguide, is studied. Properties of plasmons and magnetoplasmons in graphene layers and GNR are reviewed. The surface plasmon amplification by stimulated emission of radiation with the net amplification of surface plasmons in doped GNR is described. The minimal population inversion per unit area needed for the net amplification of plasmons in a doped GNR is reported. The various applications of graphene for photonics and optoelectronics are reviewed. The tunability of the photonic and plasmonic properties of various graphene structures by doping achieved by applying the gate voltage is discussed.

3.1 Introduction

Very attractive properties of a two-dimensional (2D) electron system were experimentally observed in graphene, which is a 2D honeycomb lattice of the carbon atoms that form the basic planar structure in graphite [1–3]. Since the band structure of graphene has unusual properties compare to the electronic properties of a two-dimensional electron gas in a semiconductor quantum well (QW) [4], graphene

O.L. Berman (✉) • R.Ya. Kezerashvili
Physics Department, New York City College of Technology, The City University
of New York, Brooklyn, NY 11201, USA
e-mail: oberman@citytech.cuny.edu

Yu.E. Lozovik
Institute of Spectroscopy, Russian Academy of Sciences, 142190 Troitsk,
Moscow Region, Russia

MIEM at National Research University HSE, 109028 Moscow, Russia

© Springer International Publishing Switzerland 2015
A. Korkin et al. (eds.), *Nanoscale Materials and Devices for Electronics,
Photonics and Solar Energy*, Nanostructure Science and Technology,
DOI 10.1007/978-3-319-18633-7_3

became the object of many recent interesting experimental and theoretical studies [1–3, 5–13]. Graphene is a 2D gapless semiconductor with effectively massless electrons and holes, which have been described as Dirac-fermions [1, 2, 14]. The specific electronic properties of graphene in a magnetic field have been analyzed [15–18].

It was demonstrated that in infrared (IR) and at above IR wavelengths the transparency of graphene is universal and defined by the fine structure constant [19]. The space-time dispersion of graphene optical conductivity was studied in [20] and the optical properties of graphene were analyzed in [21, 22].

A difference between electrons in graphene and ultrarelativistic particle physics is caused by the fact that in the Dirac equation for graphene, the Fermi velocity of electrons in graphene, which is 300 times lower than the speed of light, replaces the speed of light in vacuum. The Dirac equation for the electrons in graphene is valid only in the laboratory frame of reference, because the effective Dirac equation for graphene was derived using the symmetry of the system from the Galilean-invariant Schrödinger equation.

The unique physical properties of graphene lead to the possibilities of its numerous applications. An extremely limiting small size of a possible graphene-based element of nanoelectronics or optoelectronics can be reached in one dimension, since graphene is one atom thick. Even pristine, undoped graphene may posses a high electron mobility even at room temperature, and its thermal conductivity is much higher than that of copper. Furthermore, the intrinsic strength of graphene is 200 times greater than the strength of steel. Due to these remarkable properties, graphene is a promising material for making coatings for solar batteries and screens, new composite nanomaterials, ultracapacitors, and nanoelectronic elements.

An interesting property of graphene is the possibility of changing its transport characteristics due to adsorption of molecules, which allows design of graphene-based supersensitive nanosensors.

Moreover, an important property of graphene is related to the possibility of easily controlling the density of electrons or holes by using external control gates or external chemical doping. We will discuss in this chapter the possibilities of tunability of the photonic band structure of graphene-based photonic crystals and plasmonic quantum generator, surface plasmon amplification by stimulated emission of radiation (spaser) by gate voltage doping.

The detailed microscopic theory of graphene monolayer and multilayers spectroscopy was developed [22]. The reflectance from a gate doped graphene monolayer has been determined for the infrared region using the intraband Drude–Boltzmann conductivity and for higher frequencies using the interband absorption. At low temperatures and high carrier densities, the reflectance from multilayers has a sharp decrease with a subsequent plateau.

In this chapter we discuss the optical properties of graphene and graphene nanoribbon (GNR), resulting in the proposed novel photonic and plasmonic devices based on graphene and GNR. The chapter is organized in the following way. In Sects. 3.2 and 3.3 we present the dielectric function of graphene and GNR and discuss the universal optical absorption in graphene. The description of 1D and 2D

graphene-based photonic crystals is presented in Sect. 3.4. In Sect. 3.5 we consider
the localization of the electromagnetic waves on defects in the 1D and 2D photonic
crystals. In Sect. 3.6 the properties of plasmons and magnetoplasmons in graphene
layers and GNR are reviewed. The THz GNR-based spaser is presented in Sect. 3.7.
The applications of graphene for photonics and optoelectronics are reviewed in
Sect. 3.8. Finally, the conclusions based on the results presented in this chapter
follow in Sect. 3.9.

3.2 The Dielectric Function of Graphene and GNR

The uniqueness of graphene-based photonics and plasmonics devices is related to
the dielectric function of graphene. The dielectric function $\varepsilon_g(\omega)$ of graphene
multilayer system separated by dielectric layers with dielectric constant ε_0 and
thickness d is given by [21, 22]

$$\varepsilon_g(\omega) = \varepsilon_0 + \frac{4\pi i \sigma_g(\omega)}{\omega d}, \tag{3.1}$$

where $\sigma_g(\omega)$ is the dynamical optical conductivity of the doped graphene. In the
framework of the random-phase approximation (RPA) [23, 24], the dynamical
optical conductivity of graphene was obtained from the Kubo formula [25] in a
complex form consisting of interband and intraband contributions [26–28]. The
optical conductivity of graphene consists of intraband and interband contributions

$$\sigma_g(\omega) = \sigma_{intra}(\omega) + \sigma_{inter}(\omega), \tag{3.2}$$

where $\sigma_{intra}(\omega)$ and $\sigma_{inter}(\omega)$ represent the intraband and interband terms, correspondingly, analyzed in [29].

For high frequencies at temperature T, the dynamical optical conductivity is
given by the expression [21, 22]

$$\sigma_g(\omega) = \frac{ke^2}{4\hbar} \left[\eta(\hbar\omega - 2\mu_0) + \frac{i}{2\pi} \left(\frac{16k_BT}{\hbar\omega} \ln\left[2\cosh\left(\frac{\mu_0}{2k_BT} \right) \right] \right. \right.$$
$$\left. \left. - \ln\frac{(\hbar\omega + 2\mu_0)^2}{(\hbar\omega - 2\mu_0)^2 + (2k_BT)^2} \right) \right]. \tag{3.3}$$

Here e is the charge of an electron, $v_F = \sqrt{3}a_0 t/(2\hbar) \approx 10^8$ cm/s is the Fermi
velocity of electrons in graphene, where $a_0 = 2.46$ Å is a lattice constant and the
value of the overlap integral between the nearest carbon atoms is $t \approx 2.71$ eV [30],
$k = 9 \times 10^9$ N × m^2/C^2 is the Coulomb constant, k_B is the Boltzmann constant,
and μ_0 is the chemical potential determined by the electron concentration

$n_0 = (\mu_0/(\hbar v_F))^2/\pi$, which is controlled by doping or gate voltage. The optical conductivity of graphene given by Eq. (3.3) is valid if the damping rate is much less than the term caused by the spatial dispersion: $\gamma = \tau^{-1} \ll \omega$ and $\gamma \ll q v_F$, where q is the magnitude of the wave vector, $\gamma = \tau^{-1}$ is the electron damping rate in graphene. In the optical region, the spatial dispersion of conductivity can be neglected compared with the frequency dependence, and the dynamical conductivity can be used for a study of the graphene optical properties [21]. The optical visibility of graphene deposited on the underlying substrate was theoretically studied in [31]. The graphene transmittance spectra were experimentally observed [19], and the dynamical conductivity was found in agreement with the theoretical prediction [22].

The zero band gap in graphene is caused by the symmetry of the layer, particularly by the identical environment of the two carbon atoms in the graphene unit cell [32]. A band gap in graphene can appear and can be controlled by applying external electric or magnetic field, deformation, or doping by adatoms on the graphene sheets (e.g., by hydrogen, oxygen, or other noncarbon atoms) [33, 34]. Another method to open a band gap is related to the reduction of the dimensionality of graphene from 2D to 1D by cutting graphene into sufficiently narrow ribbons, known as GNRs. Electronic properties of GNR depend on the direction of cutting of the GNR from the extended graphene analogously to carbon nanotubes because this direction defines the effective boundary conditions at the GNR edges. The GNRs have been produced by different methods: cutting graphene applying lithographic techniques [35–38], "unzipping" carbon nanotubes [39, 40] and direct synthesis [41]. For an essential gap $\geq 0.5\,\mathrm{eV}$, which can allow room temperature operation of plasmonic devices with GNR, the GNRs should have a constant width of 2–3 nm. The electronic properties of GNRs are determined by their size and geometry [42, 43], because the edges of the ribbons parallel to different symmetry directions are described by different boundary conditions. There are zigzag edges GNRs ("zigzag ribbons") and armchair edges GNRs ("armchair ribbons") shown in Fig. 3.1. In graphene the carbon atoms crystallize in a honeycomb structure whose primitive lattice vectors are $\mathbf{a} = a_0(1,0)$ and $\mathbf{b} = a_0(1/2, \sqrt{3}/2)$. The lattice is bipartite and there are two atoms per unit cell, denoted by A and B, located at $(0,0)$ and at $d = a_0(0, 1/\sqrt{3})$.

The quasiparticle energies and band gaps of armchair and zigzag GNRs were obtained using a first-principles many-electron Green's function approach within the GW approximation [44]. It was found that the band width is larger and the effective mass is smaller for carriers in zigzag GNRs than for carriers in armchair GNRs. This difference between the electronic properties of zigzag and armchair GNRs is caused by the fact that the energy of quasiparticle states near the band gap in zigzag GNRs is wave vector sensitive.

The results of the numerical calculations of the dielectric function for the zigzag GNR were presented in [45]. For the armchair GNR the dielectric function $\varepsilon(q_x, \omega) \equiv \varepsilon_{00}(q_x, \omega, \beta, \mu_g)$ in the one-band approximation within the RPA is given by [45]

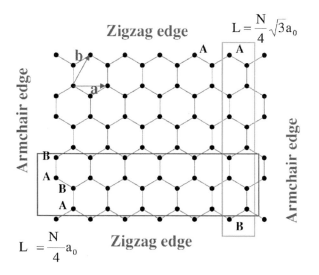

Fig. 3.1 The lattice structure of a graphene sheet. The primitive lattice vectors are denoted by **a** and **b**. *Top* and *bottom* are zigzag edges, *left* and *right* are armchair edges. Atoms enclosed in the *vertical* (*horizontal*) *rectangle* represent the unit cell used in the calculation of nanoribbons with zigzag (armchair) edges. The length of the nanoribbons, L, as function of the number of atoms, N, in the unit cell is also indicated. This figure is taken from [43]

$$\varepsilon_{00}(q_x, \omega, \beta, \mu_g) = 1 - V_{0,0}(q_x)\Pi_{0,0}(q_x, \omega, \beta, \mu_g), \tag{3.4}$$

where $V_{0,0}(q_x)$ is the Coulomb matrix element, and $\Pi_{0,0}(q_x, \omega)$ is the polarizability that can be approximated by

$$\Pi_{0,0}(q_x, \omega, \beta, \mu_g) = -\frac{g_s}{\pi} \frac{\hbar v_F q_x}{\hbar^2(\omega(\omega + i\gamma) - (v_F q_x)^2)} f_1(q_x, \beta, \mu_g) \tag{3.5}$$

with

$$f_1(q_x, \beta, \mu_g) = \frac{1}{\hbar \beta v_F}\left[-\beta \hbar v_F q_x + 2\ln\left(\frac{1 + e^{-\beta \mu_g}}{1 + e^{-\beta(\hbar v_F q_x + \mu_g)}}\right)\right], \tag{3.6}$$

where $g_s = 2$ is the spin degeneracy factor, $\beta = 1/(k_B T)$, γ is the damping rate for the GNR, and μ_g is the chemical potential of GNR controlled by the doping. Since the distribution of the doping density across the width of a GNR is quite non-uniform [46, 47], the one-dimensional carrier density n_0 along a GNR was used. The chemical potential of GNR can be calculated as $\mu_g = \pi \hbar v_F n_0/2$, where n_0 is the one-dimensional carrier density [45, 48]. The one-dimensional carrier density is $n_0 = 2k_F/\pi$ [45, 48], where k_F is the Fermi wave vector. In [45] $n_0 = 2.8 \times 10^7$ m^{-1} -5.6×10^7 m^{-1} which corresponded values of Fermi wave vector in the range of $k_F a_0 = 0.0108$–0.0216.

The Coulomb matrix element $V_{0,0}(q_x)$ is given by [45]

$$V_{0,0}(q_x) = \int_0^1 du \int_0^1 du' V(q_x W|u - u'|), \qquad (3.7)$$

where W is the width of GNR (in the y direction), and the one-dimensional Fourier transform of the Coulomb interaction has the form [49]

$$V(q_x|y - y'|) = \frac{2ke^2}{\varepsilon_g} K_0(q_x|y - y'|). \qquad (3.8)$$

In Eq. (3.8) $\varepsilon_g = 2.5$ is the static dielectric constant of graphene, $K_0(y)$ is the zeroth-order modified Bessel function of the second kind.

Substituting Eq. (3.8) for $V(q_x W|u - u'|)$ into Eq. (3.7), one obtains

$$V_{0,0}(q_x) = \frac{2ke^2}{\varepsilon_g} \int_0^1 du \int_0^1 du' K_0(q_x W|u - u'|). \qquad (3.9)$$

Finally, by substituting $\Pi_{0,0}(q_x, \omega, \beta, \mu_g)$ from Eq. (3.5) and $V_{0,0}(q_x)$ from Eq. (3.9) into Eq. (3.4), and taking into account Eq. (3.6) one can obtain the dielectric function for the armchair GNR at frequencies $\omega \gg v_F k$.

Let us mention that the dielectric function of graphene given by Eqs. (3.1) and (3.3) depends on the chemical potential of graphene μ_0 and the dielectric function of the armchair GNR given by Eq. (3.4) depends on the chemical potential μ_g of GNR. Since μ_0 and μ_g are determined by the electron concentration, that is tunable by the gate voltage [50, 51], the photonic band structure of graphene-based and GNR-based photonic crystals and frequency spectrum of plasmons in graphene and GNR determined by the dielectric function, also can be tuned by the gate voltage. The general expressions for the dielectric function of zigzag GNR were also presented in [45].

The specific properties of the band structure of graphene results in interesting features of its dielectric response: including the singularity in the low-frequency range and other highly unusual optical properties [20–22], and a weak damping of quasiparticles in graphene. These properties are very helpful for making graphene-based photonic crystals, which unlike metals has a photonic gap in the far-infrared spectral region hardly blurred by damping [52–54] and for creation of graphene-based spasers [55].

3.3 Universal Optical Absorbtion in Graphene

It was demonstrated that the optical opacity of suspended graphene is defined only by the fine structure constant, the parameter that describes the coupling between light and relativistic electrons, and that is traditionally associated with quantum electrodynamics rather than condensed matter physics [19]. This is valid for IR and larger wavelengthes for energies less than 1 eV, where the deviation from the linear isotropic dispersion law for electrons and holes is small. The calculation of the absorption of light by a graphene membrane uses the Dirac Hamiltonian $\widehat{H}$ in the presence of a vector potential $\mathbf{A}$ [19]:

$$\widehat{H} = v_F \widehat{\sigma} \cdot (\widehat{\mathbf{p}} + e\mathbf{A}), \tag{3.10}$$

where the operator $\widehat{\sigma}$ is written in terms of the Pauli matrices as $\widehat{\sigma} = (\sigma_x, \sigma_y)$, and $\widehat{\mathbf{p}}$ is the momentum operator. Taking the term $ev_F\widehat{\sigma} \cdot \mathbf{A}$ as a perturbation, in the framework of time-dependent perturbation theory the transition rate for the electronic transitions induced by light absorption is given by Fermi's golden rule [19]. Using this transition rate the fraction of transmitted light T was obtained as

$$T = 1 - \pi\alpha \simeq 0.977, \tag{3.11}$$

where $\alpha = ke^2/(\hbar c)$ is the fine structure constant, and c is the speed of light in vacuum.

The reason why perturbation theory works so well is because the final answer is controlled by the small dimensionless parameter α. The origin of the optical properties being defined by the fundamental constants is related to the two-dimensional nature and gapless electronic spectrum of graphene [19]. The experimental measurements [19] of the optical transmittance T and reflectance R are in good agreement with Eq. (3.11).

There are some predictions that due to its high transmittance, few-layer sufficiently doped graphene can potentially replace ITO (indium tin oxide) as a transparent conductor for applications in solar cells [56–58] and touch screens [59–61].

3.4 The Photonic Crystals Based on Graphene: The Properties and Advantages

Motivated by the unique optical properties of graphene, many graphene-based photonic and optoelectronic applications have been developed recently [29, 32, 59]. Here we describe photonic crystals based on graphene.

Photonic crystals are formed by structures where the dielectric constant varies periodically in space [62]. The frequency spectrum of the electromagnetic wave in photonic crystals is characterized by a band structure. The solution of Maxwell's equations with a periodic dielectric constant, resulting in photonic band-gap structures, is similar to the solution to Schrödinger's equation for a periodic potential, resulting in the electron energy band-gap structures in solids. Electromagnetic waves penetrate a photonic crystal similar to Bloch waves of electrons in a regular crystal. The width of the photonic band gap (PBG) depends on the geometrical parameters of the photonic crystal and the contrast of the dielectric constants of the constituent elements [63, 64]. There have been studies of photonic band structure of photonic crystals with different materials used for the corresponding constituent elements including dielectrics, semiconductors, and metals [63–70]. Photonic crystals with superconducting elements have been studied in [71–78]. The properties of photonic crystals provide an opportunity to manipulate the emission, propagation, and distribution of light [79, 80] and photonic crystals can be used as frequency filters. The properties of photonic crystals were reviewed in [63, 64, 81].

Photonic crystals are different from regular solid crystals in the following way. While Schrödinger's equation describes regular solid crystals via the scalar wave function, Maxwell's equations for photonic band-gap crystals describe the propagation of electromagnetic waves via the electric or magnetic fields, which are vector components of the transverse electromagnetic waves.

For the first time one-dimensional and two-dimensional graphene-based photonic crystals were proposed in [54] and [52], respectively. A one-dimensional photonic crystal can be formed by an array of periodically located parallel stacks of alternating graphene and dielectric stripes (SAGDS) embedded into a background dielectric medium. The graphene stripes are placed one under the other with the dielectric stripes placed between them. A 1D photonic crystal with graphene stripes separated by dielectric layers with a thickness d is shown in Fig. 3.2. The properties of the one-dimensional photonic crystals formed by semiconductors were analyzed in [82].

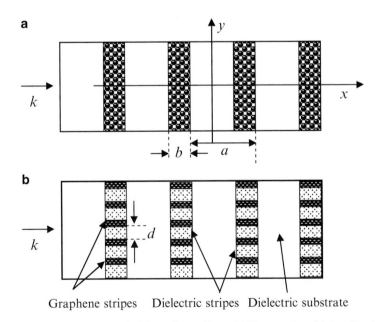

Graphene stripes   Dielectric stripes   Dielectric substrate

Fig. 3.2 The 1D photonic crystal with graphene stripes. (**a**) The *top view*; (**b**) the *side view*. The material of the dielectric stripes between graphene stripes can be the same as the material of the dielectric substrate. Figure is reprinted from [54]

A two-dimensional graphene-based photonic crystal can be fabricated using multilayer graphene formed by a stack of alternating graphene and dielectric layers (SAGDL). A 2D photonic crystal is periodic along two of its axes and homogeneous along the third axis. Below we are considering two types of 2D graphene-based photonic crystals. One type is a periodic array of cylindrical holes etched in the multilayer graphene. The cylindrical holes can be filled by the same dielectric as the dielectric placed between the graphene layers. The other type is a periodic array of cylindrical rods (pillars) of alternating graphene and dielectric discs. The dielectric between graphene discs can be the same as the material of the dielectric medium. We refer to these structures as a graphene hole photonic crystal and a graphene rod photonic crystal, respectively.

We consider polarized electromagnetic waves with the electric field **E** perpendicular to the plane of graphene stripes for 1D photonic crystal and perpendicular to the plane of the graphene layers and parallel to the axes of the cylindrical holes for 2D photonic crystal. The wave equation for the electric field in a dielectric media has the form [83]

$$\Delta \mathbf{E}(r, t) - \frac{\varepsilon(r, t)}{c^2} \frac{\partial^2 \mathbf{E}(r, t)}{\partial t^2} = 0, \tag{3.12}$$

where $\varepsilon(\mathbf{r}, t)$ is the dielectric constant of the media.

3.4.1 Graphene-Based One-Dimensional Photonic Crystal

Let us discuss the calculations of the photonic band structure of a graphene-based 1D photonic crystal. Looking for solutions of Eq. (3.12), with harmonic time variation of the electric field, i.e., $\mathbf{E}(\mathbf{r}, t) = \mathbf{E}(\mathbf{r})e^{i\omega t}$, and considering the propagation of the wave in the x-direction along the plane of graphene stripes and perpendicular to the graphene-dielectric boundaries one obtains from Eq. (3.12)

$$\frac{\partial^2 E_z(x)}{\partial x^2} + \frac{\omega^2 \varepsilon(x, \omega)}{c^2} E_z(x) = 0. \tag{3.13}$$

The dielectric constant of the 1D periodic structure is given by

$$\varepsilon(x, \omega) = \begin{cases} \varepsilon_0, \text{ for } & -\frac{1}{2}(a - b) + na < x < \frac{1}{2}(a - b) + na, \\ \varepsilon_g(\omega), \text{ for } & \frac{1}{2}(a - b) + na < x < \frac{1}{2}(a + b) + na, \end{cases}$$

where a is the period of the 1D array of graphene stripes, b is the width of the graphene stripes, and n is an integer. By introducing the filling factor ff the relation between a and b can be written as $b = aff$.

The eigenfrequencies corresponding to the electromagnetic wave penetrating in the photonic crystal shown in Fig. 3.2 are found by solving the wave equation (3.13). This wave equation is mathematically similar to the Schrödinger equation for an electron moving in a one-dimensional rectangular periodic potential barrier described by the Kronig–Penney model. The eigenenergies of the Schrödinger equation corresponding to the Kronig–Penney model are given by [84]:

$$\cos{(ka)} = \cosh(\beta b)\cos{[\alpha a(1-ff)]} + \frac{\alpha^2 - \beta^2}{2\alpha\beta}\sinh(\beta b)\sin{[\alpha a(1-ff)]},$$

(3.14)

where the wave vector k is in the range $0 \le k \le 2\pi/a$, and α and β are defined as

$$\alpha = \frac{\sqrt{\varepsilon_0}}{c}\omega,$$

$$\beta = \frac{\sqrt{\varepsilon_g(\omega)}}{c}\omega.$$

(3.15)

Applying Eqs. (3.15) and (3.14) can be written in a form

$$\cos{(ka)} = \cosh\left(\sqrt{\varepsilon_g(\omega)}\frac{aff}{c}\omega\right)\cos\left[\sqrt{\varepsilon_0}\frac{(1-ff)}{c}\omega\right]$$
$$+ \frac{\varepsilon_0 - \varepsilon_g(\omega)}{2\sqrt{\varepsilon_0\varepsilon_g}}\sinh\left(\sqrt{\varepsilon_g(\omega)}\frac{aff}{c}\omega\right)\sin\left[\sqrt{\varepsilon_0}\frac{a(1-ff)}{c}\omega\right].$$

(3.16)

The eigenfrequencies of the 1D photonic crystal as a function of the wave vector k can be obtained by substituting the dielectric constant of the multilayer graphene given by Eq. (3.1) into Eq. (3.16). The solutions of Eq. (3.16) for ω as functions of k provide the frequency band structure for a 1D photonic crystal formed by periodically located parallel SAGDS embedded into a background dielectric medium. The PBG Ω is also determined from the solutions of Eq. (3.16), which demonstrates that the PBG depends on the filling factor ff and the period of the 1D array of graphene stripes a, as well as on thickness d of the dielectric stripes that separate the graphene stripes. The dependence of the PBG on the thickness d of the dielectric layer, which separates the graphene stripes, can be found from (3.16), since the dielectric function $\varepsilon_g(\omega)$ (3.1) of a graphene multilayer system separated by dielectric layers defined in Eq. (3.1) depends on d.

Using Eq. (3.16) the band structure for 1D graphene-based photonic crystal was calculated. For simplicity, the same material was considered for a background dielectric medium and dielectric stripes between the graphene stripes. As the dielectric material we consider SiO_2 with the dielectric constant $\varepsilon_0 = 4.5$. The results of the calculations of the dispersion relation of the photonic crystal are presented in Fig. 3.3. The photonic band structure is calculated for different

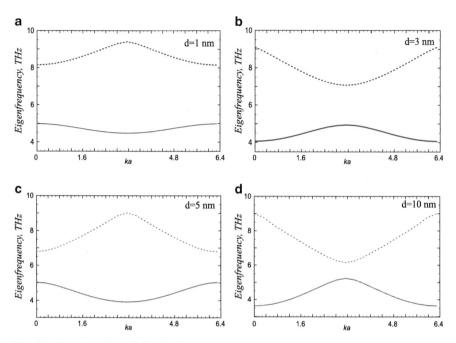

Fig. 3.3 The dispersion relation for the 1D graphene-based photonic crystal with $a = 25$ μm and filling factor $f = 0.3927$ for different thicknesses of the dielectric between graphene stripes. Figures are reprinted from [54]

distances between graphene layers d. In our calculations we assume $n_0 = 10^{11}$ cm^{-2}, which corresponds to the chemical potential for the electrons in graphene $\mu = 3.525 \times 10^{-21}$ J, temperature $T = 300$ K, the period of the 1D graphene stripes array $a = 25 \times 10^{-6}$ m, and the filling factor $ff = 0.3927$. According to the results of the calculations, the photonic band structure almost does not depend on ε_0 due to the fact that $\varepsilon_0 \ll |\varepsilon_g|$. The results of our calculations demonstrate the strong dependence of the photonic band structure on the thickness d of the dielectric that separates the graphene stripes. At $d = 1$ nm and $d = 5$ nm the distance between the lower and the upper dispersion curves is larger in the middle than at the edges. At $d = 3$ nm and $d = 10$ nm the distance between the lower and the upper dispersion curves is larger at the edges than in the middle.

For 1D graphene-based photonic crystal for the first time proposed in [54], the wave equation for the propagating electromagnetic wave was solved in the framework of the Kronig–Penney model. The frequency band structure was obtained analytically as a function of the filling factor, the period of the 1D array of graphene stripes and the thickness of the dielectric between the graphene stripes [54] generated the further studies of the 1D graphene-based photonic crystals. The photonic band structure for a 1D graphene-based photonic crystal for different polarizations were obtained, using the Bloch theorem and the discontinuity in the derivative of the electric field component at the boundaries between each graphene sheet and dielectric layer [85]. The transmission properties of the 1D graphene-based photonic crystal were investigated using transfer matrix method in [86]. It was demonstrated that the photonic band structure is almost omnidirectional and insensitive to the polarization [86]. A scaling study of the photonic band structure of the 1D graphene-based photonic crystal using the transfer matrix method was presented in [87].

3.4.2 Graphene-Based Two-Dimensional Photonic Crystal

In this section we analyze the calculations of the photonic band structure of the graphene-based 2D photonic crystal. Let us consider a SAGDL, that is a multilayer graphene. Using the multilayer graphene a two-dimensional photonic crystal that is periodic along two of its axes and homogeneous along the third axis can be fabricated. Let us consider the 2D graphene hole and graphene rod photonic crystals. The periodic arrays of holes or rods can be arranged in any of the five 2D lattice types as given by the crystallographic restriction theorem. As an example, we study a square lattice for a graphene hole crystal and a graphene rod crystal. The square lattice of graphene rods is surrounded entirely by air and holes are filled by air as well. It is evident that the rods of the graphene rod crystal can be embedded in any dielectric including the dielectric that separates the graphene layers. Also the holes in the graphene hole crystal can be filled by any suitable dielectric including the dielectric that separates graphene layers. Both the graphene hole and graphene rod photonic crystals have discrete translational symmetry in the xy plane: the square lattice has symmetries with respect to the x and y axes and with respect to both diagonals, i.e., every $\pi n/4$ with n an integer. They are homogeneous in the z direction. Therefore, in the z direction the modes are oscillatory with no restrictions on the wave vector k_z.

We consider polarized electromagnetic waves with the electric field $\mathbf{E}$ perpendicular to the plane of the graphene layers and parallel to the axes of the cylindrical holes. In this case the wave vector $\mathbf{k}$ is parallel to the plane of the graphene layers and perpendicular to the axes of the cylindrical holes or rods. The wave equation for the electric field in an ideal lattice of a graphene hole or rod crystal can be obtained from Eq. (3.12).

By expanding the electric field in terms of the Bloch waves inside a photonic crystal, one obtains from the wave equation the system of equations for Fourier components of the electric field [64, 67]:

$$(k + G)^2 E_k(G) = \frac{\omega^2(k)}{c^2} \sum_{G'} \varepsilon(\mathbf{G} - \mathbf{G}') E_{\mathbf{k}}(\mathbf{G}'), \qquad (3.17)$$

which presents the eigenvalue problem for finding the photon dispersion $\omega(\mathbf{k})$, and $\varepsilon_{in} \equiv \varepsilon_g(\omega)$ and $\varepsilon_{out} = \varepsilon_d$ for the rod crystal, while $\varepsilon_{in} = \varepsilon_d$ and $\varepsilon_{out} = \varepsilon_g(\omega)$ for the hole crystal. In Eq. (3.17) the coefficients of the Fourier expansion for the dielectric constant are given by

$$\varepsilon(\mathbf{G} - \mathbf{G}') = \varepsilon_{out}\delta_{\mathbf{GG}'} + (\varepsilon_{in} - \varepsilon_{out})M_{\mathbf{GG}'}, \tag{3.18}$$

where $M_{\mathbf{GG}'}$ for the geometry related to the square lattice for the graphene rod and hole crystals is

$$M_{\mathbf{GG}'} = \begin{cases} 2ff\dfrac{J_1(|\mathbf{G} - \mathbf{G}'|r)}{(|\mathbf{G} - \mathbf{G}'|r)} & \mathbf{G} \neq \mathbf{G}' \\ ff & \mathbf{G} = \mathbf{G}' \end{cases} . \tag{3.19}$$

In Eq. (3.19) $\mathbf{G}$ is the reciprocal lattice vector, J_1 is the Bessel function of first order, and $ff = S/A$ is the filling factor of the 2D photonic crystal that is defined as the ratio between the area S occupied by the scatterer (cylinder) and the area A of the unit cell.

Equation (3.18) for the graphene rod crystal can be rewritten as

$$\varepsilon(\mathbf{G} - \mathbf{G}') = \varepsilon_d\delta_{\mathbf{GG}'} + (\varepsilon_g - \varepsilon_d)M_{\mathbf{GG}'}, \tag{3.20}$$

while for the graphene hole crystal we have

$$\varepsilon(\mathbf{G} - \mathbf{G}') = \varepsilon_g\delta_{\mathbf{GG}'} + (\varepsilon_d - \varepsilon_g)M_{\mathbf{GG}'}. \tag{3.21}$$

For the graphene hole crystal we consider the 2D square lattice formed by the array of holes in the multilayer graphene system that consists of the SAGDL. The holes are filled by the same dielectric as the dielectric placed between the graphene layers, hence $\varepsilon_{in} \equiv \varepsilon_d = \varepsilon$. For the graphene rod crystal we consider the 2D square lattice formed by the cylinders consisting of a stack of the alternating graphene and dielectric discs and embedded in a dielectric medium. The dielectric discs consist of the same dielectric as the material where these rods are embedded into, hence $\varepsilon_{out} \equiv \varepsilon_d = \varepsilon$.

The calculations for the photonic band structure and transmittance spectrum for the 2D graphene hole and rod photonic crystal with the array of the dielectric cylinders within the multilayer graphene system arranged in a square lattice with the filling factor $ff = 0.3927$ are performed for the thickness of the dielectric layer between graphene layers $d = 10^{-3}$ μm, the period of photonic crystal lattice constant $a = 25$ μm. Thus, the lattice frequency is $\omega_a = 2\pi c/a = 7.54 \times 10^{13}$ rad/s. In the calculations below the chemical potential μ that was determined by the electron concentration $n_0 = 10^{11}$ cm^{-2}. As the dielectric material we consider SiO$_2$ with a dielectric constant $\varepsilon = 4.5$.

The results of the plane wave calculation for the photonic band structure and transmittance spectrum were performed using the plane wave expansion and finite-difference time-domain (FDTD) method [88], correspondingly, for the 2D graphene-based photonic crystals and are presented in Figs. 3.4 and 3.5, respectively.

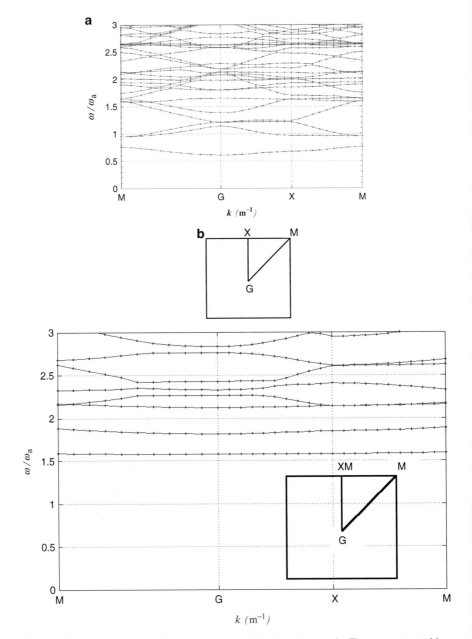

Fig. 3.4 The band structure of the graphene-based 2D photonic crystals. The array arranged in a square lattice. The filling factor *ff* = 0.3927. M, G, X, M are the points of symmetry in the first (*square*) Brillouin zone. The *insert* shows the first Brillouin zone of the 2D photonic crystal with a square lattice. (**a**) The graphene rod photonic crystal. (**b**) The graphene hole photonic crystal. Figure 4a is reproduced from [52]

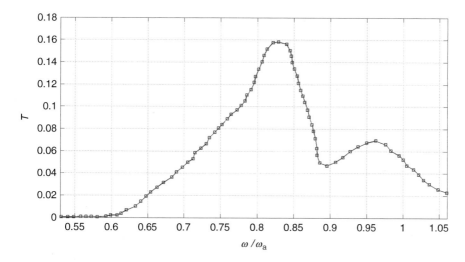

Fig. 3.5 The transmittance T spectrum of the graphene-based 2D rod photonic crystal. Figure is reproduced from [52]

The plane wave computation has been employed for an extended photonic crystal, and FDTD calculation of the transmittance has been performed for five graphene layers. The first band gap for graphene hole photonic crystal shown in Fig. 3.4b corresponds to the frequencies lower than $1.55\omega_a$. The second gap in Fig. 3.4b (at the point G) corresponds approximately to $1.6 < \omega/\omega_a < 1.79$, and the frequency at the top of the second band gap is given by $\omega = 1.79\omega_a$. For comparison, in Fig. 3.5 the calculations are presented for the transmittance spectrum for the graphene rod crystal. A band gap is clearly apparent in the range $0 < \omega/\omega_a < 0.6$ and $0.75 < \omega/\omega_a < 0.95$. The first gap originates from the electronic structure of the doped graphene, which prevents absorbtion at $\hbar\omega < 2\mu$. The photonic crystal structure manifests itself in the dependence of the lower photonic band on wave vector k. In contrast, the second gap $0.75 < \omega/\omega_a < 0.95$ is caused by the photonic crystal structure and dielectric contrast.

Comparison of Fig. 3.4a with Fig. 3.4b shows the lowest forbidden photonic gap for the 2D graphene hole photonic crystal results in a band gap almost three times larger than that of the 2D graphene rod photonic crystal. This leads to the conclusion that graphene hole photonic crystal can be used as an optical frequency filter with almost three times wider frequency region that the 2D graphene rod photonic crystal. The transmittance spectrum for the graphene-based hole photonic crystal can be obtained analogously to graphene-based rod photonic crystal.

The graphene-based photonic crystals are novel types of one-dimensional and two-dimensional photonic crystals. The photonic band structure of both types of 2D graphene photonic crystals can be modified two ways: by engineering the crystal lattice, choosing the dielectric and the radius of the cylinder during fabrication and by gate voltage doping of the graphene layers. During fabrication the band gap of 1D and 2D photonic crystals can be controlled by the type and thickness of the dielectric that separates the graphene layers, as well as by the filling factor and lattice constant. Let us mention that the band structures of the 1D and 2D graphene-based photonic crystals are tunable by gate voltage [50, 51] due to the dependence of the conductivity of graphene on the chemical potential μ that is determined by the electron concentration. In the gate controlled 1D and 2D graphene-based photonic crystals, this allows one to tune the photonic band structure. The tunability of 1D graphene-based photonic crystal by doping was supported by the calculation of the dependence of the frequency of the PBG as a function of the chemical potential in the range from 0 up to $0.5\,\text{eV}$ [86]. The tunability of photonic band structure of 1D graphene-based photonic crystal by external magnetic fields in the range from 1 T up to $14\,\text{T}$ was analyzed in [89].

3.5 The Localization of the Electromagnetic Waves in the Graphene-Based Photonic Crystals with the Defects

The common purpose of fabrication of defects in photonic crystals is creation of waveguides by localization and guiding the electromagnetic wave [90]. While electromagnetic waves with frequencies within the PBG cannot propagate through the crystal, they can be confined to the defect regions. This localization of the electromagnetic waves by the defects results in the appearance of discrete photonic states inside the band gap. Below we consider the localization of the electromagnetic wave on the defect in the 1D graphene-based photonic crystal. The same approach can be developed for 2D graphene-based photonic crystals.

Let us consider a defect in an array of SAGDS embedded in a background dielectric medium. This defect is formed by one empty space or a "1D vacancy" due to the absence of a SAGDS in one position, where it should be placed due to the periodicity. This position would be filled by the dielectric. This extra dielectric stipe contributes to the dielectric contrast that results by adding the term $-\omega^2/c^2(\varepsilon_g(\omega) - \varepsilon_0)\gamma(|x - x_0|)E_z(x)$ to the r.h.s. in Eq. (3.13). Here $\gamma(|x - x_0|) = 1$ for $|x - x_0| \le b$, and $\gamma(|x - x_0|) = 0$ for $|x - x_0| > b$, where x_0 corresponds to the coordinate in the middle of the 1D defect, which is the coordinate of the middle of the missing graphene stripe. The wave equation of the electric field for the 1D photonic crystal with the defect was obtained [54]:

$$\frac{\partial^2 E_z(x)}{\partial x^2} + \frac{\omega^2}{c^2}\left(\varepsilon(x, \omega) - (\varepsilon_g(\omega) - \varepsilon_0)\gamma(|x - x_0|)\right)E_z(x) = 0. \qquad (3.22)$$

Equation (3.22) describes the periodic array of SAGDS with the defect formed by one missing SAGDS.

The wave equation (3.22) has to be solved to obtain the eigenfrequency corresponding to the electromagnetic wave localized at the defect formed by a background dielectric medium due to the absence of a single SAGDS. This wave equation is similar to the Schrödinger equation describing the electron in the periodic potential energy in the presence of the potential energy of a defect placed at the point x_0 given in [54]. Applying a two-band model [91], we can reduce Eq. (3.22) to Dirac-type equations, which can be solved as shown in [90], and for the frequency corresponding to the localized mode in the photonic crystal with the defect, and the wave equation for the electric field in the 1D graphene-based photonic crystal is given by [54]

$$
-\frac{4k_0^2 c^4}{3}\frac{d^2 E_z(x)}{dx^2} - 2\omega^2 \Omega^2 (\varepsilon_g(\omega) - \varepsilon_0)\gamma(|x - x_0|)E_z(x) = (\omega^4 - \Omega^4)E_z(x),
$$

(3.23)

where $k_0 = 2\pi/a$ is the vector of the 1D reciprocal lattice. This Klein–Gordon type equation has the eigenvalue $\omega^4 - \Omega^4$. In Eq. (3.23) Ω is the width of the forbidden band (photonic gap) in the spectrum of the electromagnetic waves. The continuity of the electric field and its derivative at the points $x = x_0 + b/2$ and $x = x_0 - b/2$ results in the transcendental equation determining the spectrum of the even states [54]:

$$
\sqrt{\omega^2 \Omega^2 (\varepsilon_g(\omega) - \varepsilon_0)^2 - \omega^4 + \Omega^4}\,\tan\left(\sqrt{\omega^2 \Omega^2 (\varepsilon_g(\omega) - \varepsilon_0)^2 - \omega^4 + \Omega^4 b^2/c^2}\right)
$$
$$
= \sqrt{|\omega^4 - \Omega^4|},
$$

(3.24)

as well as of the odd states:

$$
\sqrt{\omega^2 \Omega^2 (\varepsilon_g(\omega) - \varepsilon_0)^2 - \omega^4 + \Omega^4}\,\cot\left(\sqrt{\omega^2 \Omega^2 (\varepsilon_g(\omega) - \varepsilon_0)^2 - \omega^4 + \Omega^4 b^2/c^2}\right)
$$
$$
= -\sqrt{|\omega^4 - \Omega^4|}.
$$

(3.25)

The frequency corresponding to the photonic mode localized by the defect is obtained by solving Eqs. (3.24) and (3.25) with respect to the frequency ω.

We applied Eqs. (3.24) and (3.25) to calculate the frequencies corresponding to the electromagnetic wave localized by the defect in the photonic crystal formed due to the absence of a single SAGDS [54]. For the photonic crystal with the defect at $d = 1$ nm we have $\nu = 1.79$ THz, at $d = 3$ nm we have $\nu = 1.85$ THz, at $d = 5$ nm we have $\nu = 3.91$ THz, at $d = 10$ nm we have $\nu = 3.62$ THz. All these frequencies are located inside the PBG. The localized frequency ν does not depend on momentum, because our approach is based on the Luttinger–Kohn model [92, 93]. The frequency localized by the defect is just a constant in the photonic band structure, which can be controlled by the thickness d of the dielectric stripes that separate the graphene stripes, as well as by the filling factor f and the period of the 1D array of graphene stripes a. The frequency corresponding to the field localization due to the defect in the 1D graphene-based photonic crystal can be tuned by the applied gate voltage, because the optical conductivity of graphene depends on the chemical potential μ_0.

3.6 Plasmons and Magnetoplasmons in Graphene Layers and GNRs

Surface plasmons are collective oscillations of electrons at the surface of a material. In spite of the fact that noble metals (e.g., Au and Ag) are considered as the best available plasmonic materials, the devices fabricated from the noble metals are characterized by large ohmic losses and nontunability once the geometry of the structure is fixed. Graphene has been recently proposed as a novel plasmonic material at infrared frequencies [94–100] and can be considered as a terahertz metamaterial [101]. Surface plasmon polaritons (SPPs) are formed due to the coupling of the photon in the infrared or terahertz region to surface plasmons in graphene. The SPPs have many interesting properties [96, 102] such as extreme confinement, tunability via electrical gating or chemical doping, and low losses resulting from long lifetime with hundreds of optical cycles. Interesting properties of plasmon polaritons in a cavity with embedded graphene were studied in [103, 104]. These features cause graphene plasmonic devices to be an attractive alternative to traditional metal plasmonic devices. Graphene is promising and perspective for photonic metamaterials [101], light harvesting [105], optical biosensing [106, 107], and transformation optics [108]. Let us also mention that the plasmon frequency spectrum can be tuned by the applied gate voltage due to the dependence of the optical conductivity of graphene on the chemical potential μ_0.

3.6.1 Plasmons in Graphene Layers and GNRs

Plasmons are self-sustaining oscillations of a carrier system, caused by the long-range nature of the electron–electron Coulomb interaction. The plasmon modes correspond to the zeros of the dynamical dielectric function at the corresponding frequency ω and wave vector q. The long-wavelength plasma oscillations are essentially fixed by the particle number (or current) conservation. The long-wavelength plasmon dispersion ω for monolayer graphene (MLG) within the RPA is given by [13]

$$\omega(q \to 0) = \left(\frac{ke^2 v_\mathrm{F} q}{\varepsilon_g \hbar} \sqrt{\pi n g_s g_v} \right)^{1/2}, \tag{3.26}$$

where n is the electron concentration in the conduction band, and $g_v = 2$ is the valley degeneracy factor.

An important property of MLG plasmon dispersion is that it is nonclassical, because $\hbar$ appears explicitly in Eq. (3.26) in the long-wavelength limit. This quantum nature of long-wavelength MLG plasmons is a direct consequence of its linear Dirac-like energy-momentum dispersion, which has no classical analogy [109].

The results of the numerical calculations of the frequency spectrum of plasmons for the zigzag GNR were represented in [45]. For the armchair GNR the plasmon frequency ω can be obtained in the one-band approximation within the RPA at $\gamma = 0$ from the condition $\text{Re}[\varepsilon(q_x, \omega)] = 0$ using Eqs. (3.4) and (3.5) [55]:

$$\omega^2 = v_F^2 q_x^2 - \frac{V_{0,0}(q_x) f_1(q_x, \beta, \mu_g) g_s v_F q_x}{\pi \hbar}. \tag{3.27}$$

In the undoped case, only metallic armchair GNR supports a propagating plasmon mode [45]. This mode propagates undamped, since the chirality of the wave functions prevents it from decaying into particle–hole pairs. However, the undoped zigzag GNR does not support plasmon excitations, because a plasmon decays into a broad continuum of particle–hole excitations with low wave vectors [45]. Doped GNRs are characterized by the properties similar to the properties of semiconductor nanowires, including a plasmon mode dispersion law proportional to $q_x^{3/2} \sqrt{-\ln(q_x W)}$ with a coefficient that vanishes when the doping vanishes, except for the metallic armchair case, and a static dielectric response, which diverges at $q_x = 2k_F$ [45]. The plasmons in doped zigzag GNRs are essentially the same as the plasmons in doped armchair GNRs and standard semiconducting nanowires: the single particle spectrum around the Fermi energy is represented by a narrow line of particle–hole excitations, above which a plasmon excitation should appear with dispersion $q_x \sqrt{-\ln(q_x W)}$ [45].

The frequency spectrum of plasmons in graphene and GNR can be tuned by the applied gate voltage due to the dependence of the plasmon frequency of graphene given by Eq. (3.26) on the electron concentration in the conduction band and the dependence of the plasmon frequency of GNR given by Eq. (3.27) on the chemical potential of the GNR μ_g. The tunability of the spectrum of plasmons by gate voltage was analyzed experimentally in the voltage range from $-20\,\text{V}$ up to $+30\,\text{V}$ [50, 51].

3.6.2 Magnetoplasmons in Graphene Layers

The frequency spectrum of plasmon excitations in graphene can also be controlled by a perpendicular external magnetic field [110]. The electrons were considered in a single graphene layer in the xy-plane in the perpendicular magnetic field **B** parallel to the positive z axis. The Zeeman splitting is neglected, and we assume valley energy degeneracy, describing the eigenstates by two pseudospins [16, 111]. The electrons in a single graphene layer in the perpendicular magnetic field are described by an effective 2×2 matrix Hamiltonian $\hat{H}_0$ whose diagonal elements are zero and whose off-diagonal elements are $\hat{\pi}_x \pm i\hat{\pi}_y$ where $\hat{\pi} = -i\hbar\nabla + e\mathbf{A}$.

Choosing $\mathbf{A} = (0, Bx, 0)$, the eigenfunctions of $\hat{H}_0$ are labeled by $\alpha = \{k_y, n, s(n)\}$, where $n = 0, 1, 2, \ldots$ is the Landau level index, k_y is the electron wave vector in the y-direction, and $s(n)$, which is defined by $s(n) = 0$ for $n = 0$ and $s(n) = \pm 1$ for $n > 0$, labels the conduction (+ 1) and valence (− 1 and 0) band, respectively. The eigenfunction $\psi_\alpha(\mathbf{r})$ is given by a spinor $\psi_\alpha(x, y)$ with components given by [111]

$$\psi_\alpha^{(1)} = \frac{C_n}{\sqrt{L_y}} e^{ik_y y} s(n) i^{n-1} \Phi_{n-1}(x + l_H^2 k_y),$$

$$\psi_\alpha^{(2)} = \frac{C_n}{\sqrt{L_y}} e^{ik_y y} i^n \Phi_n(x + l_H^2 k_y),$$
(3.28)

where $l_H = \sqrt{\hbar/eB}$ is the magnetic length, and L_y is a normalization length. We have $C_n = 1$ for $n = 0$, $C_n = 1/\sqrt{2}$ for $n > 0$ and

$$\Phi_n(x) = (2^n n! \sqrt{\pi} l_H)^{-1/2} e^{-(x/l_H)^2/2} H_n(x/l_H),$$
(3.29)

where $H_n(x)$ is the Hermite polynomial. The eigenenergies are given by $\varepsilon_\alpha = s(n)\varepsilon_n = s(n)(\hbar v_F/l_H)\sqrt{2n}$, for which successive levels are not equally separated.

The dynamic dielectric function in RPA [23] is given by $\varepsilon(q, \omega) = 1 - V_c(q)\Pi(q, \omega)$, where q is the in-plane wave vector, $V_c(q) = 2\pi e^2/(\varepsilon_s q)$ is the 2D Coulomb interaction in the momentum representation and the 2D polarization function is [110]

$$\Pi(q, \omega) = \frac{g_s g_v}{2\pi l_H^2} \sum_{n=0}^{\infty} \sum_{n'=0}^{\infty} \sum_{s(n), s'(n')} \frac{f_{s(n)n} - f_{s'(n')n'}}{\hbar\omega + \varepsilon_{s(n)n} - \varepsilon_{s'(n')n'}} F_{s(n)s'(n')}(n, n', q), \quad (3.30)$$

where $f_{s(n)n}$ is the Fermi–Dirac function, $F_{ss'}(n, n')$ arises from the overlap of eigenstates and is given by

$$F_{ss'}(n, n', q) = C_{n_1}^2 C_{n_2}^2 \left[-\frac{q^2 l_H^2}{2}\right]^{n_1-n_2} \frac{1}{|(n_1 - n_2)!|^2}$$

$$\times \left(s_1(n_1)s_2(n_2) \left|\frac{(n_1 - 1)!}{(n_2 - 1)!}\right| + \left|\frac{n_1!}{n_2!}\right|\right).$$
(3.31)

The magnetoplasmon dispersion relation for a single graphene layer was obtained from the solutions of the dispersion equation $\varepsilon(q, \omega) = 0$. The highest valence band is full and all others empty at $T = 0$ K. Transitions to the lowest five Landau levels in the conduction and valence bands were the only single-particle excitations included in the calculations [110]. The solution of the dispersion equation for a single layer of graphene when the imaginary part of the plasmon

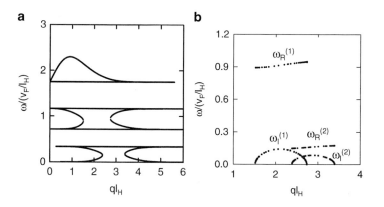

Fig. 3.6 Magnetoplasmon excitation energy as a function of wave vector, in units of l_H^{-1}, in a single graphene layer. (**a**) Real frequency solution. (**b**) The real and imaginary parts of the frequency satisfying the dispersion equation. Figure is reproduced from [110]

frequency is zero is represented in Fig. 3.6a. In this case, the magnetoplasmons are self-sustaining oscillations except when they enter the particle–hole mode region where they undergo a loss due to Landau damping. In Fig. 3.6b, the solutions of the dispersion equation were plotted, when the frequency is complex for which the real part that is linear in q and exists only where the magnetoplasmon in Fig. 3.6a has negative group velocity. The real and imaginary parts of the frequency are denoted by ω_R and $\omega_I > 0$, respectively. The loss is described by the finite imaginary part of the frequency (see Fig. 3.6b). Thus, a nonzero imaginary part of the frequency for a single graphene layer in a magnetic field was obtained [110]. The nonzero imaginary part for collective excitations for a 2D electron gas in semiconductors has been established, for several layers of semiconductor [112–115].

The negative group velocity for $ql_H > 1$ is caused by the magnetic field [116]. The fact that in [110] only the five lowest Landau levels in the conduction and valence bands were included in the calculations was justified in the following way. The calculations were performed with the same parameters as in calculating Fig. 3.6, and the summation was performed over a larger number of Landau levels in the conduction and valence bands. Qualitatively, the results were the same. The main differences were that the number of single-particle excitation lines and the frequency of the highest modes were increased compared to Fig. 3.6. However, the lower collective mode branches did not change significantly.

A magnetoplasmon instability appears in a single graphene layer even without the application of an in-plane current driving the charge carriers [110]. This instability is described by the finite imaginary part in the frequency of the collective excitations in Fig. 3.6. A magnetoplasmon instability in a bilayer semiconductor without an-in-plane current appears only in a very small region of wave vector compared to the Fermi wave vector. This difference in the spectrum of magnetoplasmons in graphene structures compared to layered semiconductors is caused by the screening properties of the dielectric function in graphene [117, 118]

and 2D semiconductors [116]. The frequency spectra of magnetoplasmons also
were calculated for a pair of parallel layers, a graphite bilayer, and a superlattice of
graphene layers in a perpendicular magnetic field [110].

3.7   The THz Spaser Based on GNR

The progress in nanoscience and nanotechnology during the past decade has led to
great interest in studying nanoscale optical fields. The surface plasmon amplification by stimulated emission of radiation was proposed in [119]. This effect also was
studied in [120, 121]. It was proposed also to increase the sensitivity of apertureless
near field optical microscopy and spaser-based intracavity spectroscopy [122]. Spaser generates coherent high-intensity fields of selected surface
plasmon (SP) modes that can be strongly localized on the nanoscale. The properties
of localized plasmons are analyzed in [123–126]. The spaser consists of an active
medium formed by two-level systems [semiconductor quantum dots (QDs) or
organic molecules] and a plasmon resonant nanosystem where the surface plasmons
are excited. The emitters transfer their excitation energy by radiationless transitions
through near fields to a resonant plasmon nanosystem.

In the same way that a laser generates stimulated emission of coherent photons, a
spaser generates stimulated emission of SPs in resonating metallic nanostructures
adjacent to a gain medium. While a spaser is the nanoplasmonic counterpart of a
laser, it does not emit photons. A spaser is similar to the conventional laser. The
laser has two principal elements: the resonator or cavity that supports photonic
modes and the gain active medium that is population inverted and supplies energy
to the lasing modes [120]. In a spaser photons are replaced by SPs and the resonant
cavity is replaced by a nanoparticle, which supports the plasmonic modes [127]. The
energy source for the spasing mechanism is formed by an active gain medium that is
excited externally. This excitation field can be characterized by optical frequencies
and be unrelated to the spaser's operating frequency; for example, a spaser can
operate in the infrared, while the excitation of the gain medium can be achieved
using a UV pulse. In a spaser loss is compensated by optical gain, and the
amplification of propagating SPs is maintained.

Various theoretical and experimental studies have been devoted to metal-based
spasers, where the excitation of SPs was considered in different metallic
nanostructures of different geometric shapes. A spaser consisting of the nanosystem
formed by the V-shaped silver nanoinclusion embedded in a dielectric host
containing the PbS and PbSe QDs was analyzed in [119]. A spaser consisting of a
silver spherical nanoshell on a dielectric core with a radius of 10–20 nm, and
surrounded by two dense monolayers of nanocrystal QDs was studied in [127]. The
propagation of SPs along the bottom of the groove (channel) in the metal surface
was considered in [128]. It was assumed that the linear chain of QDs at the bottom
of the channel excite the coherent SPs. According to [128], the gain can exceed the
loss and plasmonic lasing in a ring or linear channel of the silver surface surrounded

by a linear chain of CdSe QDs can occur at realistic values of the system parameters. The spaser with SPs excited in the metal sphere surrounded by the two-level quantum dot was studied theoretically in [129–133]. The theory for the spaser formed by the spherical gain core, containing two-level systems, coated with a metal spherical plasmonic shell was developed in [134]. The spaser consisting of 44 nm diameter nanoparticles with the gold spherical core surrounded by a dye-doped silica shell was studied experimentally in [135, 136]. In this experiment the emitters were fabricated from the dye-doped silica shells instead of QDs. A two-dimensional array of plasmonic resonators supporting coherent current excitations with high quality factor can be considered as a planar source of spatially and temporally coherent radiation [137]. This structure is formed by a gain medium slab supporting a regular array of silver asymmetric split-ring resonators. The experimental studies of the spaser formed by 55 nm-thick gold film with the nano-slits located on the silica substrate surrounded by PbS QDs were reported in [138]. Room temperature spasing of SPs at 1.46 μm wavelength was achieved by sandwiching a gold-film plasmonic waveguide between optically pumped InGaAs quantum-well gain media [139].

Since plasmons can be excited also in graphene, and damping in graphene is much less than in metals [20–22], the application of a GNR surrounded by semiconductor QDs as the nanosystem for the spaser was proposed in Ref. [55]. Plasmons in graphene can become an alternative to plasmons in noble metals, because they exhibit much tighter confinement and relatively long propagation distances, with the advantage of being highly tunable via electrostatic gating [96]. Besides, the graphene-based spaser can work in THz frequency region. The frequency spectrum of oblique terahertz plasmons in GNR arrays was obtained [140]. Graphene-based spaser meets the new technological needs, since it works at the IR frequencies, while the metal-based spaser works at the higher frequencies.

The system under consideration in [55] was the GNR, which is the stripe of graphene at $z = 0$ in the plane (x, y), that is infinite in the x direction and has the width W in the y direction. This GNR is surrounded by the deposited dense monolayer of nanocrystal quantum dots with the dielectric constant ε_d at $z < 0$ and $z > 0$. Due the laser pumping of the QDs, the resonant nonradiative transmission occurs by creating a surface plasmon localized in the GNR. The amplification by QDs can exceed absorption of the surface plasmon in the GNR [55]. As a result an increase of intensity of the surface plasmon field was demonstrated. In other words, the competition between the gain and the loss of the surface plasmon field in the GNR results in favor of the gain.

The expression for the minimal population inversion per unit area N_c, needed for the net amplification of SPs in a doped GNR, was derived in [55]. This expression was obtained from the condition that for the regime of the plasmon amplification the rate $\partial \bar{U} / \partial t$ of the transfer of the average energy of the QDs is greater than the heat released per unit time $\partial Q / \partial t$ due to the absorption of the energy of the plasmon field in the GNR.

The excitation resulting in the generation of plasmons in the GNR comes from the transitions in the QDs between the excited and ground states. Therefore, the rate of the transfer of the average energy $\overline{U}$ of the QDs characterized by the dipole moment is given by

$$\frac{\partial \overline{U}}{\partial t} = \frac{\omega}{2} \int \mathrm{Im}\left(\vec{E} \cdot \vec{P}^{\,*}\right) dV, \tag{3.32}$$

where the relation between the polarization of QDs $\vec{P}$ and electric field of the GNR plasmon $\vec{E}$ has the form [128]

$$\vec{P} = -ik\frac{\tau_p|\mu|^2 n}{\hbar}\vec{E}, \tag{3.33}$$

where $k = 9 \times 10^9$ N $\times$ m^2/C^2, n is the difference between the concentrations of the quantum dots in the excited and ground states per unit of volume, τ_p is the inverse line width, and μ is the average off-diagonal element of the dipole moment of a single QD.

Substituting Eq. (3.33) into Eq. (3.32), the rate of the transfer of the average energy of the QDs can be obtained

$$\frac{\partial \overline{U}}{\partial t} = \omega k\frac{\tau_p|\mu|^2}{2\hbar}\int n|\vec{E}|^2 dV. \tag{3.34}$$

The distances between the quantum dots was assumed to be small, so their effect on a plasmon is the same as that of a continuous (constant) gain distribution along the GNR. The 2D GNR was considered at $z = 0$, and it was assumed to be infinite in the x direction. Also the GNR has the width W in the y direction and is surrounded by the monolayer of uniformly distributed quantum dots. Therefore, for n we have $n = N_0\eta(y, -W/2, W/2)\delta(z)$, where N_0 is the difference between the numbers of the excited and ground state quantum dots per unit area, and $\eta(y, -W/2, W/2) = 1$ at $-W/2 \leq y \leq W/2$, $\eta(y, -W/2, W/2) = 0$ at $y < -W/2$ and $y > W/2$.

Taking into account the spatial dispersion of the dielectric function in the GNR [43, 45], the following expression for the rate of the heat $\partial Q/\partial t$ released due to the absorption of the energy of the plasmon field in the GNR was used [83, 123]

$$\frac{\partial Q}{\partial t} = \frac{\omega}{2} \int \mathrm{Im}\varepsilon(\omega, q_x)\eta(y, -W/2, W/2)|\vec{E}|^2 dV$$

$$= \frac{\omega}{2}\mathrm{Im}\varepsilon(\omega, q_x) \int_{-\infty}^{+\infty} dx \int_{-W/2}^{+W/2} dy \int_{-\infty}^{+\infty} dz|\vec{E}(x, y, z)|^2, \tag{3.35}$$

where $\mathrm{Im}\varepsilon(\omega, q_x)$ is the imaginary part of the dielectric function $\varepsilon \equiv \varepsilon(\omega, q_x)$ of the GNR.

The plasmons in a GNR are excited due to the radiation caused by the transitions from the excited state to the ground state on the QDs. Therefore, according to the conservation of energy, the regime of the amplification of the plasmons in the GNR is established, if the rate of the transfer of the average energy $\partial \overline{U}/\partial t$ of the QDs is greater than the heat released rate $\partial Q/\partial t$ due to the absorption of the energy of the plasmon field in the GNR:

$$\frac{\partial \overline{U}}{\partial t} > \frac{\partial Q}{\partial t}. \tag{3.36}$$

Substituting Eqs. (3.34) and (3.35) into Eq. (3.36), we get

$$k\frac{\tau_p|\mu|^2 N_0}{\hbar} \int_{-W/2}^{+W/2} dy \int_{-\infty}^{+\infty} dx |\vec{E}\,(x,y,0)|^2 \; > \mathrm{Im}\,\varepsilon(\omega,q_x) \int_{-\infty}^{+\infty} dx \int_{-W/2}^{+W/2} dy \int_{-\infty}^{+\infty} dz |\vec{E}\,(x,y,z)|^2. \tag{3.37}$$

From Eq. (3.37), one can obtain the condition for the difference between the surface densities of the quantum dots in the excited and ground state corresponding to the amplification of plasmons:

$$N_0 > N_c = \frac{\mathrm{Im}\varepsilon(\omega,q_x) \displaystyle\int_{-\infty}^{+\infty} dx \int_{-W/2}^{+W/2} dy \int_{-\infty}^{+\infty} dz\, |\vec{E}\,(x,y,z)|^2}{k\dfrac{\tau_p|\mu|^2}{\hbar} \displaystyle\int_{-\infty}^{+\infty} dx \int_{-W/2}^{+W/2} dy |\vec{E}\,(x,y,0)|^2}, \tag{3.38}$$

where N_c is the critical density of the QDs required for the amplification of the plasmons. The evaluation of the integrals in Eq. (3.38) requires the electric field of a plasmon in a GNR [55]

$$|\vec{E}\,(x,y,0)|^2 = E_0^2\left(2q_x^2\cos^2(q_y y) + q_y^2\right), \tag{3.39}$$

$$|\vec{E}\,(x,y,z)|^2 = E_0^2 e^{-2\alpha|z|}\left(2q_x^2\cos^2(q_y y) + q_y^2\right), \tag{3.40}$$

where $\alpha = \sqrt{q_x^2 + q_y^2}$ and for the armchair GNR we have $q_{ym} = 2\pi/(3a_0)((2M + 1 + m)/(2M + 1))$ at the width $W = (3M + 1)a_0$ [43], where m is the integer. We use $m = 1$. The substitution of Eqs. (3.39) and (3.40) into Eq. (3.38) results in

$$N_0 > N_c = \frac{\hbar \mathrm{Im}\varepsilon(\omega, q_x)}{\alpha k \tau_p |\mu|^2}. \tag{3.41}$$

Using Eqs. (3.4) and (3.5) one can find $\mathrm{Im}\varepsilon(q_x, \omega)$ for the armchair GNR:

$$\mathrm{Im}\varepsilon(q_x, \omega) = -\frac{V_{0,0}(q_x)f_1(q_x, \beta, \mu_g)g_s v_F q_x \omega \gamma}{\pi \hbar \left((\omega^2 - v_F^2 q_x^2)^2 + \omega^2 \gamma^2 \right)}. \tag{3.42}$$

The plasmon frequency ω can be obtained from Eq. (3.5).

The critical density N_c was calculated using Eq. (3.41) [55]. N_c is a function of the wave vector q_x, the GNR width W, temperature T, and electron concentration n_0 determined by the doping.

The typical energy corresponding to the transition between the ground and excited electron states for the PbS and PbSe QDs produced with the radii from 1 to 8 nm can be 0.7 eV. Thus, we used $\tau_p \approx 5.9$ fs, and $|\mu| = 1.9 \times 10^{-17}$ esu $= 19$ Debye (1 Debye $= 10^{-18}$ esu, 1 Debye $= 3.33564 \times 10^{-30}$ C · m) [141]. The typical frequency corresponding to the transition between the ground and excited electron states for the PbS and PbSe QDs, which is $f \approx 170$ THz, matches the resonance with the plasmon frequency in the armchair GNR [45]. Therefore, the PbS and PbSe QDs can be used for the spaser considered in [55]. The damping in graphene $\gamma = \tau^{-1}$ determined by the dissipation time that τ is assumed to be either $\tau = 1$ ps or $\tau = 10$ ps or $\tau = 20$ ps [142–145]. The optical magneto-absorption experiments for graphene in magnetic field demonstrate that the Landau level broadening allows to estimate $\tau \approx 20$ ps [142]. The values of τ used in [55] are close to the experimental quantities. In the calculations in [55] for the 1D doping electron density n_0 the values of the same order of magnitude as in [45] were used.

The dependencies of the critical density of the QDs that is required for the amplification, N_c on the wave vector q_x, the width of the nanoribbon W, the frequency f at the fixed temperature T, obtained using Eq. (3.41) are presented in Fig. 3.7 for the parameters defined in Fig. 3.7. As it was demonstrated in Fig. 3.7a, N_c decreases as q_x and n_0 increase. At q_x larger than 0.4 nm^{-1} there is almost no difference between the values of N_c corresponding to the different 1D doping electron densities n_0, and for large q_x N_c converges to approximately 15 μm^{-2}. According to Fig. 3.7b, N_c decreases as q_x and τ increase. This means that higher damping corresponds to higher N_c. It can be seen in Fig. 3.7b that starting with $q_x \approx 1.0$ nm^{-1}, N_c depends very weakly on q_x, converging to some constant values that depend on the value of τ. It follows from Fig. 3.7c that N_c increases as W increases and decreases as q_x increases. When W increases, the values of N_c depend on q_x. As shown in Fig. 3.7d, N_c increases as f and τ decrease. According to Fig. 3.7d, starting with $f \approx 160$ THz, N_c depends very weakly on the frequency and converges to some constant values that depend on the value of τ. The dependence of the plasmon frequency f on the width of the nanoribbon W, for the different wave

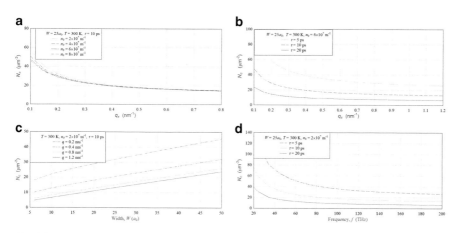

Fig. 3.7 The dependencies of the critical density of the QDs N_c on the wave vector q_x for the different 1D doping electron densities n_0 (**a**), on the wave vector q_x for the different dissipation time τ corresponding to the damping (**b**), on the width of the nanoribbon W at the different wave vector q_x (**c**), on the frequency f at the different dissipation time τ (**d**). All results are obtained at the fixed temperature T. Figures are reproduced from [55]

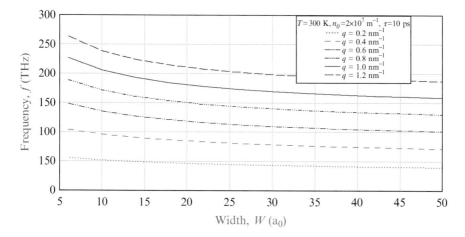

Fig. 3.8 The dependence of the plasmon frequency f on the width of the nanoribbon W, for the different wave vectors q_x at the fixed dissipation time τ corresponding to the damping, temperature T and 1D doping electron density n_0. Figure is reproduced from [55]

vectors at the fixed dissipation time corresponding to the damping, temperature and 1D doping electron density obtained using Eq. (3.27) is presented in Fig. 3.8. It follows from Fig. 3.8 that as the plasmon frequency f increases as q_x increases and the width of the nanoribbon W decreases. If in Eq. (3.41), the imaginary part of the dielectric function does not depend on the width W, N_c would not depend on W. However, due to the complicated dependence of $\text{Im}[\varepsilon(\omega, q_x)]$ on W through $V_{0,0}(q_x)$

given by Eq. (3.9), this dependence exists. According to Fig. 3.7b, d, N_c decreases when the damping time τ increases.

The advantages of the GNR-based spaser are a wide frequency generation region from THz up to IR, and small damping—the low threshold for pumping, with the possibility of control by the gate.

The critical density of the QDs N_c and frequency spectrum of plasmons in GNR can be tuned by the applied gate voltage due to the dependence of the plasmon frequency in the GNR given by Eq. (3.27) and $\text{Im}[\varepsilon(q_x, \omega)]$ given by Eq. (3.42) on the chemical potential of the GNR μ_g. Since N_c is the increasing function of W as can be seen in Fig. 3.7c, it is better to use the GNR rather than the graphene layer for the graphene-based spaser, since the GNR requires for the amplification regime a lesser density of QDs than the graphene layer.

3.8 Applications of Graphene for Photonics and Optoelectronics

Since graphene is characterized by high mobility and optical transparency, as well as flexibility, robustness, and environmental stability, there can be many applications of graphene for photonics and optoelectronics, where the combination of unique optical and electronic properties of graphene can be fully used, even in the absence of a band gap. Moreover, the linear dispersion of the Dirac electrons enables ultrawideband tunability [59]. The applications of graphene for photonics and optoelectronics include various devices, ranging from solar cells and light-emitting devices to touch screens, photodetectors, and ultrafast lasers. The multiple functions of signal emission, transmission, modulation, and detection can be realized using graphene due to its specific optical properties. The ability to integrate graphene photonics onto the silicon platform to afford broadband operation in light routing and amplification allows use graphene for fabrication of devices such as polarizers, modulators, and photodetectors [29].

Let us mention that there are two types of a band gap in a semiconductor: a "direct" band gap and an "indirect" band gap. If the k-vectors corresponding to the minimal-energy state in the conduction band and the maximal-energy state in the valence band are the same, the gap in this semiconductor is called a "direct gap." If these k-vectors are different, the gap is called an "indirect gap". The band gap is called "direct" if the momenta of electrons and holes are the same in both the conduction band and the valence band, and an electron can directly emit a photon. For an "indirect" gap, a photon cannot be emitted because the electron must transfer large momentum to a phonon and the last process is described as second order in perturbation theory. And thus the process has essentially a smaller probability. While silicon has an indirect gap, graphene has a direct gap. The energy gap in graphene between the valence and conduction bands can be changed in a wide range by an electric field [146–148]. This allows one to optimize the performance of the device by tuning the optical properties of graphene.

Various silicon-based photonic components ranging from passive devices to modulators, detectors, and light amplifiers and sources have been developed. However, all of the photonics components cannot be silicon-based, when there is a demand for broadband data transferring. Intrinsic silicon, which has an indirect band gap, is not a broadband optical material [149]. Other drawbacks of silicon include low electrooptic coefficient, low light emission efficiency, and high propagation losses due to scattering off the side walls of the waveguide. Therefore, the fabrication of hybrid materials which can combine the advantages of silicon and graphene as a second broadband photonic material is the most promising way to develop the novel materials for photonics and optoelectronics devices.

The recent developments in graphene photonics include a graphene waveguide, broadband polarizer, graphene modulator [150], graphene photodetector, surface plasmon enhanced photodetector, etc [29]. Most of these photonic devices exhibit broadband performance, which is related to the unique electronic structure in graphene. The broadband operation of these devices ranges from visible to near-infrared wavelength [29].

Due to its unique properties, graphene can be used in the wide range of optical control schemes. The fabrication of silicon-based integrated optical circuits, which are designed to provide multiple functions of light creation, routing, modulation, computing, and detection, can be enabled by the broadband optical opacity and tunable dynamical conductivity of graphene. Perhaps, progress in the direct deposition of graphene onto silicon could allow integration of graphene in hybrid silicon photonic circuits.

3.9 Conclusions

In this chapter we provided a review of the optical properties of graphene, the universal optical absorbtion in graphene, the properties of 1D and 2D photonic crystal based on graphene, the localization of the electromagnetic waves in the 1D photonic crystal with defects, plasmonics and magnetoplasmonics in graphene layers and GNR, THz spaser based on GNRs, and a brief overview of various applications of graphene for photonics and optoelectronics.

Graphene-based photonic crystals have the following advantages. They can be used as frequency filters and waveguides for a wide spectral region from infrared to THz. Furthermore, the graphene-based photonic crystals can perform very well over a wide temperature range including room temperature. Graphene-based photonic crystals similar to metallic and superconducting photonic crystals are characterized by a low frequency PBG. The photonic band structure and the localized mode in the graphene-based photonic crystals can be controlled by changing the thickness of the dielectric layers between the graphene discs and by the doping. The sizes of the graphene-based photonic crystals can be much larger than the sizes of metallic photonic crystals due to the small dissipation of the electromagnetic wave. GNR-based spasers have a wider frequency generation region from THz up to IR, smaller damping—for lower threshold for pumping compared to spasers based on metal nanostructures. The graphene-based photonic crystals and spasers can be tuned by gate voltage doping, which makes them different from metallic and dielectric photonic crystals and metallic spasers.

Motivated by the unique optical properties of graphene, various photonic and optoelectronic devices can be developed from hybrid materials, which can combine the advantages of silicon and graphene as second broadband photonic material.

Acknowledgements This research was supported in part by PSC CUNY grant: award # 64197-00 42. Yu.E.L. was supported by Program of Basic Research of HSE.

References

1. K.S. Novoselov, A.K. Geim, S.V. Morozov, D. Jiang, Y. Zhang, S.V. Dubonos, I.V. Grigorieva, A.A. Firsov, Science **306**, 666 (2004)
2. I.A. Luk'yanchuk, Y. Kopelevich, Phys. Rev. Lett. **93**, 166402 (2004)
3. Y. Zhang, J.P. Small, M.E.S. Amori, P. Kim, Phys. Rev. Lett. **94**, 176803 (2005)
4. T. Ando, A.B. Fowler, F. Stern, Rev. Mod. Phys. **54**, 437 (1982)
5. K.S. Novoselov, A.K. Geim, S.V. Morozov, D. Jiang, M.I. Katsnelson, I.V. Grigorieva, S.V. Dubonos, Nature (London) **438**, 197 (2005)
6. Y. Zhang, Y.-W. Tan, H.L. Stormer, P. Kim, Nature (London) **438**, 201 (2005)
7. K. Kechedzhi, O. Kashuba, V.I. Fal'ko, Phys. Rev. B **77**, 193403 (2008)
8. M.I. Katsnelson, Europhys. Lett. **84**, 37001 (2008)
9. A.K. Geim, Science **324**, 1530 (2009)
10. A.H. Castro Neto, F. Guinea, N.M.R. Peres, K.S. Novoselov, A.K. Geim, Rev. Mod. Phys. **81**, 109 (2009)
11. N.M.R. Peres, Rev. Mod. Phys. **82**, 2673 (2010)
12. D.S.L. Abergel, V. Apalkov, J. Berashevich, K. Ziegler, T. Chakraborty, Adv. Phys. **59**, 261 (2010)
13. S. Das Sarma, S. Adam, E.H. Hwang, E. Rossi, Rev. Mod. Phys. **83**, 401 (2011)
14. S. Das Sarma, E.H. Hwang, W.K. Tse, Phys. Rev. B **75**, 121406(R) (2007)
15. K. Nomura, A.H. MacDonald, Phys. Rev. Lett. **96**, 256602 (2006)
16. C. Tőke, P.E. Lammert, V.H. Crespi, J.K. Jain, Phys. Rev. B **74**, 235417 (2006)
17. V.P. Gusynin, S.G. Sharapov, Phys. Rev. B **71**, 125124 (2005)
18. V.P. Gusynin, S.G. Sharapov, Phys. Rev. Lett. **95**, 146801 (2005)
19. R.R. Nair, P. Blake, A.N. Grigorenko, K.S. Novoselov, T.J. Booth, T. Stauber, N.M.R. Peres, A.K. Geim, Science **320**, 1308 (2008)
20. L.A. Falkovsky, A.A. Varlamov, Eur. Phys. J. B **56**, 281 (2007)
21. L.A. Falkovsky, S.S. Pershoguba, Phys. Rev. B **76**, 153410 (2007)
22. L.A. Falkovsky, J. Phys. **129**, 012004 (2008)
23. D. Pines, *Elementary Excitations in Solids* (Benjamin, New York, 1963)
24. D.A. Kirzhnits, D.M. Brink, *Field Theoretical Methods in Many-Body Systems* (Oxford Pergamon Press, Oxford, 1967)
25. G.D. Mahan, *Many-Particle Physics* (Plenum, New York, 1990)
26. V.P. Gusynin, S.G. Sharapov, Phys. Rev. B **73**, 245411 (2006)
27. V.B. Gusynin, S.G. Sharapov, J.P. Carbotte, Phys. Rev. Lett. **96**, 256802 (2006)

28. T. Stauber, N.M.R. Peres, A.K. Geim, Phys. Rev. B **78**, 085432 (2008)
29. Q. Bao, K.P. Loh, ACS Nano **6**, 3677 (2012)
30. V. Lukose, R. Shankar, G. Baskaran, Phys. Rev. Lett. **98**, 116802 (2007)
31. D.S.L. Abergel, A. Russell, V.I. Fal'ko, Appl. Phys. Lett. **91**, 063125 (2007)
32. P. Avouris, Nano Lett. **10**, 4285 (2010)
33. D. Haberer, D.V. Vyalikh, S. Taioli, B. Dora, M. Farjam, J. Fink, D. Marchenko, T. Pichler, K. Ziegler, S. Simonucci, M.S. Dresselhaus, M. Knupfer, B. Büchner, A. Grüneis, Nano Lett. **10**, 3360 (2010)
34. D.A. Abanin, A.V. Shytov, L.S. Levitov, Phys. Rev. Lett. **105**, 086802 (2010)
35. Z. Chen, Y.-M. Lin, Ph. Avouris, Phys. E **40**, 228 (2007)
36. M.Y. Han, B. Özyilmaz, Y. Zhang, P. Kim, Phys. Rev. Lett. **98**, 206805 (2007)
37. Y.M. Lin, V. Perebeinos, Z. Chen, Ph. Avouris, Phys. Rev. B **78**, 161409R (2008)
38. X.L. Li, X.R. Wang, L. Zhang, S.W. Lee, H.J. Dai, Science **319**, 1229 (2008)
39. D.V. Kosynkin et al., Nature (London) **458**, 872 (2009)
40. J. Cai, P. Ruffieux, R. Jaafar, M. Bieri et al., Nature (London) **466**, 470 (2010)
41. E. McCann, Phys. Rev. B **74**, 161403(R) (2006)
42. M. Ezawa, Phys. Rev. B **73**, 045432 (2006)
43. L. Brey, H.A. Fertig, Phys. Rev. B **73**, 235411 (2006)
44. L. Yang, C.-H. Park, Y.-W. Son, M.L. Cohen, S.G. Louie, Phys. Rev. Lett. **99**, 186801 (2007)
45. L. Brey, H.A. Fertig, Phys. Rev. B **75**, 125434 (2007)
46. P.G. Silvestrov, K.B. Efetov, Phys. Rev. B **77**, 155436 (2008)
47. S. Thongrattanasiri, I. Silveiro, F. Javier García de Abajob, Appl. Phys. Lett. **100**, 201105 (2012)
48. D.R. Andersen, H. Raza, Phys. Rev. B **85**, 075425 (2012)
49. Q.P. Li, S. Das Sarma, Phys. Rev. B **43**, 11768 (1991)
50. J. Chen et al., Nature (London) **487**, 77 (2012)
51. Z. Fei et al., Nature (London) **487**, 82 (2012)
52. O.L. Berman, V.S. Boyko, R.Ya. Kezerashvili, A.A. Kolesnikov, Yu.E. Lozovik, Phys. Lett. A **374**, 4784 (2010)
53. O.L. Berman, V.S. Boyko, R.Ya. Kezerashvili, Yu.E. Lozovik, Electromagnetic wave propagation in two-dimensional photonic crystals, chapter, in *Wave Propagation*, ed. by A. Petrin (InTech, Vienna, 2011), pp. 83–104
54. O.L. Berman, R.Ya. Kezerashvili, J. Phys. **24**, 015305 (2012)
55. O.L. Berman, R.Ya. Kezerashvili, Yu.E. Lozovik, Phys. Rev. B **88**, 235424 (2013)
56. L.G. De Arco, Y. Zhang, C.W. Schlenker, K. Ryu, M.E. Thompson, C. Zhou, ACS Nano **4**, 2865 (2010)
57. Y. Wang, X. Chen, Y. Zhong, F. Zhu, K.P. Loh, Appl. Phys. Lett. **95**, 063302 (2009)
58. Y. Wang, S.W. Tong, X.F. Xu, B. Özyilmaz, K.P. Loh, Adv. Mater. **23**, 1475 (2011)
59. F. Bonaccorso, Z. Sun, T. Hasan, A. Ferrari, Nat. Photon. **4**, 611 (2010)
60. D.S. Hecht, L. Hu, G. Irvin, Adv. Mater. **23**, 1482 (2011)
61. S. Bae, H. Kim, Y. Lee, X. Xu, J.S. Park, Y. Zheng, J. Balakrishnan, T. Lei, H.R. Kim, Y.I. Song, Nat. Nanotechnol. **5**, 574 (2010)
62. E. Yablonovitch, Phys. Rev. Lett. **58**, 2059 (1987)
63. J.D. Joannopoulos, R.D. Meade, J.N. Winn, *Photonic Crystals: The Road from Theory to Practice* (Princeton University Press, Princeton, 1995)
64. J.D. Joannopoulos, S.G. Johnson, J.N. Winn, R.D. Meade, *Photonic Crystals: Molding the Flow of Light*, 2nd edn. (Princeton University Press, Princeton, 2008)
65. Z. Sun, Y.S. Jung, H.K. Kim, Appl. Phys. Lett. **83**, 3021 (2003)
66. Z. Sun, H.K. Kim, Appl. Phys. Lett. **85**, 642 (2004)
67. A.R. McGurn, A.A. Maradudin, Phys. Rev. B **48**, 17576 (1993)
68. V. Kuzmiak, A.A. Maradudin, Phys. Rev. B **55**, 7427 (1997)
69. Yu.E. Lozovik, S.L. Eiderman, M.V. Bogdanova, M. Willander, Laser Phys. **18**, 417 (2008)

70. S. Belousov, M. Bogdanova, A. Deinega, S. Eyderman, I. Valuev, Yu. Lozovik, I. Polischuk, B. Potapkin, B. Ramamurthi, T. Deng, V. Midha, Phys. Rev. B **86**, 174201 (2012)
71. H. Takeda, K. Yoshino, Phys. Rev. B **67**, 073106 (2003)
72. H. Takeda, K. Yoshino, Phys. Rev. B **67**, 245109 (2003)
73. H. Takeda, K. Yoshino, A.A. Zakhidov, Phys. Rev. B **70**, 085109 (2004)
74. O.L. Berman, Yu.E. Lozovik, S.L. Eiderman, R.D. Coalson, Phys. Rev. B **74**, 092505 (2006)
75. Yu.E. Lozovik, S.I. Eiderman, M. Willander, Laser Phys. **17**, 1183 (2007)
76. O.L. Berman, V.S. Boyko, R.Ya. Kezerashvili, Yu.E. Lozovik, Phys. Rev. B **78**, 094506 (2008)
77. O.L. Berman, V.S. Boyko, R.Ya. Kezerashvili, Yu.E. Lozovik, Laser Phys. **19**, 2035 (2009)
78. A.N. Poddubny, E.L. Ivchenko, Yu.E. Lozovik, Solid State Commun. **146**, 143 (2008)
79. S.Y. Lin, J. Moreno, J.G. Fleming, Appl. Phys. Lett. **83**, 380 (2003)
80. S.Y. Lin, J.G. Fleming, I. EI-Kady, Appl. Phys. Lett. **83**, 593 (2003)
81. K. Busch, G. von Freymann, S. Linden, S.F. Mingaleev, L. Tkeshelashvili, M. Wegener, Phys. Rep. **444**, 101 (2007)
82. B. Schulkin, L. Sztancsik, J.F. Federici, Am. J. Phys. **72**, 1051 (2004)
83. L.D. Landau, E.M. Lifshitz, *Electrodynamics of Continuous Media*, 2nd edn. (Pergamon Press, Oxford, 1984)
84. R. de L. Kronig, W.G. Penney, Proc. R. Soc. A (London) **130**, 499 (1930)
85. H. Hajian, A. Soltani-Vala, M. Kalafi, Opt. Commun. **292**, 149 (2013)
86. A. Madani, S.R. Entezar, Phys. B **431**, 1 (2013)
87. Z. Arefinia, A. Asgari, Phys. E **54**, 34 (2013)
88. A. Taflove, *Computational Electrodynamics: The Finite-Difference Time-Domain Method* (Artech House, Norwood, 1995)
89. D. Jahani, A. Soltani-Vala, J. Barvestani, H. Hajian, J. Appl. Phys. **115**, 153101 (2014)
90. O.L. Berman, V.S. Boyko, R.Ya. Kezerashvili, Yu.E. Lozovik, Phys. Rev. B **78**, 094506 (2008)
91. L.V. Keldysh, Sov. Phys. JETP **18**, 253 (1964)
92. J.M. Luttinger, W. Kohn, Phys. Rev. **97**, 869 (1955)
93. W. Kohn, in *Solid State Physics*, vol. 5, ed. by F. Seitz, D. Turnbull (Academic, New York, 1957), pp. 257–320
94. S. Mikhailov, K. Ziegler, Phys. Rev. Lett. **99**, 016803 (2007)
95. M. Jablan, H. Buljan, M. Soljačić, Phys. Rev. B **80**, 245435 (2009)
96. F.H.L. Koppens, D.E. Chang, F.E. García de Abajo, Nano Lett. **11**, 3370 (2011)
97. G.W. Hanson, J. Appl. Phys. **103**, 064302 (2008)
98. A.Y. Nikitin, F. Gurcía-Vidal, L. Martín-Moreno, Phys. Rev. B **84**, 161407 (2011)
99. Yu.E. Lozovik, Phys. Usp. **55**, 1035 (2012)
100. S.J. Zalyubovskiy, M. Bogdanova, A. Deinega, Y. Lozovik, A.D. Pris, K.H. An, W.P. Hall, JOSA **29**, 994 (2012)
101. L. Ju, B. Geng, J. Horng, C. Girit, M. Martin, Z. Hao, H.A. Bechtel, X. Liang, A. Zettl, Y.R. Shen et al., Nat. Nanotechnol. **6**, 630 (2011)
102. J. Christensen, A. Manjavacas, S. Thongrattanasiri, F.H.L. Koppens, F.J. García de Abajo, ACS Nano **6**, 431 (2012)
103. O.V. Kotov, Yu.E. Lozovik, Phys. Lett. A **375**, 2573 (2011)
104. O.V. Kotov, Yu.E. Lozovik, Fullerenes Nanotubes Carbon Nanostruct. **20**, 563 (2012)
105. S. Thongrattanasiri, F.H.L. Koppens, F.J.G. de Abajo, Phys. Rev. Lett. **108**, 047401 (2012)
106. L. Wu, H. Chu, W. Koh, E. Li, Opt. Express **18**, 14395 (2010)
107. O.V. Kotov, M.A. Kol'chenko, Yu.E. Lozovik, Opt. Express **21** 13533 (2013)
108. A. Vakil, N. Engheta, Science **332**, 1291 (2011)
109. S. Das Sarma, E.H. Hwang, Phys. Rev. Lett. **102**, 206412 (2009)
110. O.L. Berman, G. Gumbs, Yu.E. Lozovik, Phys. Rev. B **78**, 085401 (2008)
111. Y. Zheng, T. Ando, Phys. Rev. B **65**, 245420 (2002)

112. P. Bakshi, J. Cen, K. Kempa, J. Appl. Phys. **64**, 2243 (1988)
113. J. Cen, K. Kempa, P. Bakshi, Phys. Rev. B **38**, 10051 (1988)
114. K. Kempa, P. Bakshi, J. Cen, H. Xie, Phys. Rev. B **43**, 9273 (1991)
115. K. Kempa, J. Cen, P. Bakshi, Phys. Rev. B **39**, 2852 (1989)
116. K.W. Chiu, J.J. Quinn, Phys. Rev. B **9**, 4724 (1974)
117. E.H. Hwang, S. Das Sarma, Phys. Rev. B **75**, 205418 (2007)
118. T. Ando, J. Phys. Soc. Jpn. **75**, 074716 (2006)
119. D.J. Bergman, M.I. Stockman, Phys. Rev. Lett. **90**, 027402 (2003)
120. M.I. Stockman, J. Opt. **12**, 024004 (2010)
121. I.E. Protsenko, Phys. Usp. **55**, 1040 (2012)
122. Yu.E. Lozovik, I.A. Nechepurenko, A.V. Dorofeenko, E.S. Andrianov, A.A. Pukhov, Phys. Lett. A **378**, 723 (2014)
123. V.M. Agranovich, V.L. Ginzburg, *Crystal Optics with Spartial Dispersion, and Excitons* (Springer, Berlin, 1984)
124. Yu.E. Lozovik, A.V. Klyuchnik, The dielectric function and collective oscillations inhomogeneous systems, in *The Dielectric Function of Condensed Systems*, ed. by L.V. Keldysh, D.A. Kirzhnitz, A.A. Maradudin (Elsevier, Amsterdam, 1987), p. 299
125. A.V. Zayats, I.I. Smolyaninov, A.A. Maradudin, Phys. Rep. **408**, 131 (2005)
126. D.K. Gramotnev, S.I. Bozhevolnyi, Nat. Photon. **4**, 83 (2010)
127. M.I. Stockman, Nat. Photon. **2**, 327 (2008)
128. A.A. Lisyansky, I.A. Nechepurenko, A.V. Dorofeenko, A.P. Vinogradov, A.A. Pukhov, Phys. Rev. B **84**, 153409 (2011)
129. E.S. Andrianov, A.A. Pukhov, A.V. Dorofeenko, A.P. Vinogradov, A.A. Lisyansky, Opt. Lett. **36**, 4302 (2011)
130. E.S. Andrianov, A.A. Pukhov, A.V. Dorofeenko, A.P. Vinogradov, A.A. Lisyansky, Opt. Express **19**, 24849 (2011)
131. E.S. Andrianov, A.A. Pukhov, A.V. Dorofeenko, A.P. Vinogradov, A.A. Lisyansky, Phys. Rev. B **85**, 035405 (2012)
132. E.S. Andrianov, A.A. Pukhov, A.V. Dorofeenko, A.P. Vinogradov, A.A. Lisyansky, Phys. Rev. B **85**, 165419 (2012)
133. A.P. Vinogradov, E.S. Andrianov, A.A. Pukhov, A.V. Dorofeenko, A.A. Lisyansky, Phys. Usp. **55**, 1046 (2012)
134. D.G. Baranov, E.S. Andrianov, A.P. Vinogradov, A.A. Lisyansky, Opt. Express **21**, 10779 (2013)
135. A. Noginov, G. Zhu, V.P. Drachev, V.M. Shalaev, in *Nanophotonics with Surface Plasmons*, ed. by V.M. Shalaev, S. Kawata (Elsevier, Amsterdam, 2007)
136. M.A. Noginov, G. Zhu, A.M. Belgrave, R. Bakker, V.M. Shalaev, E.E. Narimanov, S. Stout, E. Herz, T. Suteewong, U. Wiesner, Nature (London) **460**, 1110 (2009)
137. N.I. Zheludev, S.L. Prosvirnin, N. Papasimakis, V.A. Fedotov, Nat. Photon. **2**, 351 (2008)
138. E. Plum, V.A. Fedotov, P. Kuo, D.P. Tsai, N.I. Zheludev, Opt. Express **17**, 8548 (2009)
139. R.A. Flynn, C.S. Kim, I. Vurgaftman, M. Kim, J.R. Meyer, A.J. Mäkinen, K. Bussmann, L. Cheng, F.-S. Choa, J.P. Long, Opt. Express **19**, 8954 (2011)
140. V.V. Popov, T.Yu. Bagaeva, T. Otsuji, V. Ryzhii, Phys. Rev. B **81**, 073404 (2010)
141. M.I. Stockman, *Ultrafast Phenomena in Semiconductors VII*. SPIE Proceedings, vol. 4992, ed. by K.-T.F. Tsen, J.-J. Song, H. Jiang (Society of Photo Optical, 2003), pp.60–74
142. P. Neugebauer, M. Orlita, C. Faugeras, A.L. Barra, M. Potemski, Phys. Rev. Lett. **103**, 136403 (2009)
143. M. Orlita, M. Potemski, Semicond. Sci. Technol. **25** 063001 (2010)
144. A.A. Dubinov, V.Ya. Aleshkin, V. Mitin, T. Otsuji, V. Ryzhii, J. Phys. **23**, 145302 (2011)
145. N.K. Emani, T.-F. Chung, X. Ni, A.V. Kildishev, Y.P. Chen, A. Boltasseva, Nano Lett. **12**, 5202 (2012)
146. A.B. Kuzmenko, I. Crassee, D. van der Marel, P. Blake, K.S. Novoselov, Phys. Rev. B **80** 165406 (2009)

147. K.F. Mak, C.H. Lui, J. Shan, T.F. Heinz, Phys. Rev. Lett. **102**, 256405 (2009)
148. Y. Zhang, T.T. Tang, C. Girit, Z. Hao, M.C. Martin, A. Zettl, M.F. Crommie, Y.R. Shen, F. Wang, Nature (London) **459**, 820 (2009)
149. W.B. Jackson, N.M. Am. Phys. Rev. B **25**, 5559 (1982)
150. J. Gosciniak, D.T.H. Tan, Sci. Rep. **3**, 1897 (2013)

Chapter 4
Materials Challenges for Concentrating Solar Power

Dominic F. Gervasio, Hassan Elsentriecy, Luis Phillipi da Silva,
A.M. Kannan, Xinhai Xu, and K. Vignarooban

Abstract A heat transfer fluid (HTF) is a major component in the system for concentrating solar power systems (CSP) to make electricity. The HTF carries thermal energy from the solar concentrator to a steam generator. Currently hydrocarbon oils or alkali-nitrate-based eutectic molten-salt mixtures are used as the HTF in CSP systems, but these materials have limited operating temperature range, which limits efficiency. Hydrocarbons are limited to 250 °C and alkali-nitrate salts are stable only below 600 °C. Using abundant inexpensive materials to make an HTF which is stable to 1,300 °C and compatible with a metal housing, like a Hastelloy nickel alloy, is desired. Design rules are given which tell how the desired goals can be met, which leads to mixing abundant ionic chloride salts, like NaCl and KCl, which boil at temperatures higher than 1,400 °C, with low-melting (~200 °C) covalent metal halides, such as $AlCl_3$ or $ZnCl_2$, to give low-melting (m.p. < 250 °C) eutectic mixtures, which are stable at high temperatures. To have negligible corrosion of the metals which house the eutectic, the component eutectic should have more negative reduction potentials than metals in the salt housing. Accordingly, the ternary K–Na–Zn chloride molten-salt mixtures in the alloy metal housing should be stable. However, corrosion of the metal housing is seen, especially at higher temperatures. The corrosion rates of housing alloys in molten salt in the presence of or excluding air have been experimentally determined at different temperatures. Indications are that the corrosion of the metal is not due to the salt itself but dissolved impurities like water and oxygen.

D.F. Gervasio (✉) • H. Elsentriecy • L.P. da Silva • X. Xu
Department of Chemical and Environmental Engineering, University of Arizona,
1133 E James Rogers Way, Room 108 Harshbarger Building, Tucson, AZ 85721, USA
e-mail: dominic.gervasio@arizona.edu

A.M. Kannan
Fulton School of Engineering, Arizona State University, Mesa, AZ, USA

K. Vignarooban
Fulton School of Engineering, Arizona State University, Mesa, AZ, USA

Department of Physics, Faculty of Science, University of Jaffna, Jaffna 40000, Sri Lanka

© Springer International Publishing Switzerland 2015
A. Korkin et al. (eds.), *Nanoscale Materials and Devices for Electronics,*
Photonics and Solar Energy, Nanostructure Science and Technology,
DOI 10.1007/978-3-319-18633-7_4

4.1 Introduction

4.1.1 Concentrating Solar Power

Electrical power-generating plants which are driven by concentrating solar power (CSP) have been attracting more and more interest from all over the world in recent years [1]. A CSP system converts solar-to-thermal-to-electrical energy by using mirrors to concentrate sunbeams on a collector containing a heat transfer fluid (HTF) bringing it to an elevated temperature (typically between 300 and 550 °C). The HTF flows to a heat exchanger where its heat expands a fluid to drive a turbine and thereby makes electricity by magnetic induction. In other words, the CSP system is like a conventional electrical power-generating system except the HTF is the source of heat.

There are a variety of mirror shapes, sun-tracking methods, and ways to provide energy, but they all drive the CSP generator the same way, converting solar-to-thermal-to-electrical energy in a Carnot process. A CSP plant typically generates between 50 and 280 MW of electrical power, but larger still plants are possible. CSP offers steady and reliable power and eliminates the problematic effects of solar transients associated with photovoltaic (PV) systems. CSP systems are being built with thermal storage to provide electrical power which can be dispatched from thermal storage 6 h after sundown. CSP is suitable for peak loads and base loads. CSP systems are readily integrated to the existing electrical power grid. Along with coal and nuclear sources, CSP offers another reliable large-scale source for supplying electrical power [2].

The CSP industry has been developing rapidly from being a nascent technology to becoming a mass-produced and mainstream energy generation solution. A CSP plant consists principally of three key sections:

1. The solar field with collectors where sunlight is converted to heat
2. The power block which converts heat to electrical energy
3. The storage system which saves heat for use after sundown

Building these sections to collect, move, and store heat presents materials challenges. The three sections are made of corrosion-resistant metal to perform under thermal, mechanical, and chemical stresses. The higher the temperature of the solar-to-thermal conversion, the greater the efficiency (approaching 40 %) and lower the cost (~6¢ per kWh) of the system but the greater the materials challenges.

A number of challenges have already been met and CSP technologies have matured considerably over the last few years thanks to innovations in solar collectors, heat transfer fluids, and thermal energy storage (TES) [3]. CSP installations were providing just 436 MW of the world's electricity generation at the end of 2008. In the USA, projects adding 7,000 MW are under planning and development and in Spain an additional 10,000 MW, which could all come online by 2017 [2]. With suitable further development, it has been estimated that CSP systems could satisfy 11.3 % of the global electricity demand by the year 2050 [4]. The thermal storage of excess solar energy, which is readily integrated in the CSP plant, gives another opportunity for advancing the implementation of CSP. The capacity of the thermal storage system determines the time electrical power is available from a CSP plant after sundown [5].

4.1.2   Hastelloy

A CSP system is more efficient and desirable as the temperature of operation increases. A major challenge from a materials standpoint is finding suitable heat transfer fluids and containers and pipes for housing the heat transfer fluid as the temperature of operation of the CSP plant is increased from 300 to 500 °C up to 800–1,000 °C or so. A number of heat transfer fluids have been suggested for use in CSP systems. These include steam, heavy oil, molten glass, liquid metals, and molten salts. Steam requires excessively large pipes; heavy petroleum oil is limited to 300 °C and has explosive vapors; molten glasses are too viscous and have low thermal conductivities; liquid metals have a limited range of operating temperatures, can alloy with pipe materials, and emit toxic vapors. Molten nitrates are limited to 550 °C. Halide salts can operate at higher temperatures but stable containers for halide salt at high temperatures (~1,000 °C) are an issue.

Metals are highly desired for containing and piping the heat transfer fluids, because metals have good mechanical strength and are tolerant to mechanical and thermal shock. Corrosion-resistant metals, mainly stainless steels, are presently used to contain the hot petroleum-based (b. p. = 300 °C) and molten nitrate-based (b. p. = 550 °C) heat transfer fluids presently used in CSP plants, where these metal pipes and containers have performed well under the thermal, mechanical, and chemical stresses found up to 550 °C in these CSP systems. The temperature of the CSP system can be increased (800–1000 °C) using chloride-based HTFs, but the long-term corrosion resistance of metals in hot molten halide salts is still uncertain.

Hastelloy (®Haynes International, Inc.) is a family of mainly nickel alloys known throughout the chemical process industry as one of the most corrosion-resistant structural materials [6]. For the most part, we will focus on Hastelloy® C-22®, which is a nickel–chromium–molybdenum–tungsten alloy, whose nominal chemical composition is shown in Table 4.1.

Table 4.1 Composition of Hastelloy C-276, C-22, and N types

Hastelloy type	Composition (Wt. %)								
	Ni	Co	Cr	Mo	W	Fe	Mn	Si	C
C-276	57	1[a]	16	16	4	5	1[a]	0.08[a]	0.01[a]
C-22	56	2.5[a]	22	13	3	3	0.5[a]	0.08[a]	0.01[a]
N	71	–	7	16	–	5[a]	0.8[a]	1[a]	0.08[a]

[a]Maximum

Hastelloy C-22 has good corrosion resistance at high temperatures in air mainly due to the spontaneous formation of a highly electrically resistive surface oxide layer, which is characteristic of most Ni–Cr–Mo alloys, including Hastelloys C-276, N, C-4, and alloy 625. The C-22 alloy has outstanding resistance to pitting, crevice corrosion, and stress corrosion cracking. C-22 has excellent resistance to oxidizing aqueous media including wet chlorine and mixtures containing nitric acid or oxidizing acids with chloride ions. C-22 alloy also offers resistance to environments where reducing and oxidizing conditions are encountered in process streams. Because of these properties, C-22 can be used in multipurpose plants where "upset" conditions are likely to occur [6].

The chemical process environments in which C-22 alloy has exhibited exceptional corrosion resistance include: strong oxidizers such as ferric and cupric chlorides, chlorine, hot contaminated solutions (organic and inorganic), formic and acetic acids, acetic anhydride, and seawater and brine solutions. C-22 alloy is available in most common product forms: plate, sheet, strip, billet, bar, wire, covered electrodes, pipe, and tubing. Applications of C-22 alloy include: pipes for tubular heat exchanger, geothermal wells, nuclear fuel reprocessing, and now possibly high-temperature CSP plants [6].

4.1.3   Molten-Salt Heat Transfer Fluids

There is a pressing need to find higher-temperature (operating temperature $>600\,^\circ$C) heat transfer fluids (HTFs) made from inexpensive naturally abundant materials for the next generation of CSP systems. A higher-temperature HTF made of molten salt potentially offers a number of advantages, including: (1) high energy density of latent heat storage, (2) eutectic or pure salt options which can provide operating temperatures up to 1,300 $^\circ$C, (3) simpler thermal energy storage (TES) and heat delivery to the power conversion system which is close to the salt melt temperature, and (4) expectations for 4 years or more of operation as demonstrated with a small TES system using stainless steel or nickel alloy (Hastelloy) containment [7, 8].

The state-of-the-art heat transfer fluids presently used in CSP systems are synthetic oil and a eutectic nitrate salt (which is a mix of sodium and potassium nitrate) which have operating temperature limits of 300 $^\circ$C and 550 $^\circ$C, respectively. These low operating temperatures limit the thermal-to-electric conversion efficiency in a CSP plant. New HTFs are needed to develop the next-generation CSP technology operating with a higher efficiency and power generation without water [9, 10]. The US Department of Energy (US DOE) has solicited researchers to find HTFs which meet targets found in Table 4.2.

Currently electrical generators are driven by steam generated from water heated by the HTF. Several molten-salt mixtures have been proposed as new high-temperature heat transfer fluids (HTFs) which meet critical criteria to advance the efficiency of CSP systems at temperatures of 700–800 $^\circ$C which must be achieved for fully utilizing advanced supercritical carbon dioxide (S-CO$_2$) Brayton cycles,

Table 4.2 US DOE targets for heat transfer fluids (HTFs)

Property	Target	Stretch target
Thermal stability (liquid)	$\geq$800 °C	=1,300 °C
Melting point	$\leq$250 °C	$\leq$0 °C
Heat capacity	$\geq$1.5 J/g/K	$\geq$3.75 J/g/K
Vapor pressure	$\leq$1 atm	
Viscosity	$\leq$0.012 Pa s @ 300 °C; $\leq$ 0.004 Pa s @ 600 °C	
Density	$\leq$6,000 kg/m^3 @ 300 °C; $\leq$ 5,400 kg/m^3 @ 600 °C	
Thermal conductivity	$\geq$0.51 W/m/K @ 300 °C; $\geq$ 0.58 W/m/K @ 600 °C	
Materials compatibility	Carbon steel (<425 °C), stainless steel (<650 °C), and nickel alloys (>800 °C). Corrosion <100 mm/year	
Materials cost	$\leq$$1/kg	

which eliminate the use of water during electrical power generation [10]. Halide salts, particularly chloride salts, are promising alternative HTFs for a high-temperature operation, because they have large natural abundance which leads to low cost.

The strategy for making chloride salts in a heat transfer fluid is to mix covalent and ionic chlorides. Covalently bonded chloride salts have low melting points, and ionically bonded chloride salts have high boiling points. Bonding between the Lewis-acid covalent metal halides and Lewis-base ionic halides induces mixing instead of segregation of the covalent and ionic salts at the molecular level, which leads to the formation of eutectic mixtures, which are suitable for giving new HTFs with low melting ($\leq$200 °C) and high boiling points (>1,000 °C) [11]. New molten halide salts have been made with such ionic chloride salts (e.g., NaCl and KCl) which boil at temperatures higher than (1,400 °C) mixed with such low-melting ($\leq$200 °C) covalent metal halides (e.g., $ZnCl_2$ and $AlCl_3$). In the proper proportions, these components do indeed form eutectic mixtures in which the covalent metal halide is stabilized by chloride from the ionic chloride. This stabilization occurs when the ionic chloride complexes with the transition metal in the covalent halide. These eutectic mixtures have been found to be low melting yielding thermally stable liquids which boil at the high temperatures with very low vapor pressures which are desired properties for heat transfer fluids in CSP systems.

Three ternary eutectic salts, $ZnCl_2$–KCl–NaCl, $AlCl_3$–NaCl–KCl, and $FeCl_3$–NaCl–KCl, have been found and studied to date. The Zn salt is the most promising as a suitable HTF [11]. Some relevant characteristics about the ternary $ZnCl_2$–KCl–NaCl are described next.

Three Zn eutectic compositions satisfy the melting point criterion: $T_{Salt\#1} = 204$ °C, $T_{Salt\#2} = 213$ °C, and $T_{Salt\#3} = 219$ °C. The phase diagram of this system can be found in [12], and the mole fraction of each salt is shown in Table 4.3 [3].

Table 4.3 Composition of molten salts #1, #2, and #3

Salt type	Melting point (°C)	Mole fraction		
		NaCl	KCl	ZnCl$_2$
# 1	204	0.134	0.337	0.529
# 2	213	0.186	0.219	0.595
# 3	229	0.138	0.419	0.443

The physical properties of eutectic zinc chloride molten salts virtually meet the DOE targets in Table 4.2 for use as a heat transfer fluid. However, the corrosion resistance of metals in these hot molten chloride salts is a pivotal issue. The ZnCl$_2$–KCl–NaCl halide eutectic molten-salt #1 mixture and C-22 alloy are used to illustrate the corrosion behavior characteristic of nickel alloys in molten chloride eutectic salts.

4.1.4 Corrosion

4.1.4.1 Corrosion Concepts

Metal pipes and storage vessels are important parts of CSP plants. Metals can contain fluids under thermal, mechanical, and chemical stresses and make the plant more robust and reliable [1] provided the metal is not corroded by the fluids. Only a few CSP plants in the world have tested high-temperature operation (up to 600 °C) [13]. Higher operating temperatures ($\geq$800 °C) increase the efficiency in the CSP plant but can also promote faster corrosion of the metal materials used to make pipes and containment vessels for molten salt [1]. When the salts in Table 4.3 contain little or no oxidizing impurities (oxygen and water), they are generally not corrosive since most metals are thermodynamically stable versus the metal ions in these salts [14].

There are three main restrictions that must be satisfied for obtaining a useful heat transfer fluid and a container to house it. (1) The HTF has to be low melting, which is achieved by mismatching the size, shape, or bonding between the positive ion and negative ion to make disorder leading to liquid salt at low temperatures. For example, AlCl$_3$ is a large covalently bonded tetrahedron and NaCl is a small ionic bonded cube. (2) At the same time, the covalent and ionic chloride salts must not segregate but must intermix at a molecular level to form a stable high-boiling liquid mixture with low vapor pressure at $T > 800$ °C, which is accomplished by using Lewis acid–base chemistry to attract the covalent Lewis-acid component to ionic Lewis-base component forming a stabilized Lewis salt, as is schematically shown below:

$$ZnCl_2 + NaCl + KCl \rightarrow Na^+K^+ZnCl_4^{-2}$$

Lewis acid + Lewis base → Lewis salt

$E_{\text{Zn, Al, Na, K}}$ (molten salts) $<<$ E_{Fe} (stainless, carbon steel) $<$ E_{Ni} (Hastelloys)

Less Noble $\longrightarrow$ **More Noble**

Scheme 4.1 Reduction potentials of metal ions for salt and metals for containers

(3) Lastly, the Lewis salt HTF must be contained in a material it does not corrode. The first guide, for choosing a metal container for a Lewis salt, is the tables of standard electrochemical potentials in aqueous electrolytes. The metal ions in salt must be less noble than the metal of the container. In this crude analysis, we assume that the effects of water as solvent and chloride on reduction potentials of all metals are the same and therefore do not affect the ordering of the reduction potentials in choosing materials for the corrosion of the metal container in contact with salts. The data found in the table of reduction potentials is summarized in Scheme 4.1.

Chloride salts in aqueous solution can break down the protective passive oxide layer on the metal surface by anion exchange, leaving bare metal open to oxidative attack and so promote corrosion reactions on metal surfaces. The corrosion of metal in molten-salt systems with oxygen and water impurities is somewhat similar to the corrosion in aqueous environments [15]. It is reasonable to expect that since molten-chloride-salt systems promote the breakdown of the passive oxide layers, there is no protective oxide layer on metals in molten chloride salts. For several years, many works have been performed and published about these molten-salt corrosion issues, some of these publications can be found in [14, 16–21].

One of the earliest studies about corrosion issues of CSP systems was published in 1985 concerning corrosion behavior of both iron-based and nickel-based alloys in nitrate–nitrite salts between 510 and 705 °C and nickel alloys with 15–20 % chromium content performed the best at high temperatures [14]. The corrosion of nickel-containing alloys in molten LiCl–KCl medium has been studied in a recent work and Inconel 600 and Inconel 690 alloys showed better corrosion resistance compared to Inconel 625 and alloy 800H at high temperatures [16]. The corrosion behavior of Inconel 718 alloy in molten salts was studied by electrochemical impedance spectroscopy recently [17]. A spectroelectrochemical study of stainless steel corrosion in NaCl–KCl melt found that the major corrosion products of stainless steel are iron, manganese, and chromium species [18]. The corrosion of Hastelloys in NaCl–KCl–ZnCl$_2$ ternary molten-salt eutectic mixtures has been reported for the first time only recently [22].

4.1.4.2 Methods for Determining Corrosion of Metal in Molten Salt

There are three methods that can be used to assess metal and alloy corrosion: (1) the gravimetric weight loss method, (2) the electrochemical Tafel extrapolation method, and (3) the electrochemical polarization resistance method. The last two methods are electrochemical corrosion tests based on converting the corrosion current to mass loss of metals and metal alloys.

With the gravimetric weight loss method, the metal needs to be immersed several weeks or even months to get a measurable weight change. During this time, the corrosive environment and therefore corrosion rate can fluctuate, so the gravimetric method gives an average corrosion rate over this long test time. On the other hand, the electrochemical tests can be performed in minutes, so these electrochemical methods allow the estimation of the instantaneous corrosion rates (ICRs) which occur during the fluctuations in corrosive environments during gravimetric testing [23]. Weight averaging the ICRs from electrochemical testing should give the average corrosion rate found by the gravimetric method. Electrochemical methods available for laboratory corrosion testing are potentiodynamic polarization, potentiostaircase voltammetry, and impedance spectroscopy [24]. Among these electrochemical methods, the steady-state potentiodynamic polarization method is the most reliable and most commonly used method for estimating corrosion rates [25–30].

The steady-state potentiodynamic polarization method has the metal of interest as a working electrode in an electrolyte representative of the corrosive environment. The working electrode is one of three electrodes in the test cell. Usually current is passed between the working and a counter electrode made of the same material as the working, as the potential of the working electrode is controlled versus a reference electrode. The reference electrode has a constant potential. The working electrode potential is controlled and slowly scanned ($\sim$0.1 mV s^{-1}) through a small voltage window ($\pm$30 mV from the open-circuit voltage, OCV) as the resulting current is recorded. From the resulting current/voltage (I/V) plot, the corrosion current, i_{corr}, can be estimated at the corrosion potential, E_{corr}. This corrosion current is used to estimate the instantaneous corrosion rate (ICR).

The corrosion rates of a number of metals in molten ternary eutectic chloride salts have been evaluated previously and recently reported [22]. The corrosion behavior of a nickel-based Hastelloy, C-22, in molten salt will illustrate how these results were obtained and the challenges of using metals to house hot molten salts. The electrochemical potentiodynamic method was used at three different temperatures (300, 500, and 800 °C) to estimate the corrosion rate of the alloy in the ternary NaCl–KCl–ZnCl$_2$ eutectic molten salt with a melting point of 204 °C.

4.2 Experimental Methods

4.2.1 Materials Preparation

4.2.1.1 Molten Salt

The salt used in this study was a Na–K–Zn chloride called salt # 1. This salt has the following % molar compositions 13.4NaCl–33.7KCl–52.9ZnCl$_2$ [11, 31], and the preparation of 100 g of salt mixture has the grams of each component as given in Table 4.4.

Salt #1 has a melting temperature of 204 °C and all components were stored in an inert atmosphere or under vacuum before loading them into the furnace where

Table 4.4 Component weight for the eutectic salt #1 mixture

Required weight of the salt (g)	NaCl (g)	KCl (g)	ZnCl$_2$ (g)
100	7.45	23.91	68.63

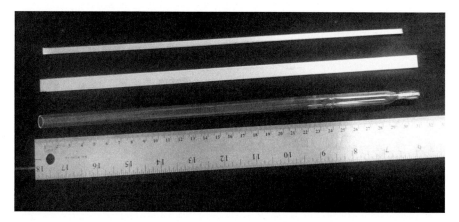

Fig. 4.1 Working, counter, and reference electrodes (from the top to the bottom of the picture)

the salt mixture was heated to above 200 °C within 20 min so there was little bulk water. There, the only source of water or oxygen in the salt came from these diffusing into the molten salt from air.

4.2.1.2 Reference Electrode

To do an electrochemical measurement of the corrosion rate of metal in molten salt at high temperatures, a reference electrode (RE) with a stable potential at each temperature is required in order to control the potential of the sample (working electrode, WE) during the measurements. The reference electrode was a quartz tube with a ceramic rod sealed at the tip of the tube making a tortuous path for ionic exchange between the molten salts around the working electrode and reference electrode during the electrochemical measurements. Proper amounts of silver (Ag), silver chloride (AgCl), and potassium chloride (KCl) powders were weighed and mixed and inserted in the quartz tube. A silver wire (about 1 mm diameter) was inserted in the salt in the quartz tube for electrical connection between the RE and the potentiostat, the device used for performing the electrochemical corrosion tests. The resulting electrode is called "a cracked junction Ag/AgCl reference electrode," which has a potential of 0.2 V versus NHE at 25 °C and shifts in potential by −0.0007 V/°C.

Figure 4.1 shows the working and counter Hastelloy electrodes and the Ag/AgCl reference electrode. All of them have similar length, approximately 30 cm.

4.2.1.3 Hastelloy

The Hastelloy C-22 (thickness: 0.6 mm, from Haynes International Inc., Kokomo, IN, USA [32]) was evaluated for its corrosion behaviors at three different temperatures 300, 500, and 800 °C. The major elements in this alloy are Ni, Co, Cr, Mo, W, and Fe. The Haynes product literature on this alloy provides the specific composition along with trace elements (the composition is shown in Table 4.1). The Hastelloy samples (WE and counter electrode, CE) immersed in the molten salt for electrochemical tests were polished by a 600 grit SiC paper, cleaned with distilled water, cleaned with acetone, and then dried before immersion in the molten salt.

4.2.2 Experimental Electrochemical Setup

The experimental setup for conducting potentiodynamic scans of Hastelloy as the working electrode in molten salt at high temperatures is shown in Fig. 4.2a, b. The high quartz cell holds the working, counter, and reference electrodes in the molten-salt mixture. In order to eliminate contamination in the three-electrode system, the counter electrode material was the same alloy used as the working electrode in every measurement and the cracked junction separated the reference electrode electrolyte from the electrolyte in the cell. The working electrode in the salt mixture and reference and counter electrodes were connected to a BioLogic VSP potentiostat/galvanostat, which computer controlled potential and recorded measured current.

As shown in the experimental setup in Fig. 4.2a, all the corrosion measurements were made with the reference electrode at the operating temperature. The temperature of the electrochemical quartz cell was controlled by the digital controller of Thermolyne 48000 furnace (Fig. 4.2b). Figure 4.2b shows on the left side the overview of the setup, with working, counter, and reference electrodes inside of the furnace, and on the right side the connection of three electrodes to the VSP modular 5 channels potentiostat/galvanostat.

4.2.2.1 Potentiodynamic Method

The geometric area of the working electrode was 2.0 cm^2 and the area of the counter electrode was 4 cm^2. The potentiodynamic polarization experiments were conducted using a VSP modular 5 channels potentiostat/galvanostat system at a scan rate of 0.3 mV s^{-1}. The sample (WE) was polarized ±30 mV versus the open-circuit voltage (OCV) versus the reference electrode, starting at the negative limit and scanning in the positive direction. The potential of the WE was determined versus the Ag/AgCl reference electrode. The potentiodynamic scans for corrosion rate estimation were conducted in air at 300, 500, and 800 °C.

4.2.2.2   Data Analysis

As shown in Fig. 4.3, the corrosion current density values are determined from tangent lines drawn on the anodic and cathodic portions of the current density versus potential curves and intersecting at the corrosion potential, using the EC-Lab software in the VSP modular 5 channels potentiostat/galvanostat.

The corrosion rates (CR) are estimated by using the formula derived from Faraday's law, as given by the American Society for Testing and Materials (ASTM) standards G59 and G102 [33, 34]:

$$CR = K_1[(i_{corr} \cdot EW)/\rho] \qquad (4.1)$$

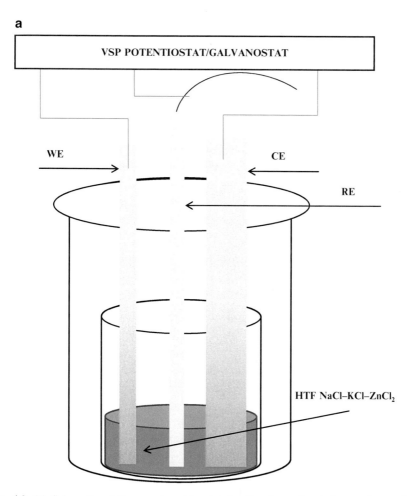

a

VSP POTENTIOSTAT/GALVANOSTAT

WE

CE

RE

HTF NaCl–KCl–ZnCl$_2$

Fig. 4.2 (**a**) Schematic of the experimental setup for the electrochemical corrosion testing. (**b**) Picture of the cell in oven and connections for electrochemically measuring the corrosion rate of C-22 Hastelloy in NaCl–KCl–ZnCl$_2$ eutectic molten salt at different high temperatures

b

Fig. 4.2 (continued)

Fig. 4.3 Method to
determine corrosion
current density

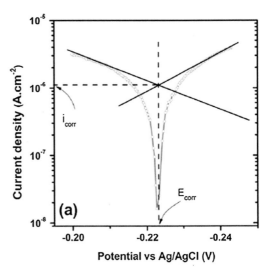

where CR is the corrosion rate in µm/year, $K_1 = 3.27$ in µm g µA/cm/year, i_{corr} is the corrosion current density in µA/cm^2, and EW and ρ are the equivalent weight and density (g cm^{-3}) of the Hastelloy, respectively [35]. The current density is from the apparent (projected) area, not real area, that is, no correction is made for the roughness of the surface which was finished by abrading with 600 grit SiC paper before corrosion measurements, as described above.

The expression, $EW \times K_1/\rho$, yields a constant value of 9.8 when the known values for the material properties of Hastelloy C-22 are inserted, and resulting constant is used in Eq. (4.1) to find the corrosion rates of Hatelloy from the experimentally measured values of i_{corr}.

4.2.3   Gravimetric Immersion Method

In all cases, the weight loss for a metal alloy coupon was measured after immersing it in eutectic molten chloride salt with a composition of 13.4 mol % NaCl, 33.7 mol % KCl, and 52.9 mol % $ZnCl_2$. This was done for the metal in the molten salt in a container open to and sealed from atmospheric air. Different coupons of Hastelloy were immersed in a quartz vessel containing molten salt from the same batch of eutectic chloride molten salt and heated at different temperatures in the presence of air and absence of air. The weight loss was measured for each coupon that has a different time of immersion. CR values were calculated from the weight loss values by using the following formula, assuming uniform dissolution:

$$CR \ (\text{in } \mu\text{m per year}) = (365)10^4 \left[WL/(\rho AT) \right] \tag{4.2}$$

where WL is the weight loss in gram, ρ is density (g cm^{-3}) of the alloy, A is the total immersed area in cm^2, and T is the immersion duration in days. All the data were collected by conducting three trials for each sample and the average CR values were calculated for comparison.

4.3   Results and Discussion

There are three methods that are commonly used to assess metal and alloy corrosion: (1) the gravimetric weight loss method, (2) the electrochemical Tafel extrapolation method, and (3) the electrochemical polarization resistance method [25].

The electrochemical polarization resistance method is a sensitive method that takes minutes and can be repeatedly used on the same sample to detect small changes in relative corrosion rates of an alloy sample in real time, semicontinuously and rapidly [36, 37]. Besides monitoring the temporal evolution of corrosion rate, the electrochemical polarization resistance method conveniently allows the screening of the relative corrosion rates as a function of changes in different parameters affecting corrosion, such as temperature and pressure and when the salt is equilibrated with different chemical environments (when the salt is equilibrated with air, dry and humid inert gas, etc.). The electrochemical method is considered more convenient than the gravimetric method, because electrochemical measurement allows hundreds of determinations of metal corrosion rate in the time it takes to do one gravimetric test. Therefore, the electrochemical corrosion methods were chosen to screen the corrosion rate in a variety of conditions. Corrosion rates were obtained by conducting steady-state potentiodynamic scans in a narrow voltage range to allow repeated use of the working electrode. The working electrode can be reused because there is no change in the surface composition or properties of the working electrode when its potential is no more than ± 30 mV from the value of the open-circuit potential (OCP) of the working versus the reference electrode. If a wider voltage range is used (e.g., ± 250 mV from the open-circuit voltage as is done during the Tafel extrapolation method), then the surface changes and the electrode cannot be reused for a corrosion test unless the electrode is removed from the cell and its surface reground and polished and equilibrated with the cell [25]. As reported in the literature, using a very low scan rate such as 0.1–0.3 mV s^{-1} is an important requirement to insure the polarization is essentially at steady state and capacitive contribution to the total current is negligible [26]. In recent experiments and publications, a small shift in E_{corr} (in the rage of 10–40 mV) has been observed, and a similar change in E_{corr} has already been reported in an ASTM standard by performing 33 repeated potentiodynamic polarization scans of the same corrosive system [22, 35].

4.3.1 Electrochemical Corrosion Testing

Figure 4.4 shows the polarization curves for Hastelloy C-22 in salt #1 at 300 °C.
    Operating conditions are shown in Table 4.5 for the polarization of Hastelloy C-22 in the K–Na–Zn chloride eutectic molten salt at 300 °C. The corrosion

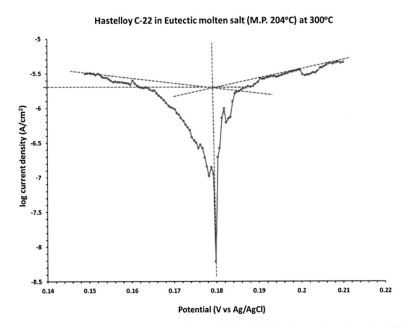

Fig. 4.4 Polarization curve for C-22 Hastelloy in ZnCl$_2$–KCl–NaCl eutectic salt in air at 300 °C

Table 4.5 General data and operating conditions at 300 °C

Surface area of working electrode	2 cm^2
Surface area of counter electrode	4 cm^2
Hastelloy sample type	C-22
Operation temperature	300 °C
Scan rate	0.3 mV s^{-1}
Salt (ZnCl$_2$–KCl–NaCl) melting point	204 °C
Atmosphere	Air

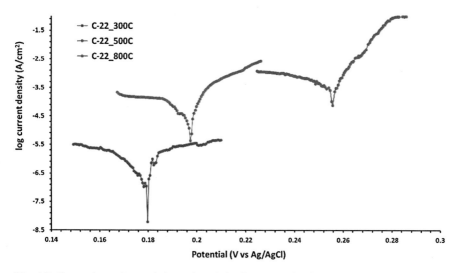

Fig. 4.5 Comparison of potentiodynamic polarization curves for C-22 Hastelloy in ZnCl$_2$–KCl–NaCl eutectic salt in air at different temperatures

potential (E_{corr}) and corrosion current density (i_{corr}) values from the potentiodynamic scans were estimated by drawing tangent lines on the anodic and cathodic portions of the current density versus potential curves and intersecting at the corrosion potential, as illustrated in Fig. 4.3 in the previous section above. The potentiodynamic curves conducted for Hastelloy C-22 in the same salt composition (salt #1, composition in Table 4.4) at 300, 500, and 800 °C, respectively, are given in Fig. 4.5.

Except for temperature, the operating conditions for curves taken at 300, 500, and 800 °C are identical and are as given in Table 4.5 The corrosion potential, the corrosion current density, and the calculated corrosion rate (CR) at the three temperatures are summarized in Table 4.6.

The corrosion rate values given in Table 4.6 are presented as a bar graph in Fig. 4.6.

Table 4.6 Corrosion potential, current density, and corrosion rate comparison for C22 in K–Na–Zn chloride molten salt in air

Temperature (°C)	E_{corr} (V)	i_{corr} (µA/cm^2)	CR (µm/year)
300	0.18	2.00	19.6
500	0.198	112.20	1,099.6
800	0.255	316.23	3,099.0

Fig. 4.6 C-22 Hastelloy corrosion rate comparison in ZnCl$_2$–KCl–NaCl eutectic salt at different temperatures

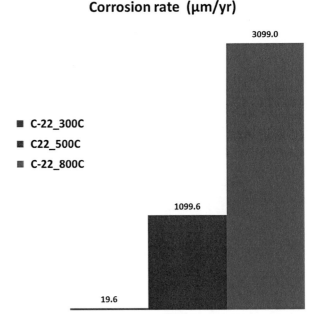

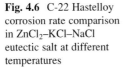

As shown in Fig. 4.6, the corrosion rates increased with increasing temperature. The corrosion rate at 800 °C is nearly 158 times more than the values at 300 °C and about three times more compared to corrosion rate at 500 °C.

Using the above analysis, the corrosion current densities have been found for Hastelloy C-276, C-22, and N as well as stainless steel 304 at 250 and 500 °C [22]. The ratio of these current densities is proportional to the ratio of the metal corrosion rates. These current densities are listed in Table 4.7.

Clearly Hastelloys C-22 and C-276 are the most stable metals under these conditions.

The effect of soaking time was studied and in general the corrosion rate seem to decelerate as a fixed amount of metal is soaked longer in a fixed amount of molten salt as is evident from Fig. 4.7a, b.

It is noteworthy that there is good agreement between the electrochemical and gravimetric test methods. The reason for the deceleration of corrosion rate may be due to a modification of the metal surface composition to a more noble alloy. This is the subject of ongoing work.

Table 4.7 Corrosion currents for 4 metal alloys in 3 aerated molten salts at 250 and 500 °C

Sample number	Alloy types	Corrosion current density (μA/cm^2) NaCl–KCl–ZnCl$_2$					
		13.4–33.7–52.9 mol% Salt #1		18.6–21.9–59.5 mol% Salt #2		13.8–41.9–44.3 mol% Salt #3	
		250 °C	500 °C	250 °C	500 °C	250 °C	500 °C
1	C-276	1.0	3.8	1.4	4.3	1.9	12.9
2	C-22	1.5	4.3	1.8	4.4	2.6	14.4
3	N	3.7	16.1	4.8	17.6	4.8	29.2
4	SS-304	2.2	38.1	2.3	16.9	2.7	36.3

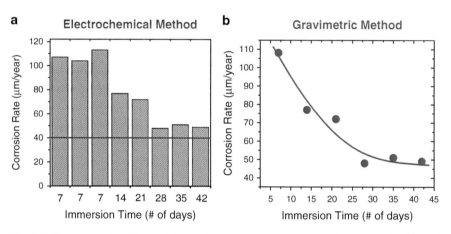

Fig. 4.7 Corrosion rate with metal immersion time in aerated molten salt as determined by (**a**) electrochemical and (**b**) gravimetric methods for Hastelloy C-276 in aerated NaCl–KCl–ZnCl$_2$ (13.4–33.7–52.9 mol%) eutectic salt mixture at 500 °C

4.3.2 Immersion Testing in Sealed Cells

Table 4.8 shows a preliminary result for the estimation of the corrosion rate of Hastelloy C-22 when the metal alloy was immersed in molten salt with virtually no air for 840 h.

The average corrosion rate is 40 μm/year for Hastelloy C-22 immersed in anaerobic salt which is ~80 times lower than for C-22 in the same salt in air (3,100 μm/year; see Table 4.6). Clearly dissolved oxygen and/or water from air accelerates the corrosion of metal in molten salt, and elucidating the role of these impurities in the mechanism of metal corrosion in molten salts is of paramount importance and is the subject of ongoing work in meeting the materials challenges for concentrating solar power.

Table 4.8 Measured results for C-22 corrosion in anaerobic K–Na–Zn chloride salt #3 when immersed for 840 h at 800 °C in ~40 g salt

Coupon	1	2	3	4
m_0 (mg)	806.4	1,070.1	1,380.1	1,679.1
M_f (mg)	802.1	1,065.0	1,373.1	1,672.1
$\Delta m/m_0$ (%)	0.5	0.5	0.5	0.4
Unit Δm (mg/cm^2)	2.9	2.6	1.5	0.8
CR (µm/year)	43	38	42	35

4.3.3 Discussion

Hastelloy C-22 has a substantial lower corrosion rate at 500 °C when compared in similar experiments made at the same temperature with other alloys such as Hastelloy N, as reported in [22]. Hastelloy N has five times higher corrosion rate than Hastelloy C-22, 150 µm/year versus 30 µm/year, respectively. It has also been reported that the higher the nickel content in the alloy, the lower the corrosion rate; however, pure nickel is poorly resistant to corrosion in molten salts [38]. The corrosion rate values at 300 °C and higher temperatures are important to know, because using Hastelloy to contain these ternary chloride salts will most likely be used when these salts are used for heat transfer in CSP applications. The corrosion rate will determine the service life of the pipes and containers.

On the outside of the pipe exposed to air and water, chromium can be used as a protective surface layer to improve the corrosion resistance and prevent higher metal corrosion at high temperatures due to exposure to oxygen and water from air [39]. The low silicon content in C-22 can also be a reason for the high corrosion resistance in this alloy, as there are some studies that attribute a poor corrosion resistance to alloys containing high levels of silicon in molten salts. Such silicon-containing alloys are Nicrofer-3718 and RA-330 [14, 38]. In the literature, some publications express that nickel-based alloys with around 20 % Cr content can perform well in molten-salt-corrosive environments [15, 40]. In addition, it is reported that 15–20 % of chromium-containing nickel alloys perform as the best corrosion-resistant alloys in molten nitrate–nitrite salts at high temperatures, which was determined using traditional weight loss studies. On the other hand, iron alloys perform poorly even with considerable chromium content [14]. The Hastelloy C-22, used in this present study, has approximately 22 % Cr content (see Table 4.1), and that can explain the low corrosion rate of this alloy; however investigation of the effect of Cr is still underway to determine the mechanism by which Cr might (or might not) prevent corrosion on the inside of the pipe in direct contact with molten salt. Alloys with cobalt and with refractory metals have been found to tend to increase the resistance of alloys to corrosion in hot molten salts. This corrosion resistance was followed by measuring the solubility of chromium into the chloride salt. When no refractory metals or cobalt was used, the Cr content in the salt was found to be around 0.1–0.2 %. However, alloying 12 % refractory metals or 18 % cobalt yielded 10–20 times less Cr in the salt, because stable spinel structure with a refractory or cobalt forms and acts as a barrier, slowing the diffusion of Cr out of the base alloy, and inhibits the diffusion of oxide into the alloy; thus nickel alloys with Mo, W, and Co are recommended [41].

4.4   Conclusions

The ternary $NaCl-KCl-ZnCl_2$ halide systems produce network-forming molten salts that exhibit a set of promising thermophysical properties and low corrosion to Hastelloy metal (demonstrated with $NaCl-KCl-ZnCl_2$) for application as heat transfer fluids (HTFs) at temperatures of 800 °C or above. These new ternary eutectic chlorides are formed by the complexation between a Lewis acid like $ZnCl_2$ and a Lewis base like NaCl and/or KCl forming a stabilized "Lewis salt" ternary halide salt system, like $Na-K-ZnCl_4$. The ternary chloride salts are eutectics which have low melting (~200 °C) and high boiling (>1,000 °C) points and act as good thermal transfer fluids (HTFs) for concentrating solar power (CSP) as high-grade heat in the collectors of solar radiation. However, a deeper physical understanding of the unusual dynamic properties and composition–property relationship in relation to the network-forming character of the HTFs is required before deployment in the CSP field. In particular, the mechanism of the resistance of container alloys to corrosion in the ternary chloride salts must be elucidated and optimized through compositional refinement of the salts and/or metal of the containers without compromising the critical physical properties of the HTF.

Hastelloys are a set of high nickel alloys known to be stable in air at high temperatures (~1,000 °C), because the Hastelloys spontaneously form a durable insulating layer which inhibits corrosions on the outside of a container in air. It is important to understand the corrosion behavior of Hastelloys exposed to molten chloride salt on the inside, to find if these chloride salts are suitable for long-term use as heat transfer fluids in next-generation electrical power-generating plants driven by CSP. To date, a number of Hastelloy and ternary eutectic chloride salt combinations have been studied and the ternary salts appear suitable as a heat transfer fluid, and from the corrosion standpoint, that is, on the salt side, the ternary halide salts appear compatible with Hastelloy containers, provided impurities, like water and oxygen from air, are excluded from the salt.

We have shown here how the corrosion rate of the Hastelloy C-22 in several ternary eutectics in aerated molten chloride salts is determined using the potentiodynamic method. We have also shown the determination of the corrosion rate of Hastelloy C-22 in anaerobic molten chloride salt by the gravimetric immersion method at high temperatures. The average corrosion rate is 40 μm/year for Hastelloy C-22 immersed in anaerobic salt at 800 °C which is 80 times lower than for Hastelloy C-22 in the same salt under the same conditions but in air. Clearly oxygen and/or water from air dissolved in molten salt accelerate the corrosion of the metals. Elucidating the role of these and other impurities in the molten salt on the mechanism of metal corrosion at high temperatures is of paramount importance if these salts are to be used as a heat transfer fluid to improve CSP operation and is the subject of ongoing work in meeting the materials challenges for concentrating solar power.

Acknowledgment   The authors would like to thank the US Department of Energy for the financial support of this work through DOE MURI award number DE-EE00059.

References

1. A. Modi, C.D. Pérez-Segarra, Thermocline thermal storage systems for concentrated solar power plants: one-dimensional numerical model and comparative analysis. Sol. Energ. **100**, 84–93 (2014)
2. C. Richter, S. Teske, R. Short, *Concentrating Solar Power: Global Outlook 2009: Why Renewable Energy is Hot* (Greenpeace International, SolarPaces, ESTELA, Amsterdam, The Netherlands, 2009)
3. A. Fernández-García, E. Zarza, L. Valenzuela, M. Pérez, Parabolic-trough solar collectors and their applications. Renew. Sustain. Energy Rev. **14**(7), 1695–1721 (2010)
4. International Energy Agency, *Technology Roadmap: Concentrating Solar Power* (OECD/IEA Publishing, France, 2010)
5. D. Barlev, R. Vidu, P. Stroeve, Innovation in concentrated solar power. Sol. Energ. Mater. Sol. Cells **95**(10), 2703–2725 (2011)
6. Haynes international, *Corrosion-Resistant alloys, Hastelloy C-22 alloy* (Haynes International, Inc., Kokomo, IN, 2002)
7. High operating temperature fluids funding opportunity announcement, #DE-FOA-0000567: Multidisciplinary University Research Initiative (MURI) of the US Department of Energy, 2012.
8. S. Qiu, M. White, R. Galbraith, Phase change salt thermal energy storage with integral pool boiler for dish stirling solar power, *SunShot Concentrating Solar Power Program Review*, (2013), pp. 103–106.
9. R.I. Dunn, P.J. Hearps, M.N. Wright, Molten-salt power towers: newly commercial concentrating solar storage. Proc. IEEE **100**(2), 504–515 (2011)
10. J. Stekli, Thermal Energy Storage and the United States Department of Energy's SunShot Initiative, in *SolarPACES2011* (SolarPACES, Granada, Spain, 2011)
11. P.W. Li, C.L. Chan, Q. Hao, P.A. Deymier, K. Muralidharan, D.F. Gervasio, M. Momayez, S. Jeter, A.S. Teja, A. M. Kannan, Halide and oxy-halide eutectic systems for high performance high temperature heat transfer fluids, *SunShot Concentrating Solar Power Program Review*. April 2013, pp. 85–86. Accessed 10 July 2014, from http://www.nrel.gov/docs/fy13osti/58484.pdf
12. C. Robelin, P. Chartrand, Thermodynamic evaluation and optimization of the (NaCl + KCl + MgCl2 + CaCl2 + ZnCl2) system. J. Chem. Thermodyn. **43**, 377–391 (2011)
13. A. Gil, M. Medrano, I. Martorell, A. Lazaro, P. Dolado, B. Zalba, L.F. Cabeza, State of the art on high temperature thermal energy storage for power generation. Part 1—Concepts, materials and modellization. Renew. Sustain. Energy Rev. **14**(1), 31–55 (2010)
14. J.W. Slusser, J.B. Titcomb, M.T. Heffelfinger, B.R. Dunbobbin, Corrosion in molten nitrate–nitrite salts. J. Met. **37**, 24–27 (1985)
15. G. Sorell, The role of chlorine in high temperature corrosion in waste-to-energy plants. Mater. High Temp. **14**, 137–150 (1997)
16. A. Ravi Shankar, A. Kanagasundar, U. Kamachi Mudali, Corrosion of nickel-containing alloys in molten LiCl–KCl medium. Corros. Sci. Sect. NACE Int. **69**, 48–57 (2012)

17. J.L. Trinstancho-Reyes, M. Sanchez-Carrillo, R. Sandoval-Jabalera, V.M. Orozco-Carmona, F. Almeraya-Calderon, J.G. Chacon-Nava, J.G. Gonzalez-Rodriguez, A. Martinez-Villafane, Electrochemical impedance spectroscopy investigation of alloy Inconel-718 in molten salts at high temperature. Int. J. Electrochem. Sci. **6**, 419–431 (2011)
18. A.V. Abramov, I.B. Polovov, V.A. Volkovich, O.I. Rebrin, T.R. Griffiths, I. May, H. Kinoshita, Spectroelectrochemical study of stainless steel corrosion in NaCl–KCl melt. ECS Trans. **33**(7), 277–285 (2010)
19. C.S. Ni, L.Y. Lu, C.L. Zeng, Y. Niu, Electrochemical impedance studies of the initial-stage corrosion of 310S stainless steel beneath thin film of molten $(0.62Li, 0.38 K)_2 CO_3$ at 650 °C. Corros. Sci. **53**, 1018–1024 (2011)
20. J.W. Ambrosek, *Molten chloride salts for heat transfer in nuclear systems*, Ph.D. Dissertation in Nuclear Engineering and Engineering Physics, University of Wisconsin-Madison, (2011)
21. D.F. Williams, *Assessment of Candidate Molten Salt Coolants for the NGNP/NHI Heat-Transfer Loop* (Oak Ridge National Laboratory, Oak Ridge, TN, 2006)
22. K. Vignarooban, P. Pugazhendhi, C. Tucker, D. Gervasio, A.M. Kannan, Corrosion resistance of Hastelloys in molten metal-chloride heat-transfer fluids for concentrating solar power applications. Sol. Energ. **103**, 62–69 (2014)
23. L. L. Wang, S. I. Martin, R. B. Rebak, 2006, Methods to calculate corrosion rates for alloy 22 from polarization resistance experiments, in *Proceedings of ASME Pressure Vessels and Piping Division Conference*, Vancouver, BC, July 23–27, 2006
24. A.C. Ciubotariu, L. Benea, M.L. Varsanyi, V. Dragan, Electrochemical impedance spectroscopy and corrosion behavior of Al_2O_3–Ni nano composite coatings. Electrochim. Acta **53**, 4557–4563 (2008)
25. M.K. Hsieh, D.A. Dzombak, R.D. Vidie, Bridging gravimetric and electrochemical approaches to determine the corrosion rate of metals and metal alloys in cooling systems: bench scale evaluation method. Ind. Eng. Chem. Res. **49**, 9117–9123 (2010)
26. X.L. Zhang, Z.H. Jiang, Z.P. Yao, Z.D. Wu, Effects of scan rate on the potentiodynamic polarization curve obtained to determine the Tafel slopes and corrosion current density. Corros. Sci. **51**, 581–587 (2009)
27. A. Poursaee, Potentiostatic transient technique, a simple approach to estimate the corrosion current density and Stern–Geary constant of reinforcing steel in concrete. Cem. Concr. Res. **40**, 1451–1458 (2010)
28. L. Wang, Y. Chao, Corrosion behavior of $Fe_{41}Co_7Cr_{15}Mo_{14}C_{15}B_6Y_2$ bulk metallic glass in NaCl solution. Mater. Lett. **69**, 76–78 (2012)
29. Y. Zou, J. Wang, Y.Y. Zheng, Electrochemical techniques for determining corrosion rate of rusted steel in seawater. Corros. Sci. **53**, 208–216 (2011)
30. D. Inman, Corrosion in fused salts, in *Corrosion*, 2nd edn., Metal/Environment Interaction, vol. 1, ed. by L.L. Shreir (Newnes-Butterworths, Boston, MA, 1976).
31. C.W. Bale, E. Belisle, P. Chartrand, S.A. Decterov, G. Eriksson, K. Hack, I.H. Jung, Y.B. Hang, J. Melancon, A.D. Pelton, C. Robelin, S. Petersen, FactSage thermochemical software and databases–recent developments. CALPHAD **33**, 295–311 (2009)
32. *Corrosion Resistant Alloys*. Haynes International Inc. Accessed July 19, 2014, from http://www.haynesintl.com/CRAlloys.htm
33. ASTM. Standard practice for preparing, cleaning, and evaluating corrosion test specimens, ASTM Standard G1-03. *2005 Annual Book of ASTM Standards*, vol. 03.02, (ASTM, Philadelphia, PA, 2005)
34. ASTM. Standard test method for corrosivity of water in the absence of heat transfer (weight loss method), ASTM D2688-05. *2005 Annual Book of ASTM Standards*, vols. 11.01 and 11.02, (ASTM: Philadelphia, PA, 2005)
35. ASTM International, vol. 03.02, Standards G 5, G 48, G 59, G 61 and G 102 (ASTM International, West Conshohocken, PA, 2003).
36. D.A. Jones, *Principles and Prevention of Corrosion*, 2nd edn. (Prentice-Hall, Upper Saddle River, NJ, 1996)

37. S. Keysar, D. Hasson, R. Semiat, D. Bramson, Corrosion protection of mild steel by a calcite layer. Ind. Eng. Chem. Res. **36**, 2903 (1997)
38. High temperature corrosion and materials, Molten salt corrosion. *High-Temperature Corrosion and Materials Applications*, Chapter 15, ed. by George Y. Lai, (ASM International, Materials Park, OH, 2007), pp. 409–421.
39. R.V. Dennis, L.T. Viyannalage, A.V. Gaikwad, T.K. Rout, S. Banerjee, Graphene nanocomposite coatings for protecting low-alloy steels from corrosion. Am. Cer. Soc. Bull. **92**, 18–24 (2013)
40. G. Michel, P. Berthod, S. Mathieu, M. Vilasi, P. Steinmetz, Chromium deposition on cobalt-based alloys by pack-cementation and behavior of the coated alloys in high temperature oxidation. Open. Corros. J. **4**, 27–33 (2011)
41. I.V. Oryshich, O.S. Kostyrko, Influence of molybdenum, tungsten and cobalt on the corrosion of high-temperature strength nickel alloys in molten salts. Met. Sci. Heat Treat. **27**(9-10), 740–746 (1985) (English Translation of Metallovedenie i Termicheskaya Obrabotka)

Chapter 5
Atomistic Simulations of Electronic and Optical Properties of Semiconductor Nanostructures

Marek Korkusinski

Abstract We present a tutorial introduction to atomistic calculations of electronic and optical properties of semiconductor nanostructures. The systems of interest, self-assembled quantum dots and colloidal nanocrystals, are composed of thousands to millions of atoms, beyond the applicability of ab initio schemes. Our approach, implemented as the QNANO computational package, consists of (1) atomistic geometry optimization using the valence force-field (VFF) model, (2) calculation of single-particle states confined in a nanostructure using the tight-binding scheme, (3) computation of Coulomb interaction matrix elements using the atomistic wave functions, and (4) calculation of multi-exciton spectra and optical response functions using the configuration interaction approach coupled with Fermi's Golden Rule. We illustrate this methodology by computing the single-particle states and the exciton fine structure of a self-assembled quantum dot, and by analyzing the multiexciton generation processes in colloidal nanocrystals.

5.1 Introduction

5.1.1 Types of Quantum Dots

Nanostructured materials and devices are currently one of the main areas of focus in solid-state research. In these systems, the carrier motion is restricted in one or more dimensions, allowing to tailor the density of states and, in consequence, transport and optical properties by material engineering. In this chapter, we shall discuss the zero-dimensional nanostructures, or quantum dots, in which the carriers of one or both types (electrons or/and holes) are fully confined spatially. This results in a

M. Korkusinski (✉)
Quantum Theory Group, Security and Disruptive Technologies,
National Research Council, Ottawa, ON, Canada
e-mail: Marek.Korkusinski@nrc-cnrc.gc.ca

© Springer International Publishing Switzerland 2015
A. Korkin et al. (eds.), *Nanoscale Materials and Devices for Electronics,*
Photonics and Solar Energy, Nanostructure Science and Technology,
DOI 10.1007/978-3-319-18633-7_5

discrete single-particle energy spectrum, with gaps between states which are broadly tunable by the choice of size, shape, and material composition of the quantum dot. The discrete energy spectrum can be probed optically at well-defined photon wavelengths, or by transport at well-defined gate voltages. For in-depth reviews of properties of quantum dots of various types, the interested reader is directed to [1–3]. The brief review of different types of quantum dots, which we shall present below, is aimed to provide context for the computational methods appropriate for their theoretical treatment, focused primarily on optical properties.

The self-assembled quantum dots are fabricated by an epitaxial process and grow as a result of the mismatch of the lattice constant of the substrate and the dot material. As a result of the accumulation of strain, the dot material spontaneously forms small droplets on the surface of the substrate. The dots are fabricated in such manner in many laboratories around the world [3], however in this chapter we shall be interested particularly in samples grown at the National Research Council of Canada. The first system of interest consists of the InAs quantum dot grown on the GaAs substrate. Figure 5.1a shows the electronic micrograph of such a single dot. The lattice mismatch between InAs and GaAs is $\sim 7\%$, resulting in the structures of typical diameters of ~ 20 nm [3]. Such dots, when capped with GaAs, consist of $\sim 2 \cdot 10^6$ atoms, including a region of surrounding barrier material necessary to contain the confined single-particle states. These systems typically luminesce at ~ 0.9 μm. They can be engineered along their vertical (growth) dimension by the indium-flush technique [6], which allows to fabricate lenses, disks, or pyramids with various thicknesses. They can even be arranged in vertically coupled quantum dot molecules [7–10]. However, due to the stochastic character of the growth itself the sample contains randomly distributed dots with a variety of diameters. In order to improve the reproducibility and enable the precise positioning of single dots, one has to change the constituent materials.

A much greater control over the size, shape, and position of the InAs quantum dot is achieved by growing it on patterned InP substrates [4, 5]. Figure 5.1b shows an electron micrograph and a schematic diagram of such a system. In the growth process, the planar InP substrate is covered by a SiO_2 mask, in which rectangular openings are made using lithographic methods. The deposition of InP is further resumed, leading to the growth of the pyramid-shaped structure, depicted in Fig. 5.1b in blue. The InP growth is stopped upon formation of the pyramidal structure with a flat region on the top. At this point, the growth is switched to InAs, and the dot material is deposited preferentially on the top of the pyramid rather than on its side walls due to the marked difference of diffusion lengths on these crystallographic planes. Thus, the quantum dot forms in a well-specified spot and its shape and lateral size can be tuned by the size of the top surface of the substrate. Once the dot is formed, the InP growth is resumed to cap the sample and complete the pyramid, as depicted in Fig. 5.1b. The InAs quantum dots on the InP patterned substrate are typically larger than the InAs/GaAs dots, with diameters of about 40 nm. This is due to the smaller mismatch of lattice constants ($\sim 3\%$). As a result, typically these dots absorb and emit light with wavelengths close to ~ 1.55 μm, making them excellent candidates for telecommunication devices.

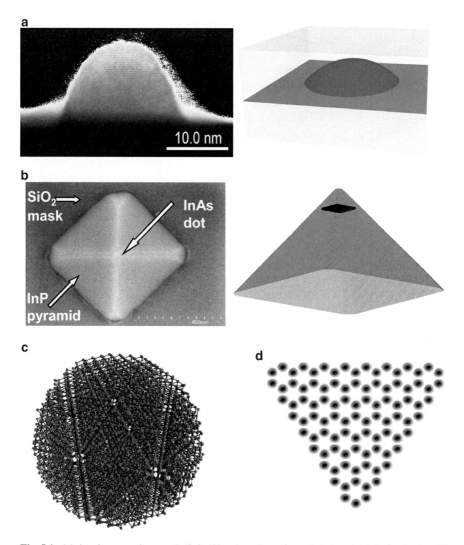

Fig. 5.1 (**a**) An electron micrograph (*left*) [3] and a schematic model view (*right*) of a single, self-assembled InAs quantum dot grown on the GaAs substrate. (**b**) An electron micrograph (*left*) [4, 5] and a schematic model view (*right*) of a single, self-assembled InAs quantum dot grown on the patterned InP substrate. (**c**) A schematic image of a wurtzite colloidal nanocrystal composed of $\sim 6,500$ atoms. (**d**) A schematic image of a triangular graphene quantum dot

However, atomistic modeling of these structures requires computational domain sizes in excess of $\sim 10^7$ atoms.

In general, the self-assembled dots described above confine both electrons and holes in a small volume. This results in the appearance of discrete single-particle energy spectra, with energy gaps of tens of meV which are additionally tunable by size, shape, and the choice of materials. We will discuss these issues in greater detail later on. The spatial localization of carriers translates into values of dipole matrix elements which are large compared with the structures of higher dimensionality, resulting in faster electron–hole recombination. All these properties make the self-assembled dots ideal candidates for active components of optical and optoelectronic devices with improved thermal stability. The dots or dot molecules can function as optical memories [11, 12]. The dots grown on patterned substrates are utilized in high-quality laser applications [13, 14]. Due to their tunability, they are also proposed as emitters of non-classical light, in the form of single photons [15–18] and entangled photon pairs [19–22]. The state of electrons and holes confined in the dot can also be controlled coherently, making these systems good candidates for optical-based qubits in quantum information [23–30]. Moreover, the ability to position the quantum dots deterministically using patterned substrates allows to engineer their strong coupling to photonic modes of an optical cavity [31]. Such a coupling can be used to modify the emission dynamics [32], to enhance the efficiency [33–35], and to guide the photons from the dot to an external optical system [36].

The second broad category of quantum dots, which we cover in this chapter, is the family of colloidal nanocrystals. These nanosized particles consist typically of 10^2–10^5 atoms, i.e., they are much smaller than the self-assembled quantum dots. A schematic picture of such a nanocrystal is shown in Fig. 5.1c. Currently, large quantities of nanocrystals of various materials and sizes and with good optical properties are fabricated in several groups worldwide in the colloidal growth process [37–49]. Interest in these systems is driven by their potential applications as biomarkers [50–52], light emitting diodes [53–55], photodetectors [56, 57], single-photon sources in quantum cryptography [58], and lasers [59, 60].

The nanocrystals are also of interest in solar cell research. This is because their optical properties can be tailored by the size and material engineering [61–65]. However, most importantly, the nanocrystals promise a highly efficient process of multiexciton generation following the absorption of a single high-energy photon [39, 47, 66–71]. This process will be discussed in detail later on in this chapter. Here let us only state that it can be described simply as the following sequence. First, a high-energy photon is absorbed by the nanocrystal creating a high-energy (highly excited) exciton. Next, this single electron–hole excitation is converted to multiple electron–hole excitations by Coulomb interactions, with the total energy of the system being conserved [69]. This conversion process is reversible, i.e., the multi-exciton states can evolve into states of smaller number of excitations by Auger processes, which limits the number of additional quasiparticles. This charge multiplication phenomenon competes with the phonon-assisted relaxation of the highly-excited exciton. Here, a part of the energy of the original photon is lost to heating the sample, while only one electron–hole pair is created in the dot. The multiexciton generation has been demonstrated in several material systems [72, 73] with efficiency up to 700 % [45, 67, 74, 75].

We conclude this brief review with a description of the novel and emerging class of nanostructures—graphene flakes, or clusters of carbon atoms organized in a single-sheet lattice. The first successful attempts to fabricate the graphene flakes composed of a single layer of atoms, including the famous exfoliation (scotch-tape) method, were carried out about a decade ago [76–79]. The resulting samples are nanoplatelets with area of $\sim 1 \, \mu m^2$. By virtue of their dimensions, their properties are close to those of the infinite two-dimensional graphene sheet, and therefore can be considered as ultimate examples of two-dimensional physics. The crystal lattice here is hexagonal (honeycomb). Moreover, the band states occupied by mobile electrons and holes are created predominantly from the spin-degenerate p_z orbitals, one per carbon atom. These simple structural properties are responsible for a number of fascinating phenomena observed in graphene sheets. Linear quasiparticle dispersion coupled with zero gap causes the electrons to behave as massless Dirac particles, which, coupled with suppression of backscattering and excellent thermal and mechanical properties of carbon, make graphene a promising material for future electronic and optoelectronic devices [80–83].

Much smaller graphene quantum dots, containing several hundred carbon atoms only, can be fabricated by colloidal methods [84–87]. Using mass spectrometry methods it is possible to establish the number of constituent atoms precisely, and by analyzing the vibronic spectra it is possible to establish the symmetry of the resulting quantum dots. At present, the optical response of ensembles of such quantum dots can be measured [84, 88], however single-dot experiments have not yet been reported.

Figure 5.1d presents a schematic view of an idealized, triangular graphene quantum dot. The structure is, again, extremely simple—carbon atoms arranged in a hexagonal lattice, with one spin-degenerate p_z orbital per atom available for mobile carriers. However, unlike in the graphene sheet, here we also deal with the properties arising from the quantum confinement, that is, the finite and small size of the system. Perhaps the most striking consequence of the shape and size engineering of those dots is the ability to open a bandgap and to tune it over a very broad range of frequencies—from UV to THz, just by tuning the size of the dot [89]. The electronic properties can be further modified by engineering the type of the edge of the flake. Structures with armchair edge exhibit an open bandgap separating the electron and hole states, while the existence of zigzag edges brings an additional band of zero-energy states, lying in the middle of the bandgap, with degeneracy dependent on the dot size [90, 91]. This electronic structure gives clear signatures in absorption [89, 92, 93]. Moreover, the zero-energy shell is only partially occupied with electrons in a charge-neutral system, which leads to nontrivial magnetic properties [94–100]. The total magnetic moment of the system can be probed and tuned using optical means [89, 101]. In this chapter we shall not consider the graphene quantum dots. The interested reader will find an extensive discussion of their properties, covering both experimental and theoretical aspects, in [82]. In the context of this chapter let us only mention that these systems are small enough to be tractable using ab-initio methods, and yet large enough for the tight-binding approaches to be still valid. This gives a unique opportunity to test different models against one another and confront their predictions with experimental results.

5.1.2   Choice of Computational Approach

In the previous section, we have reviewed the physical systems, whose electronic and optical properties we are trying to describe, understand, and model. Presently we shall choose the computational approach which is going to give us the most insight, while at the same time not being overwhelming in terms of computational time and resources.

At the most fundamental level, our samples can be understood as collections of atomic nuclei surrounded by electrons, in number of several tens per one nucleus (indium brings 49 electrons, arsenic—33, cadmium—48, while selenium—34). Even if we assume that the nuclei do not move, the complete description of the system requires handling the wave function of N electrons, i.e., it must obey Fermionic anticommutation rules, and will depend on $3N$ spatial variables and N spin degrees of freedom. Clearly, for a million-atom sample, $N \sim 10^7$. At present, this number of electrons is not tractable using any exact-diagonalization approaches. The most efficient density-matrix renormalization group schemes can presently handle several tens of electrons [102–104], that is, in our case they are barely sufficient to account for one atom.

The next approximation involves treating each nucleus and all the electrons belonging to it except for the valence electrons as a single unit (the ion). Typically, on average we have four valence electrons per atom, which applies equally to the III–V systems such as InAs (self-assembled dots material) and to the II–VI systems such as CdSe colloidal nanocrystals. Therefore, the Coulomb nuclear potential of each nucleus is replaced by the ionic pseudopotential, which is experienced locally (close to each atom) by a much smaller number of valence electrons. Even now, however, we have to account for $\sim 10^6$ electrons in a million-atom epitaxial quantum dot, making this system intractable. However, the smallest, colloidal quantum dots, consisting of $\sim 1,000$ atoms or less, may already be tractable by the density-functional approaches [105]. Such calculations can be carried out by commercially available computational packages such as SIESTA [106], VASP [107], or Quantum ESPRESSO [108], and allow to extract mainly ground-state properties of the system. Quasiparticle excitations (electrons, holes, excitons) can be built on top of these results using packages such as BerkeleyGW [109] or YAMBO [110]. The Quantum Theory group at the National Research Council of Canada has performed several SIESTA studies of colloidal nanocrystals [111, 112] as well as graphene quantum dots [96, 99]. Results of similar calculations have been recently employed in engineering of optical properties of colloidal nanocrystals [113]. It has to be mentioned, however, that these calculations, and particularly those of quasiparticle excitations, require large computational resources, both in terms of the size of the cluster and the computational time. For those reasons, at present it is not possible to apply this methodology to the epitaxial quantum dots.

It seems that maintaining the atomistic resolution of the system and accounting for multiple electrons at the same time causes our system to become intractable. Let us, therefore, consider an approximation which neglects both these elements, while still being capable of characterizing our system, albeit in a limited fashion. Such a level of approximation is supplied by the $\mathbf{k} \cdot \mathbf{p}$ approach. Here, the material is treated as a continuous (non-granular) medium characterized by certain material constants, which may change discontinuously at the interfaces between different parts of the system. Also, in this method one obtains the states of a single quasiparticle, that is, a particle dressed in interactions with all other charge carriers present in the system. The energy of this particle can fall within the conduction band, above the bandgap, and then we deal with a quasi-electron, or within the valence band, which makes it a quasi-hole. The interactions with other, "real" electrons present in the system are reflected in the material parameters, such as the bandgap, the effective masses, and the subband-mixing constants, so that our problem has been reduced to solving the Schroedinger equation for a single particle confined in a composite, but finite sample. The $\mathbf{k} \cdot \mathbf{p}$ approach was developed originally to understand the band structure of bulk semiconductors [114, 115]. Two strategies are used to adapt this model to nanostructures. On one hand, one tries to solve the continuous Schroedinger equation analytically or semi-analytically, often restricting oneself to considering only two subbands, the electron and the heavy hole. This approach works best for quantum dots of simple shapes, such as spheres, lenses, or disks, where the solution can be sought in the form of a series of appropriate spatial harmonics. In our group, this strategy has been used to understand the single-particle states of the electron [10, 116, 117], the hole [8, 118, 119], and both the electron and the hole [120–122] confined in an epitaxial quantum dot. The full eight-subband approach is also widely used to describe colloidal spherical nanocrystals [38, 123].

The second strategy is to discretize the Schroedinger equation on a grid of points encompassing the sample, that is, to rewrite the second-order differential equation in the form of finite differences. Here, typically one considers the mixing of all eight quasiparticle subbands in one Hamiltonian. This method allows to gain insight into properties of nanostructures of less regular shapes, such as pyramids [124–126], truncated pyramids [127], as well as quantum wires [128].

In spite of its simplicity and portability, the $\mathbf{k} \cdot \mathbf{p}$ model can offer only qualitative understanding of the physics of quantum dots. In order to be a reliable engineering and design tool, a computational approach has to account for the symmetry of the underlying crystal lattice and effects taking place on the atomistic scale. This is perhaps most straightforward to see in smaller nanostructures, such as graphene flakes or small colloidal nanocrystals. We have already mentioned the importance of the edge type in graphene dots. In nanocrystals, due to their small size, the surface-to-volume ratio is larger than that in epitaxial structures, and therefore the type of atoms (anions or cations) terminating the system, as well as faceting and ligands influence the resulting electronic structure to a large degree. Indeed, imperfect passivation of surfaces leads to the formation of trap states, which degrade the optical performance of the nanocrystals [41, 44, 46, 113, 129]. Such details are not straightforward to account for in a continuous-medium model.

In larger structures, the symmetry of the crystal lattice modulates the overall symmetry of the sample, leading to removal of degeneracies in their single-particle spectra [130]. Interestingly enough, by a careful construction of the quantum dot one atomistic layer by another, one can also engineer degeneracies in many-body spectra (e.g., exciton), which is crucial for entangled photon pair generation [131].

The epitaxial quantum dots, even if they are nominally grown from a single semiconductor material, are rarely pure: typically, there is interdiffusion in the regions of the barrier and well close to the interface, and atoms of the barrier material (e.g., phosphorus) can be present as random impurities in the interior of the quantum dot. With the $\mathbf{k} \cdot \mathbf{p}$ approach these processes can be accounted for by making the material parameters vary over the volume of the dot. However, the interface roughness and random impurities, when treated atomistically, lower the overall symmetry of the system even further. This randomness translates into local and random distortions of the quasiparticle wave functions, and, in consequence, influences such global properties of the sample as optical oscillator strengths or dipole moments [132].

In single-dot measurements (see e.g. [133, 134]), apart from the energy of emitted photons one also measures their polarization. As we will show in detail later on, the plane of polarization is defined by crystallographic planes of the sample rather by the shape of the dot, therefore its accurate prediction goes beyond the capabilities of the $\mathbf{k} \cdot \mathbf{p}$ model. If one further wants to obtain a strong coupling of the confined exciton to the mode of the optical cavity, in which the dot is positioned, careful engineering of the exciton polarization is crucial.

In computational practice, one encounters two approaches which allow for the atomistic resolution, and yet are formulated as a single quasiparticle problem. One of them is the empirical pseudopotential approach, in which each atom is represented as a locally confining pseudopotential. This atomic pseudopotential is, however, different than the ionic pseudopotential discussed above, in that the atoms are not stripped of their valence electrons, but remain charge-neutral objects. As a result, our quasiparticle experiences an effective potential which is a super-position of the local pseudopotentials. Typically, one solves for the eigenenergies and wave functions of the single particle in this complex confinement using the plane wave basis set. This produces a large and dense Hamiltonian (with few elements vanishing identically) which has to be diagonalized numerically.

The second method is the tight-binding approach, in which one defines a basis of atomistic orbitals (rather than pseudopotentials) and describes the motion of the electron as hopping from one orbital to another, with some hopping amplitude, which depends on the type of orbitals involved in the process and the distance between atoms. This locality of the basis set makes it possible to express the Hamiltonian of the system in terms of a very sparse matrix, with a constant (and known a priori) number of nonzero elements per row. The diagonalization of such a matrix is straightforward in the parallel computing paradigm, which allows to treat efficiently very large systems. This computational convenience comes at a price—approximately 30 hopping elements have to be supplied to describe each material, while the empirical pseudopotential approach requires only 4 constants per atom type (8 constants for binary compounds). A critical comparison of the two methods will be given in the next section.

Once the single-particle energies and wave functions are obtained, one can populate the dots with several charge carriers, for example with electron–hole pairs. This can be done typically in several ways, that is, a number of various configurations of carriers can be created. However, since the charge carriers now interact, these configurations are not proper eigenstates of the system. These eigenstates have to be computed, e.g., by exact diagonalization of the system Hamiltonian written as a matrix in the basis of the many-body configurations. Further, once these many-body states are established, we can calculate the optical response of the system, such as absorption and emission spectra, and the coherent evolution of the system initialized optically. These optical properties can also be probed experimentally. This makes it possible to formulate and test the validity of our approach, and then, once calibrated, to use our tool to design systems with tailored properties. In the following section, we shall describe in detail the theoretical framework of each of the elements of the model. The third section contains a selection of examples, which illustrate several aspects of our atomistic approach.

5.2   The Computational Model

In the previous section we have presented, in general terms, the motivation of the atomistic approach and the high-level overview of the theoretical models, which can be used in its implementation. This section contains a more detailed description of the software package QNANO, developed in the Quantum Theory Group of the National Research Council of Canada, which is designed to compute the optical properties of nanostructures atomistically in a sequence of steps [135]. We start with the definition of the sample in terms of atomic positions, and then establish the equilibrium positions of the constituent atoms to account for strain. The calculation of single-quasiparticle levels, constituting the next step, is realized in QNANO by the semiempirical tight-binding approach. The atomistic wave functions are further used to calculate the Coulomb interaction matrix elements as well as the dipole matrix elements. The next step involves calculation of many-body states in the exact diagonalization approach. The procedure is concluded by computing the absorption and emission spectra using the Fermi's Golden Rule. In this section we shall describe the theoretical foundations of each of these steps.

5.2.1   Definition of the Computational Domain

The computational procedure starts by defining the shape and material composition of the sample. In practical terms, this task is reduced to preparing a list of atoms, in which we define the three spatial coordinates of the atom and its chemical identity. Such a computational domain encompasses the nanostructure together with a region of barrier material around it, which is sufficiently large to allow such a sample to be treated as a closed system. The size of such a computational domain depends on the type of calculation that we are about to perform. We shall evaluate the relevant requirements at each computational step.

The crystal structure of the actual samples is typically based on the substrate crystal lattice, and can be identified—most typically—as zinc-blende, wurtzite, rock-salt. Clearly, the actual samples will also exhibit some amount of lattice faults (vacancies, dislocations etc.). The QNANO package does not account for these faults, as they are not straightforward to treat within the tight-binding approach. Instead, we assume that the atomic coordinates correspond to those of atoms in bulk.

As for the chemical identity of the atoms, it could be established, for instance, by simulating the epitaxial or colloidal growth using the kinetic Monte Carlo techniques [136–140]. This approach allows to obtain a very good match with the actual sample morphology [141], however it is computationally intensive.

In computational practice, the material composition of the sample is inferred from the electron microscope images, presented earlier in Fig. 5.1. However, these images show dots before they are capped with the additional barrier material. This process is known to lead to redistribution of the material and even to changes in the shape of the dot [143]. A much more reliable microscopic technique, the cross-sectional scanning tunneling microscopy (XSTM) was developed to obtain the structural information about the nanostructures after capping [143–145]. In this technique, the sample is grown and probed optically, whereupon it is cleaved. The cleaving edge may be chosen to pass through the fully encapsulated quantum dot, which exposes its cross-section as shown in Fig. 5.2. Since the quantum dot material is strained (compressed), it will attempt to lower the total elastic energy by buckling outwards, thereby creating a structure extending from the cleaving surface. The topography of this structure can be further measured by scanning tunneling microscopy techniques. Thus, the bright areas in Fig. 5.2a correspond to the quantum dot material, while the darker areas, which have not buckled, correspond to the barrier. This technique can be used to investigate the morphology of quantum dots in a variety of material systems [142, 146, 147]. It is also possible to measure the molar fractions in ternary structures, such as $In_xGa_{1-x}As$ dots [148], and the distribution of these fractions across the sample. As we can see from Fig. 5.2b, the dots are well approximated by lenses, or truncated lenses if the indium flush step was included in the growth sequence [142]. In the QNANO package it is also possible to construct samples with impurities, such as the lens-shaped structure shown in Fig. 5.2c.

For smaller nanostructures it is possible to ascertain their shapes and atomic content by less destructive means. We have already discussed the case of graphene quantum dots, whose composition can be measured by atomic mass spectroscopy, and the symmetry can be derived from infrared vibronic spectra [84–87].

The details of the size, shape, and symmetry of the epitaxial dots can also be inferred from their optical spectra. We have already mentioned the relationship between the atomistic symmetry and the fine structure of exciton levels [131]. We can also "reverse engineer" the structure from the optical spectra (absolute energy of the emission lines, their polarization, degeneracies, oscillator strength) by postulating several probable sample geometries and observing the resulting trends in the calculated optical response.

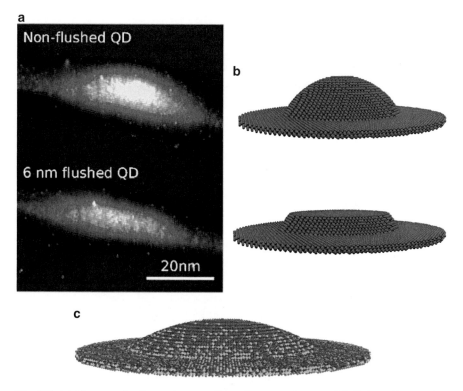

Fig. 5.2 (**a**) Cross-sectional scanning tunneling microscopy images of InGaAs self-assembled quantum dots in the GaAs barrier grown without (*top*) and with (*bottom*) indium flush step [142]. (**b**) Atomistic models of the lens-shaped and truncated lens-shaped quantum dots corresponding to the samples in panel (**a**). *Red* (*blue*) *dots* denote the indium (arsenic) atoms. (**c**) Atomistic model of the InGaAs sample with *green dots* denoting the gallium impurities

The final remark concerns the output from this "sample builder" module. As we already stated, here we obtain the list of atoms with their positions and chemical identification. However, it is also convenient to create an additional file, containing nearest-neighbor links between atoms. This information can be easily obtained just by knowing all atomic positions. This process, although trivial to implement, is however time consuming, while the information about nearest-neighbor links will play a crucial role in the two subsequent stages of computation. We choose to perform it once and save these links for convenience.

5.2.2   Atomistic Calculation of Strain

At this point our computational sample is represented by a list of atoms, whose positions are chosen to match the bulk lattice of the substrate of the nanostructure (GaAs or InP). In Fig. 5.3a we show just one cubic unit cell illustrating this situation.

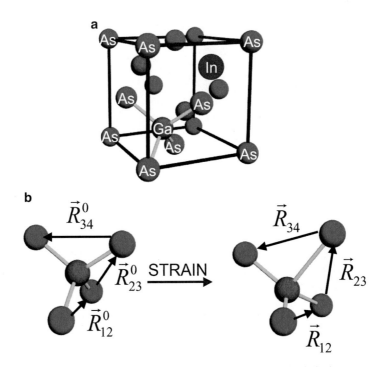

Fig. 5.3 (a) Model representation of the cubic unit cell of the zinc-blende lattice, corresponding here to GaAs (*red* and *blue spheres* denote the constituent atoms). *Green lines* denote bonds to the four nearest neighbors of one of gallium atoms. The *brown sphere* represents and indium atom, which has been substituted for one of the gallium atoms. (b) A schematic representation of the bulk tetrahedral unit (*left*) and the strained unit (*right*). Vector coordinates are used to establish strain tensor matrix elements

The blue spheres denote the arsenic atoms, while the red spheres represent the gallium atoms. One of the gallium atoms has been replaced by indium (the brown sphere). Note that this sample lattice is of the zinc-blende type, with the characteristic tetrahedral coordination. The bonds to the four nearest neighbors of one of the gallium atoms have been denoted in green. However, the lattice constant of the dot material is different than that of the substrate ($d = 0.60583$ nm for InAs as opposed to $d = 0.56532$ nm for GaAs and $d = 0.58687$ nm for InP). The atoms must, therefore, be displaced from their initial positions and their new, equilibrium distribution has to be found. This is achieved by defining the total elastic energy of the system in the valence force-field (VFF) approach in the following form [149, 150]:

$$
U = \frac{1}{2} \sum_i \left\{ \sum_{j=1}^{n.n} \frac{3\alpha_{ij}}{4(d_{ij}^0)^2} \left((\mathbf{R}_j - \mathbf{R}_i)^2 - (d_{ij}^0)^2 \right)^2 \right.
$$
$$
\left. + \sum_{j=1}^{n.n} \sum_{k<j}^{n.n} \frac{3\beta_{ijk}}{4 d_{ij}^0 d_{ik}^0} \left[(\mathbf{R}_j - \mathbf{R}_i) \cdot (\mathbf{R}_k - \mathbf{R}_i) - \cos\theta_{ijk} d_{ij}^0 d_{ik}^0 \right]^2 \right\}.
$$

(5.1)

Here the first summation is carried out over all atoms, while the inner sums are taken over the nearest neighbors of the atom i. The constants α_{ij} and β_{ijk} are material parameters, the vectors $\mathbf{R}_i$ are positions of atoms, the constants d_{ij}^0 are material-dependent bulk lattice constants, and the angle θ_{ijk} is the bulk angle between adjacent bonds. The above total energy consists of two terms, the first one describing the bond stretching, and the second one describing the bond bending. The function of both of these terms is illustrated schematically in Fig. 5.3b. The bond stretching term accounts for the changes in interatomic distances by comparing the actual bond lengths (expressed as the difference in atomic positions) to the ideal bulk bond length for a given material. This corresponds simply to changes in the length of green lines connecting atoms in the figure. The bond bending term interrogates three atoms at a time, and compares the actual angle of two adjacent bonds to the ideal bulk value (for zinc-blende materials, $\theta = 2\pi/3$). This term will therefore be activated by changes in the angles between any pair of green bonds in the figure. This quartic total elastic energy is a reasonable approximation if the differences in lattice constants, and therefore the resulting strain, are small. Higher-order formulas, including cross-terms between the bond stretching and bond-bending components, are also available [151]. The QNANO package utilizes the quartic VFF energy as in Eq. (5.1).

The values of material parameters α_{ij} and β_{ijk} are established by fitting the VFF model to the elastic constants of the bulk material [150]. The cubic materials are characterized by three independent elastic constants, c_{11}, c_{12} and c_{44} [152]. These constants connect to the VFF parameters by the following relations:

$$c_{11} + 2c_{12} = \frac{\sqrt{3}}{4d^0} (3\alpha + \beta), \tag{5.2}$$

$$c_{11} - c_{12} = \frac{\sqrt{3}}{d^0} \beta, \tag{5.3}$$

$$c_{44} = \frac{\sqrt{3}}{4d^0} \frac{4\alpha\beta}{\alpha + \beta}. \tag{5.4}$$

Note that we have three equations for two unknowns; in fact, typically it is not possible to satisfy all these relations simultaneously. The common practice in this case is either to eliminate one of the equations, or to perform least-square fitting of the VFF parameters. In Table 5.1 we give the parameters used for InAs, GaAs, and InP in the QNANO package. Note that higher-order VFF models are characterized by a larger number of parameters, which makes it possible to fit the model to the bulk elastic properties more accurately.

Since the nanostructure is composed of more than one material, a question arises as to what parameters to take at the interfaces. Since our typical systems, InAs/GaAs and InAs/InP, share one of the atomic species, this problem never arises for the bond-stretching term scaled by the parameter α. For the bond-bending term, the

Table 5.1 Elastic parameters of the valence force-field model used in the QNANO package

Material	α (dyne)	β (dyne)
InAs	35,180	5,490
GaAs	41,490	8,940
InP	39,520	6,600

parameter at the interface is taken as the harmonic average of the parameters of each material, $\beta_{INT} = \sqrt{\beta_1\beta_2}$ [126].

With the total elastic energy fully parametrized, we can now move on to calculating the equilibrium distribution of atoms. To this end we minimize the energy with respect to atomic positions using the standard conjugate gradient algorithm. Since the elastic potential is quartic, there typically exists one, stable minimum, which is reached usually rather quickly (for a two-million atom computational domain containing a quantum dot of about 100,000 atoms, we typically obtain the converged result in several hundred iterations). As for the computational resource requirements, assuming that each atom is represented by three eight-byte floating point numbers (atomic coordinates), a structure of 10^8 atoms will require approximately 2.2 GB of memory. The conjugate gradient algorithm typically requires about three such data structures to be available simultaneously. Therefore, this part of the QNANO package can be executed even on a single-CPU workstation with a reasonably large amount of memory, without the need for parallelization. As an output, we store the list of atoms with updated positions, while the nearest-neighbor connectivities typically remain unchanged.

The information encoded in the new positions of atoms can be made accessible by mapping them onto the local values of strain tensor matrix elements. This operation also allows us to make contact with the continuous-medium methods, such as $\mathbf{k} \cdot \mathbf{p}$, where these matrix elements are used directly as input to the Bir–Pikus strain Hamiltonian, which translates them into the language of electronic structure [125, 153]. Our structure is atomistic (granular), and so the definition of strain tensor matrix elements,

$$\varepsilon_{ij} = \frac{\partial u_i}{\partial x_j}, \tag{5.5}$$

cannot be applied directly (here, $\mathbf{u}$ is the displacement, $\mathbf{x}$ is the spatial coordinate, and $i,j = x,y,z$). The principle of the mapping proposed instead [154] is illustrated in Fig. 5.3b, that is, it involves comparing the characteristic dimensions of the primitive tetrahedral unit cell of a locally deformed system (right) to those of the unstrained bulk (left). These two sets of dimensions are connected in the following manner:

$$\begin{pmatrix} R_{12,x} & R_{23,x} & R_{34,x} \\ R_{12,y} & R_{23,y} & R_{34,y} \\ R_{12,z} & R_{23,z} & R_{34,z} \end{pmatrix} = \begin{pmatrix} 1+\varepsilon_{xx} & \varepsilon_{xy} & \varepsilon_{xz} \\ \varepsilon_{yx} & 1+\varepsilon_{yy} & \varepsilon_{yz} \\ \varepsilon_{zx} & \varepsilon_{zy} & 1+\varepsilon_{zz} \end{pmatrix} \begin{pmatrix} R^0_{12,x} & R^0_{23,x} & R^0_{34,x} \\ R^0_{12,y} & R^0_{23,y} & R^0_{34,y} \\ R^0_{12,z} & R^0_{23,z} & R^0_{34,z} \end{pmatrix}.$$

$$\tag{5.6}$$

From this relation one obtains the local strain tensor matrix elements by simple matrix inversion and multiplication. Note that the relative atomic positions entering the above equation are those of the corner atoms of the tetrahedron (blue atoms in Fig. 5.3b). The resulting values ε_{ij} are, however, assigned to the central atom (red in the figure) by convention.

To illustrate the points made in this section, we have performed a sample calculation of the displacements and resulting strain tensor matrix elements in a sample consisting of a small InAs disk of diameter 16 nm and height 3 nm, embedded in the center of the GaAs computational domain of dimensions 50 × 50 × 20 nm. The domain contains 2.2×10^6 atoms. Its schematic view is shown in Fig. 5.4a, rendering faithfully the relative dimensions of the elements of the system. The results of this calculation are presented in Fig. 5.4b–d. Instead of showing the diagonal strain tensor matrix elements ε_{ii}, representing the relative changes of length along the direction $i = x, y, z$, respectively, we plot combinations of these elements as follows. The hydrostatic strain, $\varepsilon_h = \varepsilon_{xx} + \varepsilon_{yy} + \varepsilon_{zz}$, represents the relative change of volume of each tetrahedral cell. As we can see from Fig. 5.4b, this strain is concentrated inside the quantum dot and is negative, i.e., on the whole, the volume of each unit cell has decreased. The biaxial strain, $\varepsilon_b = \varepsilon_{zz} - (\varepsilon_{xx} + \varepsilon_{yy})/2$, describes the anisotropy of the strain, that is, quantifies how much the strain along the Z axis differs from that in the dot plane. From Fig. 5.4c we see that inside the dot this strain is positive, that is, the dot material is decompressed along the dot height, and compressed in-plane. Finally, Fig. 5.4d shows the variation of the shearing strain ε_{yz} across the sample. This strain component is much weaker than the other two, and is concentrated predominantly at the corners of the rectangular cross-section of the dot. The significance of these strain fields will be discussed in the next section.

With the total elastic energy defined in Eq. (5.1) and the equilibrium positions of atoms found in the minimization procedure, we can in principle compute the phonon spectra of our sample. This can be done by setting up the dynamical matrix, calculated as the matrix of second partial derivatives of the total elastic energy over the atomic positions [155]. The normal modes of phonons are the eigenvectors of this matrix. This functionality, although straightforward, is not implemented in QNANO yet, as in our computational experience it was not needed. Indeed, the optical spectra such as absorption or emission are modified only slightly by the phonon modes. However, if one desires to calculate the dynamical properties of our system, such as time-resolved spectra or issues related to the coherent control by external fields, one needs to account for the coupling to the phonons, as they provide important non-radiative relaxation channels for many-electron and electron–hole systems. The significance of that relaxation channel will be discussed later on, on the example of Auger processes in colloidal nanocrystals.

In closing of this stage of the computational procedure we wish to mention that the strain can be used as an important tool in engineer of the optical properties of the nanostructures. Indeed, recent studies show that by embedding the dots in semi-conductor nanobridges, which then can be deformed in a symmetric or shearing

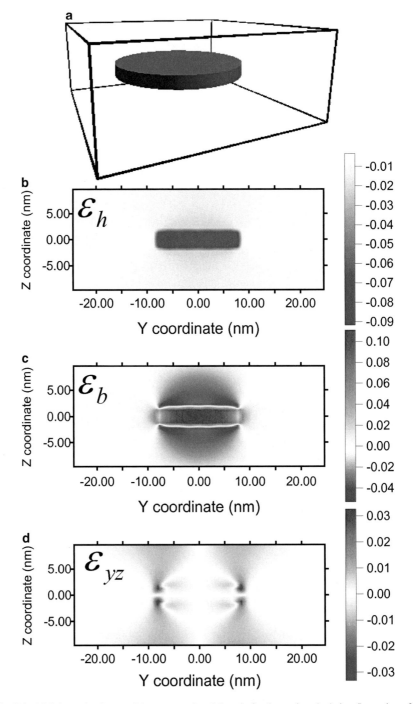

Fig. 5.4 (a) Schematic picture of the computational domain for the strain calculation. It consists of an InAs disk with the diameter of 16 nm and height of 3 nm, embedded in a GaAs domain of dimensions $50 \times 50 \times 20$ nm. Lower panels show the local values of the hydrostatic strain (**b**), the biaxial strain (**c**), and the shear strain (**d**) computed across the YZ plane intersecting the disk along its diameter

manner, one can tune the shape, orientation, and phase of the confined exciton [156], and even control the system dynamically by transferring carriers between quantum dots [157]. One can also change the ordering of light- and heavy-hole levels in the single-quasiparticle spectrum. Typically, the heavy-hole levels will occur closer to the semiconductor bandgap, and therefore will participate predominantly in the exciton states. However, in appropriately strain-engineered samples these levels can be replaced by the light-hole states, leading to the formation of the light-hole exciton [158]. The light-hole exciton is an important intermediate state in proposals for coherent conversion of photons into electron spins [159, 160].

5.2.3 Single-Quasiparticle Levels

In this section we shall describe the next step in the computational procedure of the QNANO package: the atomistic calculation of single-quasiparticle states. As we already mentioned in Sect. 5.1, in practice we can apply two computational approaches: the empirical pseudopotential method and the empirical tight-binding model. For completeness, we will describe the former briefly, while for the latter we will provide a complete description.

5.2.3.1 Empirical Pseudopotential Approach

This approach is the method of choice for calculating the electronic structure of bulk semiconductors, and has been first implemented by M. Cohen in 1960s, with subsequent significant development by Chelikovsky and Louie [105]. Its use in the calculations of the properties of nanostructures is largely due to A. Zunger and his group, and will be briefly summarized here using his formulation.

Suppose we have a single atom, e.g., indium, and we want to compute its electronic properties using the density-functional approach. The relevant Kohn–Sham Hamiltonian will in this case attain the familiar form:

$$\widehat{H}_{KS} = \frac{\widehat{\mathbf{p}}^2}{2m} - \frac{N_e e^2}{|\mathbf{r}|} + \int d^3 r_i \frac{n(\mathbf{r}_i)}{|\mathbf{r}_i - \mathbf{r}|} + V_{XC}[n(\mathbf{r})], \qquad (5.7)$$

where $N_e = 49$ is the number of electrons in the indium atom, m is the bare electron mass, $n(\mathbf{r})$ is the total electron density, and V_{XC} is the exchange-correlation potential in some approximation (local density approximation, generalized gradients, etc.). The origin was chosen on the atomic nucleus, and the Coulomb potential generated by it is represented as the second term. The third term represents the Hartree potential, describing the repulsion of our electron and all other electrons present in the system, and is manifestly a functional of the total charge density. So is the exchange-correlation potential, the last term of the Hamiltonian. With a chosen

V_{XC}, the Kohn–Sham procedure is carried out self-consistently until convergence. As a result, we obtain the eigenenergies and eigenstates of the electron dressed in interactions with all other electrons, that is the quasiparticle (Kohn–Sham) energies and wave functions. Let us, however, identify the following potential, which can be constructed once the convergence is reached:

$$\widehat{V}_{ps}(\mathbf{r}) = -\frac{N_e e^2}{|\mathbf{r}|} + \int d^3 r_i \frac{\tilde{n}(\mathbf{r}_i)}{|\mathbf{r}_i - \mathbf{r}|} + V_{XC}[\tilde{n}(\mathbf{r})], \tag{5.8}$$

where the converged charge density is denoted by $\tilde{n}(\mathbf{r}_i)$. The above equation defines the pseudopotential, that is, the effective potential that our quasiparticle perceives, which originates from all the charges present in the system. The above prescription allows, at least in principle, to generate pseudopotentials for all isolated, charge-neutral atoms. Once this is done, we can express the motion of our quasielectron in a solid simply with the Hamiltonian

$$\widehat{H}_{PS} = \frac{\widehat{\mathbf{p}}^2}{2m} + \sum_{i=1}^{N_{at}} V_{PS}^{(i)}(\mathbf{r} - \mathbf{R}_i), \tag{5.9}$$

in which the potential term is a simple superposition of all atomic pseudopotentials (not necessarily identical), whose origins are defined by the atomic coordinates $\mathbf{R}_i$. At this point we run into the problem of transferability: are the pseudopotentials generated for isolated atoms appropriate to use in a solid? Since the total charge densities on each atoms are certainly going to be affected by their environment (e.g., by the fact that bonds of various degree of covalence are formed), we cannot count on such a universal transferability. The pseudopotentials will be specific to the particular type of solid, even if we consider the same atom in different environments. We have two ways to proceed further. On one hand, one can attempt to simulate a larger domain of the system in the density-functional approach. On the other hand, one parametrizes the pseudopotentials in some fashion, and then one chooses these parameters such that the electronic properties predicted by our model match those seen in the experiment. The second technique, the empirical pseudopotential method, will be discussed in what follows.

   In order to understand what is required for the fitting procedure, we first derive the bulk band structure based on the Hamiltonian (5.9). This derivation follows closely that presented in the original papers, [161, 162]. First, we account for the fact that our bulk solid has a periodic structure, i.e., we can define the unit cells. Then,

$$\widehat{V}_{ps}(\mathbf{r}) = \sum_{i=1}^{N_{uc}} \sum_{\alpha} V_{\alpha}(\mathbf{r} - \mathbf{R}_i - \tau_{\alpha}), \tag{5.10}$$

where the index α enumerates atoms within one unit cell, and their coordinate measured from the center of the i-th unit cell is τ_{α}. There are N_{uc} unit cells in the crystal. The coordinate $\mathbf{R}_i$ defines the position of the center of the i-th unit cell relative to the global coordinate system. Owing to the periodicity of our solid, the superposition of pseudopotentials in each unit cell does not depend on the position of that cell in the crystal, that is,

$$V_\alpha(\mathbf{r} - \mathbf{R}_i - \tau_\alpha) = V_\alpha(\mathbf{r} - \tau_\alpha).$$

This periodicity allows us to express the total pseudopotential as the Fourier series

$$\widehat{V}_{ps}(\mathbf{r}) = \sum_G V_{ps}(\mathbf{G})e^{i\mathbf{G}\cdot\mathbf{r}}, \tag{5.11}$$

where the Fourier transform of our pseudopotential

$$V_{ps}(\mathbf{G}) = \frac{1}{N_{uc}\Omega} \int d^3 r V_{ps}(\mathbf{r})e^{-i\mathbf{G}\cdot\mathbf{r}},$$

and Ω is the volume of the unit cell. The reciprocal lattice vectors $\mathbf{G}$ obey the periodicity relation $e^{i\vec{G}\cdot(\vec{r}+\vec{R}_i)} = e^{i\mathbf{G}\cdot\mathbf{r}}$. By substituting the real-space form of the pseudopotential, and by using the periodicity we arrive at

$$V_{ps}(\mathbf{G}) = \frac{1}{\Omega} \sum_\alpha \int d^3 r V_\alpha(\mathbf{r} - \tau_\alpha)e^{-i\mathbf{G}\cdot\mathbf{r}}$$

noting that the above sum is composed only of the pseudopotentials of atoms belonging to the single unit cell. Typically the unit cell contains two (as in zinc-blende semiconductor) or four atoms (in wurtzite). We shall focus on the former.

In zinc-blende materials, whose cubic unit cell is shown in Fig. 5.3a, it is convenient to choose the center of the unit cell at the midpoint of the bond between the anion and the cation. Note that the actual, "primitive" unit cell is composed only of the two atoms, one at position $(-1, -1, -1)a/8$ (in our case, the corner arsenic), and one at position $(1, 1, 1)a/8$, where a is the lattice constant (in our case, the gallium atom connected to the corner arsenic with a green line). This allows to write the pseudopotential in a particularly simple form

$$V_{ps}(\mathbf{G}) = V_S(\mathbf{G})\cos(\mathbf{G}\cdot\tau) + iV_A(\mathbf{G})\sin(\mathbf{G}\cdot\tau),$$

where the vector $\tau = (1, 1, 1)a/8$, the symmetric and antisymmetric combinations of atomic pseudopotentials are

$$V_{S,A}(\mathbf{G}) = \frac{1}{\Omega}(V_1(\mathbf{G}) \pm V_2(\mathbf{G})),$$

and the pseudopotentials in the reciprocal space of individual atoms are simply

$$V_a(\mathbf{G}) = \int d^3 r V_a(\mathbf{r}) e^{-i\mathbf{G}\cdot\mathbf{r}}.$$

Note that for the materials with the diamond structure, such as carbon (diamond) or silicon, the antisymmetric pseudopotential vanishes identically.

Thus, the entire model is parametrized only by two functions, $V_S(\mathbf{G})$ and $V_A(\mathbf{G})$. Let us suppose for the moment that we know these two functions. We obtain the band structure by solving the Schroedinger equation with the Hamiltonian (5.9), but written in the reciprocal space. To this end, we note that we are really looking for the single-particle wave function in the Bloch form

$$\Psi(\mathbf{k}, \mathbf{r}) = e^{i\mathbf{k}\cdot\mathbf{r}} u_n(\mathbf{k}, \mathbf{r}), \tag{5.12}$$

where the Bloch function $u_n(\mathbf{k}, \mathbf{r})$ is indexed by the band quantum number n and is periodic both in the real space (with the real unit cell periodicity) and in the reciprocal space (with the periodicity of the reciprocal space). Due to this periodicity, we can express the Bloch function in the Fourier series as well:

$$u_n(\mathbf{k}, \mathbf{r}) = \sum_{\mathbf{G}} U_n(\mathbf{k}, \mathbf{G}) e^{i\mathbf{G}\cdot\mathbf{r}}.$$

The Schroedinger equation, after trivial calculus and recasting in the reciprocal space, takes thus the form

$$\frac{\hbar^2}{2m} |\mathbf{G} + \mathbf{k}|^2 U_n(\mathbf{k}, \mathbf{G}) + \sum_{\mathbf{G}'} V_{ps}(\mathbf{G} - \mathbf{G}') U_n(\mathbf{k}, \mathbf{G}') = E_n(\mathbf{k}) U_n(\mathbf{k}, \mathbf{G}). \tag{5.13}$$

Here, $\mathbf{k}$ is a parameter—it is the wave vector (within the unit cell) for which we want to compute the band energies $E_n(\mathbf{k})$ and Bloch functions $U_n(\mathbf{G}, \mathbf{k})$, which can then be transformed back to the real space by the inverse Fourier transform.

The last element we require is the suitable set of vectors $\mathbf{G}$. We define these vectors based on the unit vectors of the reciprocal space. As usual for the zinc-blende materials, the real space unit vectors are defined as $\mathbf{R}_1 = (1, 0, 1)a/2$, $\mathbf{R}_2 = (1, 1, 0)a/2$, and $\mathbf{R}_3 = (0, 1, 1)a/2$. This gives the reciprocal space unit vectors in the form $\mathbf{K}_1 = (1, -1, 1)2\pi/a$, $\mathbf{K}_1 = (1, 1, -1)2\pi/a$, and $\mathbf{K}_1 = (-1, 1, 1)2\pi/a$. Just as the position of any atom can be written as a superposition of the vectors $\mathbf{R}_i$, the reciprocal vectors are constructed as $\mathbf{G}_i = h_i\mathbf{K}_1 + k_i\mathbf{K}_2 + l_i\mathbf{K}_3$. The numbers (h, k, l) are nonnegative integers, and so the vectors $\mathbf{G}$ form the discrete, orthonormal, and infinite basis set for our problem. These properties allow us to write our Schroedinger equation in the matrix form, with matrix elements

Table 5.2 Empirical pseudopotential parameters for the simple calculation of bulk band structure

Element	Si	InAs (eV)
$V_S(\sqrt{3})$	−0.21	−0.22
$V_S(\sqrt{8})$	0.04	0.0
$V_S(\sqrt{11})$	0.08	0.05
$V_A(\sqrt{3})$		0.08
$V_A(\sqrt{4})$		0.05
$V_A(\sqrt{11})$		0.03

$$\langle \mathbf{G}_i|H_{PS}|\mathbf{G}_j \rangle = \frac{\hbar^2}{2m}|\mathbf{G}_i + \mathbf{k}|^2 \delta_{ij} + V_S(\mathbf{G}_i - \mathbf{G}_j) + V_A(\mathbf{G}_i - \mathbf{G}_j).$$

One further remark needs to be made about the nature of the pseudopotentials. We assume that the pseudopotentials are symmetric so that

$$V_{S,A}(\mathbf{G}_i - \mathbf{G}_j) \equiv V_{S,A}(|\mathbf{G}_i - \mathbf{G}_j|),$$

that is, they depend only on the length of the vector argument. Since the vectors $\mathbf{G}$ have discretized lengths, we will only need to ascertain the values of the pseudopotentials at discrete points of the reciprocal space. Of course, the size of the Hamiltonian matrix is determined by the number of vectors $\mathbf{G}$ that we choose to use. Typically one chooses these vectors using the energy cutoff E_C such that the basis vectors fulfill the relationship $\hbar^2|\mathbf{G}|^2/2m \leq E_C$. Moreover, by setting the values of pseudopotentials $V_{S,A}(|\mathbf{G}_i - \mathbf{G}_j|)$ to zero for sufficiently large difference $|\mathbf{G}_i - \mathbf{G}_j|$, we can obtain the matrix in a sparse form. Indeed, for diamond and zinc-blende crystals it is usually sufficient to retain nonzero values of pseudopotentials only for $|\mathbf{G}| = \sqrt{3}\frac{4\pi}{a}, \sqrt{4}\frac{4\pi}{a}, \sqrt{8}\frac{4\pi}{a}$, and $\sqrt{11}\frac{4\pi}{a}$. In Table 5.2 we give the parameters for silicon and indium arsenide taken from [162]. Note that we have omitted the values of $V_S(\sqrt{4})$ and $V_A(\sqrt{8})$, as the values of pseudopotentials are irrelevant due to the identical vanishing of the cos and sin functions for those arguments in the definition of the symmetric and antisymmetric potentials.

In Fig. 5.5 we plot the resulting band structures of silicon (a) and indium arsenide (b) close to the bandgap, obtained with the basis of 49 vectors $\mathbf{G}$.

These calculations were carried out without accounting for the spin-orbit interaction, which is the reason for the absence of the spin-orbit split-off subband in the band structure. The spin-orbit Hamiltonian, which needs to be added to H_{PS} given in Eq. (5.9), has the general form

$$H_{SO} = \sum_{i=1}^{Nuc} \sum_{\alpha} \sum_{l=0}^{\infty} V_{\alpha,l}^{SO}(\mathbf{r} - \mathbf{R}_i - \tau_\alpha)\mathbf{L} \cdot \mathbf{S} \sum_{m=-l}^{l} |l,m\rangle_{i,\alpha \ i,\alpha}\langle l,m|. \quad (5.14)$$

Fig. 5.5 Band structures for the bulk silicon (**a**), bulk InAs (**b**), and graphene (**c**) calculated with the empirical pseudopotential method without the spin-orbit interaction. The unit length a_B is the bare hydrogenic Bohr radius

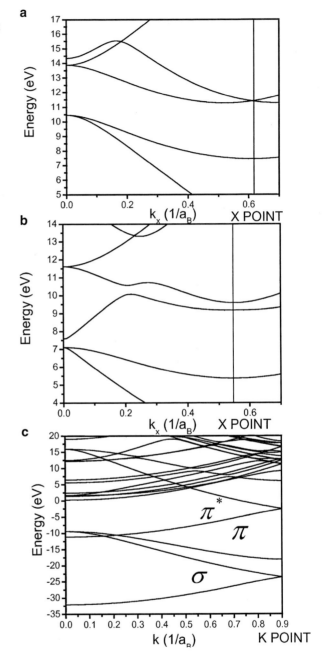

This Hamiltonian couples the electronic spin and orbital degrees of freedom, but does so locally on each atom. Indeed, in the above formula we project the extended electronic wave function onto local atomic spherical harmonics, and apply the angular-momentum dependent atomic potential $V_l^{SO}(\mathbf{r})$. As a result, our basis states can no longer be identified by just the reciprocal vector $\mathbf{G}$, but also have to be labeled by the spin number, i.e., have to be in the form $|\mathbf{G}, \sigma\rangle = \frac{1}{\sqrt{N_{uc}\Omega}} e^{i\mathbf{G}\cdot\mathbf{r}}\chi_\sigma$, where the last term denotes the spinor, which for electrons can describe the spin-up or spin-down. The spin-orbit Hamiltonian will couple the basis states with different spins, which leads to the doubling of the size of our Hamiltonian matrix. For completeness, we provide here the explicit form of the Hamiltonian matrix element in our spinor basis [151, 163]:

$$
\langle \mathbf{G}_i \sigma_i | H_{SO} | \mathbf{G}_j \sigma_j \rangle = i\frac{4\pi}{\Omega} \langle \sigma_i | \mathbf{S} | \sigma_j \rangle \frac{\mathbf{G}_i \times \mathbf{G}_j}{|\mathbf{G}_i||\mathbf{G}_j| \sin\gamma}
$$
$$
\times \sum_\alpha e^{i(\mathbf{G}_j - \mathbf{G}_i)\cdot\tau_\alpha} \sum_{l=0}^\infty (2l+1) V_{l,\alpha}^{SO}(\mathbf{G}_i, \mathbf{G}_j) P_l^1(\cos\gamma), \tag{5.15}
$$

where γ is the angle created by the vectors $\mathbf{G}_i$ and $\mathbf{G}_j$, $P_l^1(\cos\gamma)$ is the associated Legendre polynomial, and the spherical Fourier transform of the spin-orbit potential is $V_{l,\alpha}^{SO}(\mathbf{G}_i, \mathbf{G}_j) = \int dr r^2 V_{l,\alpha}^{SO}(r) j_l(|\mathbf{G}_i|r) j_l(|\mathbf{G}_j|r)$ with $j_l(x)$ being the spherical Bessel function. Typically in pseudopotential calculations of nanostructures one restricts this element to contain only terms with $l = 1$, and parametrizes the local spin-orbit pseudopotential by a Gaussian function with one adjustable parameter per atom type [151].

We have discussed the application of the empirical pseudopotential model to calculating the electronic structure of bulk solids. Let us now discuss briefly how this theory can be used to nanostructures. To make the discussion graphic, let us assume that we have parametrized the pseudopotential for the diamond bulk, neglecting the spin-orbit interaction due to the low atomic number of carbon. Typically this parametrization will consist of several numbers corresponding to the parameters for the symmetric pseudopotential listed in Table 5.2 for silicon. Can we use this parametrization, for example, to compute the electronic structure of the graphene sheet? At first look it would seem that we cannot, because diamond is coordinated tetrahedrally, while graphene is a two-dimensional, one-atom-thick hexagonal (honeycomb) lattice (a piece of such a lattice can be seen in Fig. 5.1d). Indeed, the basis of $\mathbf{G}$ vectors in the two cases will be different, and there is no guarantee that pseudopotentials at the points suggested in Table 5.2 will be relevant. The solution is to formulate the pseudopotential in the reciprocal space as a continuous function, with adjustable parameters, which can be fitted to give values relevant for any allotrope of carbon. We find such a function in [164] in the following form:

$$V_C(q) = \lambda \frac{\mu q^2 - \xi}{e^{\mu q^2 - \nu} + 1}, \tag{5.16}$$

with the following values of parameters: $\lambda = 1.781 \mathcal{R}, \mu = 0.354 a_B^{-1}, \xi = 1.424$, and $\nu = 0.938$, with $\mathcal{R}$ and a_B being the bare hydrogenic Rydberg and Bohr radius, respectively. The lattice constant of diamond is taken as $a = 0.35668$ nm, a parameter necessary for the proper normalization of the pseudopotential to the unit cell volume. Since the unit cell of graphene contains two atoms, just as is the case in the diamond lattice, the pseudopotential formalism presented above survives without changes. The only difference is visible in the basis of vectors $\mathbf{G}$, which have to be appropriate for the triangular two-dimensional lattice with lattice constant $a_g = 0.246$ nm. Moreover, since the atomic orbitals extend above the graphene layer, we cannot reduce our structure to two dimensions only. In practice, one considers a superlattice of identical graphene layers, separated from one another by a sufficiently large distance to avoid spurious coupling. The band structure calculated with this method is shown in Fig. 5.5d. There are two groups of subbands, denoted as σ and π (π^*). These subbands describe the in-plane motion of the electron, with the σ subband being created by the in-plane (covalent) bonds between carbon atoms, and the π subband being created by the p_z orbitals. The characteristic degeneracy of π and π^* subbands at the edge of the Brillouin zone (point K), and the linear dispersion close to that point are famously responsible for the Dirac Fermionic character of the mobile electrons.

In the case of both epitaxial quantum dots and colloidal nanocrystals, one typically does not change the coordination of atoms: the InAs/GaAs dots are still defined in the zinc-blende lattice, while the CdSe nanocrystals are still clusters in wurtzite lattice. However, the functional representation of the pseudopotential as in Eq. (5.16) is still necessary owing to the fact that these lattices are not of perfect bulk character. The atoms are displaced from their lattice sites due to strain. However, more importantly, the translational symmetry is broken by the existence of the nanostructure itself—the material composition varies across the computational domain. It is necessary, therefore, to define an artificial periodicity of the system, creating the entire nanostructure as a supercell, and repeating this supercell to form a superlattice in all directions. The barrier material surrounding the nanostructure should be thick enough to prevent the adjacent supercells from coupling—mechanically or quantum-mechanically, unless such coupling is a part of the design. The basis of reciprocal vectors $\mathbf{G}$ has to be chosen according to this high-level periodicity, and the total pseudopotential will be constructed as a superposition of all atomic pseudopotentials in the entire supercell. For a more detailed description of this computational procedure and examples of its use the reader is directed to [151, 165–170].

In concluding our presentation of the semiempirical pseudopotential approach, we emphasize that the functions composing the basis set are defined in the entire volume of the supercell, rather than in the immediate vicinity of a given atom. This means that interactions between a chosen atom and all atoms in the supercell are accounted for, even though they diminish rapidly with the distance (the model is "up to all neighbors"). The size of the Hamiltonian matrix is therefore defined only indirectly by the number of atoms (the size of the supercell), and has more to do with the quality of the atomic pseudopotentials: for less-soft pseudopotentials (changing rapidly close to atoms) more plane waves are needed to obtain satisfactory accuracy. It must be admitted, however, that the pseudopotentials are usually parametrized by only several material constants, as can be seen from the form (5.16). The lack of locality of basis functions, however, makes it impossible to implement this method using order-N approaches [106], which scale linearly with the number of atoms. On the other hand, obtaining such scaling is possible in the empirical tight-binding approach, to which we now turn.

5.2.3.2 Empirical Tight-Binding Approach

In deriving the tight-binding model one starts with the same form of the Hamiltonian for the single quasielectron as seen in the previous section:

$$\widehat{H}_{PS} = \frac{\widehat{\mathbf{p}}^2}{2m} + \sum_{i=1}^{N_{at}} V_{PS}^{(i)}(\mathbf{r} - \mathbf{R}_i), \tag{5.17}$$

expressed as a superposition of atomic pseudopotentials $V_{PS}(\mathbf{r} - \mathbf{R}_i)$, each centered on the atom i, whose coordinate is denoted by $\mathbf{R}_i$. The critical assumption in the tight-binding model is, however, the choice of a local basis set, composed of atomistic orbitals $\langle \mathbf{r}|i, \alpha \rangle = \phi_\alpha(\mathbf{r} - \mathbf{R}_i)$. Manifestly, each atom has its own set of these orbitals, enumerated by the index α. These orbitals do not have to be (and typically are not) orthogonal.

Using these orbitals, we express the sought wave function of the quasielectron as the linear combination (LCAO approach):

$$\Phi_n(\mathbf{r}) = \sum_{i=1}^{N_{at}} \sum_{\alpha=1}^{M} F(n, i, \alpha) \phi_\alpha(\mathbf{r} - \mathbf{R}_i), \tag{5.18}$$

where $F(n, i, \alpha)$ are numerical coefficients which we want to find in the calculation. The index n enumerates the resulting quasielectron states. In bulk it will be a composite index, denoting the subband and the value of the vector $\mathbf{k}$. In the nanostructure, on the other hand, it will just be the index of the state within the electronic structure. The index may also contain spin, although in the presence of the spin-orbit interaction the spin is not resolved as a good quantum number.

Our computational problem is therefore reduced to setting up the Hamiltonian matrix in the basis of the atomistic orbitals and diagonalizing this matrix numerically. The size of the Hamiltonian matrix is readily obtained as the product $N_{at}M$, that is, the number of atoms multiplied by the number of orbitals per atom. As we shall see, the QNANO package assumes $M = 20$, therefore we are faced with a problem of diagonalization of a matrix of order of tens of millions.

Let us define our Hamiltonian matrix elements. As is clear from the above discussion, we need the following terms:

$$\langle i, \alpha | H_{PS} | j, \beta \rangle = \int d^3 r \phi_\alpha(\mathbf{r} - \mathbf{R}_i)^* H_{PS}(\mathbf{r}) \phi_\beta(\mathbf{r} - \mathbf{R}_j). \tag{5.19}$$

If we knew the explicit forms of the pseudopotentials, by assuming some set of atomistic orbitals we could compute the above elements directly. Such attempts are being made in the form of the tight-binding-density functional theory approaches ([171], and references therein). However, the empirical tight-binding approach assumes that these elements are not calculated, but obtained by fitting to the known properties of semiconductor materials. We will discuss the fitting process later on; first let us complete the formulation of the model itself.

The essential assumption in the QNANO implementation of the model is that the Hamiltonian matrix elements $\langle i, \alpha | H_{PS} | j, \beta \rangle$ are set identically to zero whenever the distance between the atom i and j is greater than a single bond (the nearest-neighbor approximation). In tetrahedrally coordinated materials, therefore, the only nonzero elements in the Hamiltonian will comprise the onsite elements $(i = j)$ and the elements connecting our atom to its four neighbors. Thus, our Hamiltonian has only $5M$ nonzero elements per row, which in the QNANO implementation amounts to 100, irrespective of the overall number of atoms. Thus, our Hamiltonian matrix may be large, but it is extremely sparse. Here let us mention that implementations involving second neighbors do exist [172], but they present difficulties in fitting, especially in the presence of interfaces.

Note that even within the nearest-neighbor approximation a large number of empirical constants (for binary compounds and for $M = 20$, 40 onsite and 400 hopping parameters) are needed. In order to reduce their number, we take further assumptions as to the nature of our atomic orbitals. We write them in the following form:

$$\phi_\alpha(\mathbf{r}) = R_{nl}(|\mathbf{r}|) Y_{lm}(\theta, \phi) \chi_\sigma, \tag{5.20}$$

that is, as the product of the radial, angular, and spin part. For electrons we only have $\sigma = \uparrow$ or $\sigma = \downarrow$, so our set of $M = 20$ functions per atom is split into 10 orbitals with spin-up, and 10 with spin-down. Further, in our basis we include one orbital with $l = 0$ (the s orbital), three orbitals with $l = 1$ (the p orbitals), five orbitals with $l = 2$ (the d orbitals), and one additional, excited orbital with $l = 0$ (the s^* orbital). The characteristic symmetries of these orbitals are shown in Fig. 5.6. However, we choose not to define the radial component at this time.

Figure 5.6 also reveals the organization of the diagonal matrix elements of the Hamiltonian. The orbitals with the same spin are treated as local eigenstates of the pseudopotential of the i-th atom, that is, $\langle i, \alpha | H_{PS} | i, \beta \rangle = 0$ if $\alpha \neq \beta$, neglecting for now any spin-orbit coupling. Further, as it is in the hydrogen atom, we choose the onsite elements to be equal for all components m of the same angular momentum l.

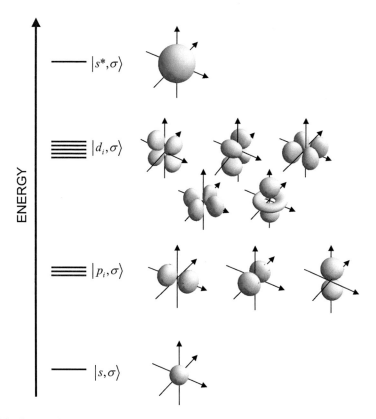

Fig. 5.6 Characteristic symmetries of the atomistic orbitals. *From the bottom*: the s orbital, three p orbitals, five d orbitals, and the excited s^* orbital

As a result, instead of 20 different onsite elements per atom we only have four, grouped as depicted symbolically on the energy scale of Fig. 5.6. We shall denote these onsite energies by $\varepsilon_{i,\alpha}$ (in binary compounds they will be different for the cation and for the anion, hence the index i).

Let us now include the spin-orbit interaction. Following the formulation of Chadi [173], we restrict this interaction to connect only the orbitals on the same atom (that is $i = j$), and work only with the three p orbitals. This is an approximation, as this interaction can also give matrix elements connecting the d orbitals. The spin-orbit interaction can connect the orbitals with the same spin, which results in the following elements $\langle x\uparrow|H_{SO}|y\uparrow\rangle = -i\Delta/3$, $\langle x\downarrow|H_{SO}|y\downarrow\rangle = +i\Delta/3$. There are also elements connecting states with opposite spins: $\langle x\uparrow|H_{SO}|z\downarrow\rangle = \Delta/3$, $\langle y\uparrow|H_{SO}|z\downarrow\rangle = -i\Delta/3$, $\langle z\uparrow|H_{SO}|x\downarrow\rangle = -\Delta/3$, $\langle z\uparrow|H_{SO}|y\downarrow\rangle = i\Delta/3$. The numerical parameter Δ is a material constant, which needs to be fitted for each atom type (i.e., two constants for a binary compound). Note that we have rotated our p orbitals from the basis with a definite quantum numbers $l = 1, m = 0, \pm 1$ to the "linear" orbitals x, y, and z. Figure 5.6 shows the symmetries of these axially resolved orbitals.

Fig. 5.7 The projection method of obtaining hopping matrix elements for a simple cubic (**a**) and zinc-blende lattice (**b**)

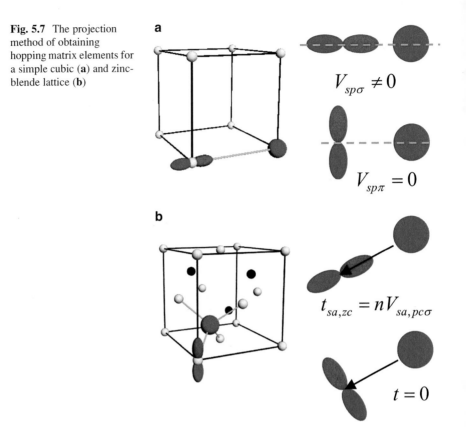

Finally, we assume that hopping from one atom to another can be carried out only without changing the spin. That is, if we unfold the composite index α using the orbital and spin quantum numbers, we have $\langle i, n_1, l_1, m_1, \sigma_1 | H_{PS} | j, n_2, l_2, m_2, \sigma_2 \rangle = 0$ whenever $\sigma_1 \neq \sigma_2$. This reduces the number of hopping elements from 400 to 100, as we only have to consider elements between 10 orbitals from each atom.

At this point, for a binary compound we require now 110 parameters, a number still prohibitively large. To further reduce it we shall employ a projection procedure originally described by Slater and Koster [174]. We illustrate this procedure on an example of hopping from an s orbital on an atom to a p orbital on its nearest neighbor. In Fig. 5.7 we discuss this procedure for the simple cubic and the zinc-blende lattice. In the case of the simple cubic lattice we have positioned our orbitals on the nearest-neighbor atoms, whose bond is denoted in green. We may reasonably expect that the hopping element between these two will be finite and equal $V_{sp\sigma}$. However, if we were to place a p_z orbital on one atom, and the s orbital on the other, the hopping element will be zero by symmetry. This is shown in the lower right-hand diagram of Fig. 5.7a: the product of the two orbitals is going to be antisymmetric about the horizontal plane, and therefore its integral must be zero. However, the zinc-blende lattice is more complicated, as the bond directions are not aligned with the coordinate axes, as evident from Fig. 5.7b. The vector connecting the two atoms (one of the green bonds) can be written as $\mathbf{d} = d_0(l\widehat{x} + m\widehat{y} + n\widehat{z})$ in terms of the Cartesian versors, with l, m, and n being directional cosines, and d_0—the bond length. Let us now project our p_z orbital onto the bond (upper right-hand diagram of Fig. 5.7b) and in a direction perpendicular to the bond (lower right-hand diagram). The hopping element in the first case, accounting for the appropriate directional cosine, can be expressed in the form $t_{sa,zc} = nV_{sa,zc}$, with the anion-cation parameter $V_{sa,zc}$ being a material constant. On the other hand, the other projection will yield zero contribution due to the symmetry. The key point of this procedure is that now we can immediately generate the hopping elements between the s orbital and the p_x, p_y orbitals simply as $t_{sa,xc} = lV_{sa,zc}$ and $t_{sa,yc} = mV_{sa,zc}$, that is, the same material constant, only premultiplied by the directional cosine, gives us three different hopping matrix elements. This procedure can be applied to all 100 hopping matrix elements, reducing the number of independent hopping material constants to 21. Thus, the tight-binding model for the ideal zinc-blende binary compound will be fully parametrized by 31 material constants.

Let us write explicitly the tight-binding Hamiltonian in the nearest-neighbor basis, obeying all the constraints listed above.

$$\widehat{H}_{\text{TB}} = \sum_{i=1}^{N_{at}} \sum_{\alpha} \varepsilon_{i,\alpha} c_{i,\alpha}^+ c_{i,\alpha} + \sum_{i=1}^{N_{at}} \sum_{\alpha,\beta} \lambda_{i,\alpha,\beta} c_{i,\alpha}^+ c_{i,\beta} + \sum_{i=1}^{N_{at}} \sum_{j}^{n.n(i)} \sum_{\alpha,\beta} t_{i,\alpha,j,\beta} c_{i,\alpha}^+ c_{j,\beta}, \quad (5.21)$$

where the spin-orbit parameters λ are expressed in terms of the material constants Δ defined above, and the hopping elements $t_{i,\alpha,j,\beta}$ are composites of the 21 material hopping constants. In the above formula, the operator $c_{i,\alpha}^+(c_{i,\alpha})$ is the creation (annihilation) operator of an electron on the orbital α of atom i.

The numerical values of the parameters are obtained by fitting the bulk band structure obtained from our model to that obtained experimentally or with more advanced computational techniques. For the $sp^3d^5s^*$ model, that is, the one that QNANO implements, the canonical example of the fitting procedure is presented in [175]. Here, the authors have used the muffin-tin potential approximation to obtain a high-quality description of the band structure of semiconductor materials throughout the entire Brillouin zone. Further, group-theoretical arguments were used to isolate the tight-binding parameters responsible for coupling of different subbands at different points of the reciprocal space, which allowed for a systematic fitting procedure of smaller sections out of the full 31-parameter set. High-quality parameter sets were given for group IV and III–V semiconductor materials. The quality of this fit is, however, limited by the quality of the underlying, more microscopic model. Often one desires to be able to fit directly to the characteristic properties of the system, such as band gaps and effective masses of the electron and the hole, measured in experiment. Here the fitting procedure is more complex, particularly if one wants to obtain good fit to the effective masses, as in these cases the convenient group-theoretical analysis is not possible. Therefore, the entire

31-parameter set has to be fitted at once. It seems that the only algorithm capable of treating this highly nonlinear optimization problem satisfactorily is the genetic algorithm. The genetic fitting has been successfully applied to silicon [176] giving a set of parameters for a simpler, sp^3s^* model. The fitting was also performed for silicon and germanium in the full 20-band model, with the algorithm augmented by semi-analytical expressions of relevant effective masses of these materials at the Γ point of the Brillouin zone [177]. The complete sets of parameters for GaAs and InAs were given in [132]. Note that this particular fitting strategy, suggested by the group from Purdue University, assumes that the onsite and spin-orbit elements of arsenic ought to be the same irrespective of whether this atom is in the GaAs or InAs material. In that respect, this approach differs from the original fitting to muffin-tin potentials as in [175]. An excellent critical discussion of different tight-binding models and different parameter sets for the $sp^3d^5ds^*$ approach is presented in [178]. It demonstrates that the inclusion of the five d orbitals is crucial to understand the optical spectra of GaAs nanocrystals, however the results produced by the two available parameter sets for the 20-orbital $sp^3d^5s^*$ model were very similar. Our recent work was focused on InAs dots embedded in GaAs material [179, 180], however we have developed our own parametrization, in general more consistent with that of [175].

Also, we have developed a parametrization of the CdSe wurtzite material, in which the approach described above has to be modified slightly. The hexagonal symmetry of the wurtzite lattice introduces a subtle symmetry breaking, not present in the zinc-blende lattices, but visible only at the distance of the third neighbor. This symmetry breaking is visible in the band structure as the so-called crystal field splitting. Since our approach is nearest-neighbor only, our model will not capture this property naturally. To account for the crystal field splitting, one typically allows the p orbitals to split into two manifolds: the p_x and p_y orbitals remain degenerate, and the p_z orbital becomes split-off; the amount of this splitting is another fitting parameter [181]. In [111] we present a full list of the $sp^3d^5s^*$ parameters appropriate for CdSe. Figure 5.8 shows the bulk band structures for InAs (a) and CdSe (b) across the Brillouin zone obtained using our parametrizations.

5.2.3.3 Incorporation of Strain

Up to now we have discussed the tight-binding calculation of electronic structure of bulk materials, in which the atoms occupy their ideal positions. However, as we have discussed at length before, in the nanostructures the atoms are displaced from their bulk coordinates due to strain. Presently we discuss how these displacements are incorporated into the tight-binding model. The model approach used in QNANO consists of two aspects. The first element is illustrated schematically in Fig. 5.9. In the previous section we have discussed the Slater–Koster rules allowing us to parametrize different hopping elements with relatively few material constants utilizing the directional cosines as scaling factors. This ideal situation is repeated in Fig. 5.9a, where the hopping matrix element is defined for an ideal bond and scaled

Fig. 5.8 Band structure of InAs (**a**) and CdSe (**b**) obtained with the 20-band tight-binding model with our parametrizations

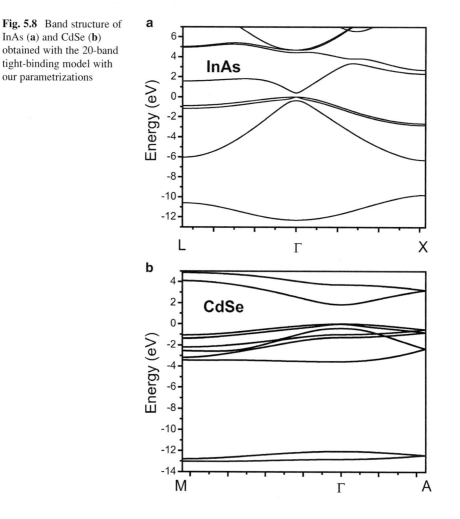

by the directional cosine n. When the atoms are displaced, the net effect on the bond is twofold: the bond can change direction and length (Fig. 5.9b). The change in direction is easily accounted for by recomputing the directional cosine. However, the change in length requires a modification of the characteristic material constant. Therefore, for the particular $p_z - s$ hopping shown in the figure, the matrix element for a strained material takes the form

$$t'_{sa,zc} = n' V_{sa,pc} \left(\frac{d}{d_0} \right)^{\eta_{sz}},\tag{5.22}$$

where d and d_0 are respectively the strained and ideal bond length. The exponent η_{sz} is treated as a material parameter and has to be fitted in the procedure described in detail below. This approach has been proposed in [175].

Fig. 5.9 Illustration of the procedure of incorporation of atomic displacements into the tight-binding model: (**a**) hopping without strain, (**b**) hopping with strain

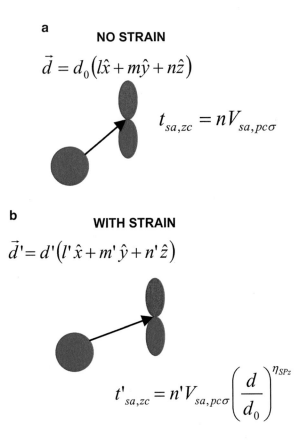

a **NO STRAIN**

$$\vec{d} = d_0\left(l\hat{x} + m\hat{y} + n\hat{z}\right)$$

$$t_{sa,zc} = nV_{sa,pc\sigma}$$

b **WITH STRAIN**

$$\vec{d}' = d'\left(l'\hat{x} + m'\hat{y} + n'\hat{z}\right)$$

$$t'_{sa,zc} = n'V_{sa,pc\sigma}\left(\frac{d}{d_0}\right)^{\eta_{SPz}}$$

The second aspect by which the displacements enter the tight-binding parameters involves the modification of diagonal (onsite) tight-binding matrix elements [182]. This modification follows from our implicit assumption of orthogonality of our atomistic orbitals—not only locally (different orbitals on the same atom), but also globally (different orbitals on different atoms). We can build such globally orthogonal orbitals out of our local, atomistic orbitals by the Loewdin procedure. If we limit ourselves to nearest neighbors only, we get

$$\langle \mathbf{r}|\mathbf{R}_i\alpha\rangle = \phi_\alpha(\mathbf{r} - \mathbf{R}_i) - \frac{1}{2}\sum_{j=1}^{nn}\sum_{\beta}S_{i\alpha,j\beta}\phi_\beta(\mathbf{r} - \mathbf{R}_j), \qquad (5.23)$$

where the ket notation describes the global basis, while the explicit real-space orbitals are local, atomistic functions, and $S_{i\,\alpha,j\,\beta}$ is the overlap matrix element.

Bearing that in mind, we can now express the bulk, unstrained diagonal matrix element formally as:

$$\langle \mathbf{R}_i \alpha | H | \mathbf{R}_i \alpha \rangle = \varepsilon_{i\alpha}^{(0)} - \sum_{j=1}^{nn} \sum_{\beta} S_{i\alpha, j\beta} t_{\mathbf{R}_i \alpha, \mathbf{R}_j, \beta}. \tag{5.24}$$

Of course, the above energy in total is never computed, it is obtained by fitting. The above formula only exposes its internal structure in terms of the local atomistic orbitals. Here, $\varepsilon_{i\alpha}^{(0)}$ is the bare, vacuum-referenced energy, that is, one that we would have if the atom i was isolated. The element t is the matrix element with which the hopping terms are parametrized; it is obtained in the fitting procedure. The overlap matrix element is, however, unknown, as we never specify the explicit form of our atomic orbitals. The key of the procedure proposed in [182] is to express this element using the extended Hueckel theory, in the following form:

$$S_{i\alpha, j\beta} = C_{i\alpha, j\beta} \frac{t_{i\alpha, j\beta}}{\varepsilon_{i\alpha}^{(0)} + \varepsilon_{j\beta}^{(0)}}, \tag{5.25}$$

where $C_{i\alpha, j\beta}$ is a material parameter. Thus, our diagonal energy has an internal structure, which can be expressed by the vacuum-referenced energies, hopping matrix elements to nearest neighbors, and certain material scaling constants.

Of course, the very same procedure can be carried out for the strained case. In our notation, in such case the relevant symbols will acquire a "prime", but let us analyze more closely which of these symbols will actually be modified. The vacuum-referenced energies will stay the same, as they pertain to the case of isolated atoms. The coefficients C will stay the same, as we choose them to be material- and not position-dependent. The position information enters through the tunneling element t, which will certainly acquire a new value. By comparing the formulas for the strained and unstrained case, we find

$$\varepsilon_{i\alpha}' = \varepsilon_{i\alpha} + \sum_{j=1}^{nn} \sum_{\beta} C_{i\alpha, j\beta} \frac{\left(t_{i\alpha, j\beta}'\right)^2 - \left(t_{i\alpha, j\beta}\right)^2}{\varepsilon_{i\alpha}^{(0)} + \varepsilon_{j\beta}^{(0)}}. \tag{5.26}$$

At this point we are essentially done, however we do not know the vacuum-referenced energies appearing in the denominator. The authors of [182] propose to use the unstrained diagonal elements $\varepsilon_{i\alpha}$ diminished by the bare Rydberg ($\Delta E = -13.6\,\text{eV}$). This is a somewhat disputable point of this theory, however any inaccuracy introduced by it will be compensated by the appropriate choice of the constants C.

Both hopping exponents η and constants C describing the diagonal energy shifts have to be obtained in a fitting procedure. To this end, we need to know (from experiment or from ab initio calculations) what the behavior of the band edges is when the bulk material is stressed. We choose to consider two cases only, the hydrostatic strain and the biaxial strain, since in both cases the translational symmetry remains intact. In the first case the bulk is stressed from all directions in the same manner, so that the unit cell of the crystal changes its volume, but is not otherwise deformed. Here, for small stresses the shifts of band edges are described by the Bir–Pikus linear laws [153]:

$$E_{CB} = E_{CB}^{(0)} + a_c(\varepsilon_{xx} + \varepsilon_{yy} + \varepsilon_{zz}), \tag{5.27}$$

$$E_{HH,LH} = E_{HH,LH}^{(0)} + a_v(\varepsilon_{xx} + \varepsilon_{yy} + \varepsilon_{zz}), \tag{5.28}$$

$$E_{SO} = E_{SO}^{(0)} + a_v(\varepsilon_{xx} + \varepsilon_{yy} + \varepsilon_{zz}), \tag{5.29}$$

where the indices CB, HH, LH, SO denote the electron, heavy and light hole and spin-orbit split-off subbands, respectively, the quantities indexed with (0) represent unstrained band edges, and the constants a_c and a_v are deformation potentials. These deformation potentials are well-known material parameters, but can also be computed using density-functional methods [183–186].

In the second case the unit cell changes its dimensions as we apply the stress in the $x - y$ plane. That is, we choose the x and y coordinates of the atoms, but allow the unit cells to expand naturally in the z direction. This expansion, and therefore the resulting vertical strain, is governed by the Poisson ratio:

$$\varepsilon_{zz} = -2\frac{c_{12}}{c_{11}}\varepsilon_{xx}, \tag{5.30}$$

where the constants c_{ij} are the elastic parameters discussed in the section devoted to strain. By using this ratio, we can calculate the corresponding z coordinates of our atoms. However, translation of this biaxial strain into the shifts of band edges is not straightforward. Typically it is necessary to perform density-functional calculations for a broad range of strains to map out this highly nonlinear dependence [184, 187, 188]. The full set of parameters obtained in this fitting procedure for GaAs and InAs can be found in [132]. In Fig. 5.10 we show the comparison of target InAs band edge behavior under the hydrostatic (a) and biaxial (b) strain to the results obtained with our model.

In closing, we note that currently more sophisticated methods of incorporating strain into the tight-binding model are being developed [189–191]. It is understood that the approach described above is appropriate only for nanostructures under diagonal stresses (hydrostatic, biaxial), but is not sufficient for shear strains. Stresses of that nature appear for nanostructures grown on substrates with orientation (111). However in what follows we restrict ourselves to discussing the samples grown on (001) substrates, in which case the above approach is sufficient.

5.2.3.4 Calculation of Nanostructures

Up to this point we have discussed the tight-binding model in the context of the properties of bulk (strained or not) with its translational symmetry. We conclude the

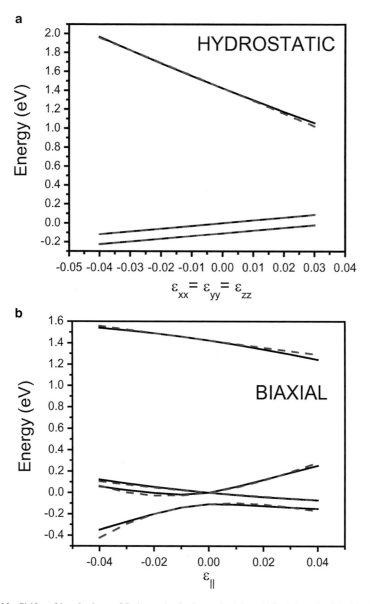

Fig. 5.10 Shifts of band edges of InAs under hydrostatic (**a**) and biaxial strain (**b**). *Black lines*: target behavior from DFT calculations, *red lines*: our fit

discussion of calculations of single-particle state by specializing our model to nanostructures.

In the case of nanostructures we have to be mindful of two issues. First, as the atom moves across the system, it will encounter interfaces, that is places where atoms of two (or more) different materials meet as nearest neighbors. In systems in which one atom is shared, such as InAs/GaAs (common arsenic) or InAs/InP (common indium) such material interfaces never pose a problem for bonds: if a bond connects the In and As atoms, it is surely of the InAs nature, with the appropriate hopping matrix elements; if the In atom is connected to P, it is of the indium phosphide type. However, the ambiguity may arise for the onsite elements of this shared atom: at the interface it may be difficult to establish whether the In atom belongs to InAs or InP. In some parametrizations [132] this issue is solved by enforcing that the onsite and spin-orbit elements of this shared atom (in this case In) be the same, and this has to be done already at the fitting step. Indeed, in such case the bulk tight-binding parameters of InAs and InP have to be fitted simultaneously. Our parametrization allows the shared atoms to be different. The interface ambiguity is solved by simple arithmetic averaging of the onsite and spin-orbit constants weighed with the type of the nearest neighbors.

Second aspect is connected to the fact that we do not have the translational symmetry any more. Therefore, our computational domain cannot be narrowed down to only several atoms making up the unit cell, but has to include all atoms (possibly millions) making up the entire system. This issue has been discussed in detail before for empirical pseudopotentials, in which case the approach of choice is to copy the supercell infinitely in all directions, thereby creating a periodic superlattice. Of course, this approach can also be realized in QNANO, simply by applying the periodic boundary conditions: the atoms from the side of our domain will have, as their nearest neighbors, the atoms from the opposite wall. Sometimes, however, this approach is not appropriate, such as for example for colloidal nanocrystals. As we shall see in the third part of this chapter, the nanocrystals are free-standing, spherical structures, which are not embedded in any extended medium. Therefore, we need to account for the fact that our computational domain has a surface. As a result of that, some of the atoms will be deprived of some of their nearest neighbors, i.e., dangling bonds will appear. The dangling bonds unfortunately create spurious subbands, whose energy is close to the energies of the states confined in the nanostructure (these "artificial impurity" bands are inside the semiconductor gap). The way of dealing with these bonds was discussed in detail in [192]. First, one finds the atoms which exhibit the dangling bonds. Then, one identifies these bonds by performing orthogonal transformation of the orbitals belonging to that atom. This is necessary, since the atomistic basis contains the orbitals s, p etc., while the actual bonds in tetrahedrally coordinated materials are sp^3 hybrids. The appropriate transformation is very simple:

$$|sp_1^3\rangle = \frac{1}{2}(|s\rangle + |p_x\rangle + |p_y\rangle + |p_z\rangle), \tag{5.31}$$

$$|sp_2^3\rangle = \frac{1}{2}(|s\rangle + |p_x\rangle - |p_y\rangle - |p_z\rangle), \tag{5.32}$$

$$|sp_3^3\rangle = \frac{1}{2}(|s\rangle - |p_x\rangle + |p_y\rangle - |p_z\rangle), \tag{5.33}$$

$$|sp_4^3\rangle = \frac{1}{2}(|s\rangle - |p_x\rangle - |p_y\rangle + |p_z\rangle). \tag{5.34}$$

After this transformation we are in the basis of hybrids. By checking the coordinates of the missing nearest neighbors, we can identify which of these hybrids are dangling. We shift the diagonal Hamiltonian energies of these hybrids up in energy, so they fall in regions outside of our scope (i.e., they become remote in energy from the states confined in the nanostructure). We complete the procedure by performing the inverse transformation back into the basis of the s and p orbitals. The whole procedure can be condensed to a precomputed set of corrections, which have to be applied to the diagonal Hamiltonian blocks of surface atoms. The form of these matrices depends on how many dangling bonds there are, and in which direction they point. This "surface passivation" procedure becomes simply encoded in the overall Hamiltonian of the system.

At this point we are ready to compute the single-article states of the electron and the hole of our atomistically defined, strained nanostructure. Here we would like to stress that our QNANO package is to this point functionally analogous to another atomistic computational tool, NEMO3D, which has been developed within the Nanohub initiative of the Purdue University by the group of Professor G. Klimeck [193]. This large-scale effort is, however, devoted to modeling the electronic transport through the nanostructured systems. In what follows we shall discuss how the elements of electronic structure computed within the atomistic tight-binding approach can be used to investigate optical properties of nanostructures.

5.2.4 States of Interacting Electrons and Holes

Upon diagonalization of the tight-binding Hamiltonian, we obtain the single-quasiparticle energy levels and wave functions. As explained in detail in the beginning of Sect. 5.2.3, these levels are to be understood in analogy to the Kohn–Sham orbitals in the self-consistent density functional approach. That is, we construct the ground state of the system by filling these states with quasielectrons up to the Fermi level. In our case, all the valence-band states are to be filled with electrons, and the conduction-band states are empty. In order to write this ground state formally, we first define the quasielectron creation and annihilation operator on our single-quasiparticle levels as c_i^+ and c_i, respectively. Here, the index i enumerates the quasiparticle states, and not the atoms. The ground state of our system is thus

$$|v\rangle = \prod_{i\in\text{VB}} c_i^+ |0\rangle, \tag{5.35}$$

where $|0\rangle$ denotes the vacuum state (the system emptied of all valence electrons).

Note that the number of the quasiparticle states is very large—in our model it is $20N_{at}$, where N_{at} is the total number of atoms in our supercell. In the case of free-standing nanostructures, such as colloidal nanocrystals, all these states are confined inside the dot, and therefore all of them are relevant for the next step. However, for epitaxial dots, only a small number of states (typically several tens) will be confined inside our quantum dot, and their wave functions will decay exponentially in the barrier. The vast majority of the states will be extended over the entire computational domain (their energies will lie within the bands of the barrier material). The properties of these states will be influenced by the boundaries of the computational domain, and therefore we do not expect that they are a true representation of the actual propagating barrier states in the real sample.

The many-body physics in our system involves creating excitations from the ground state $|v\rangle$. We will remove electrons from occupied valence states and/or add electrons onto the empty conduction states. However, in light of the above, it will only be meaningful to consider the confined quasiparticle levels, and therefore all the electrons placed deeper in the valence band (on the extended states) will never be moved. This makes the form of the ground state (5.35) cumbersome. It is convenient to define a set of quasihole creation and annihilation operators, defined as follows: $h_{i,\text{VB}}^+ = c_{i,\text{VB}}$ and $h_{i,\text{VB}} = c_{i,\text{VB}}^+$. These operators will act only in the valence band. Creating the hole on the level i is equivalent to removing an electron from that level, and conversely. In the conduction band we continue to use the quasielectron operators $c_{i,\text{CB}}^+$ and $c_{i,\text{CB}}$, and in the following we will drop the band indices CB and VB.

From the above definition we see that the quasihole state is a collective state of many electrons (with one electron missing). However, it is convenient to define the hole orbital in relation to that of the missing electron. The electronic and hole operators are respectively the Hermitian conjugates of each other. Therefore, if the orbital of an electron on the valence band orbital i has the LCAO form

$$\Phi_i^{(e)}(\mathbf{r}) = \sum_{k=1}^{Nat}\sum_{\alpha=1}^{M} F(i,k,\alpha)\phi_\alpha(\mathbf{r} - \mathbf{R}_k), \tag{5.36}$$

then the hole orbital is

$$\Phi_i^{(h)}(\mathbf{r}) = \sum_{k=1}^{Nat}\sum_{\alpha=1}^{M} F^*(i,k,\alpha)\phi_\alpha(\mathbf{r} - \mathbf{R}_k). \tag{5.37}$$

Let us now use our electron and hole creation operators to form the excitations from the ground state $|v\rangle$. The single-pair excitations will have the form:

$$|i, p\rangle = c_i^+ h_p^+ |v\rangle, \tag{5.38}$$

while the two-pair excitations are

$$|i, j, p, q\rangle = c_i^+ c_j^+ h_p^+ h_q^+ |v\rangle. \tag{5.39}$$

Note that the indices i, j refer to the confined electron orbitals, while p, q refer to the confined hole orbitals. There may be many such excitations, even in the small basis of confined states. Indeed, with 10 electron and 10 hole states we can generate 100 single-pair and 2,025 two-pair configurations.

Now we have to account for the fact that the electrons and the holes are charged quasiparticles, and therefore they interact with one another. The Hamiltonian of the interacting electron–hole system, written in the language of our electron and hole creation and annihilation operators has the form:

$$\begin{aligned}
\widehat{H}_{eh} = {} & \sum_i E_i c_i^+ c_i + \frac{1}{2} \sum_{ijkl} \langle ij|V_{ee}|kl\rangle c_i^+ c_j^+ c_k c_l \\
& + \sum_p E_p h_p^+ h_p + \frac{1}{2} \sum_{pqrs} \langle pq|V_{hh}|rs\rangle h_p^+ h_q^+ h_r h_s \\
& - \sum_{iqrl} (\langle iq|V_{ehd}|rl\rangle - \langle iq|V_{ehx}|lr\rangle) c_i^+ h_q^+ h_r c_l.
\end{aligned} \tag{5.40}$$

In the above Hamiltonian, the energies E_i and E_p are, respectively, the single-quasiparticle energies of the electron and the hole calculated in our tight-binding approach. The matrix elements $\langle ij|V_{ee}|kl\rangle$ and $\langle pq|V_{hh}|rs\rangle$ scale, respectively, the electron–electron and hole–hole Coulomb interactions. The elements $\langle iq|V_{ehd}|rl\rangle$ and $\langle iq|V_{ehx}|lr\rangle$ are, respectively, the direct and exchange electron–hole Coulomb matrix elements. In the next section we will show how we calculate these matrix elements using our atomistic quasiparticle wave functions. Here let us only state that the many-body eigenstates of the above Hamiltonian are obtained in the configuration interaction approach: we generate all possible excitations from our state $|v\rangle$ (typically restricting them to a chosen number of electron–hole pairs), we create the Hamiltonian matrix in the basis of these configurations, and we diagonalize this matrix numerically. A detailed and systematic discussion of this procedure, albeit without the electron–hole exchange term, is given in [194].

We wish to emphasize that the electron–hole exchange is not usually included in the calculations based on the $\mathbf{k} \cdot \mathbf{p}$ model of single-particle states. The interested reader may find an extensive discussion of this term and its microscopic origin in [195, 196]. The atomistic model, by virtue of its resolution, allows to treat this aspect of Coulomb interactions on equal footing with the three other terms.

5.2.5 Coulomb Matrix Elements

To complete the description of the electron–hole Hamiltonian (5.40) we now describe the procedure of calculating the Coulomb matrix elements. We choose to discuss the electron–electron element, with all other terms calculated in an analogous manner. This computational equivalence between the matrix elements of the different forms of Coulomb interactions is one of the strengths of the atomistic approach.

Having in mind the LCAO form of the electronic single particle states as in Eq. (5.36), we write down the complete Coulomb matrix element in the form:

$$
\begin{aligned}
\langle ij|V_{ee}|kl\rangle = {} & \frac{e^2}{4\pi\varepsilon_0\varepsilon} \sum_{a1}\sum_{\alpha} F^*(i,a_1,\alpha) \sum_{a2}\sum_{\beta} F^*(j,a_2,\beta) \\
& \times \sum_{a3}\sum_{\gamma} F(k,a_3,\gamma) \sum_{a4}\sum_{\delta} F(l,a_4,\delta) \\
& \times \int d^3r_1 \int d^3r_2 \frac{\phi_\alpha^*(\mathbf{r}_1-\mathbf{R}_{a1})\phi_\beta^*(\mathbf{r}_2-\mathbf{R}_{a2})\phi_\gamma(\mathbf{r}_2-\mathbf{R}_{a3})\phi_\delta(\mathbf{r}_1-\mathbf{R}_{a4})}{|\mathbf{r}_1-\mathbf{r}_2|}.
\end{aligned}
$$

$$(5.41)$$

Here the indices a_1, a_2, a_3, a_4 enumerate atoms, e is the electron charge ($e > 0$), and ε_0 and ε are respectively the vacuum and relative dielectric constants. Note that in the above expressions we have four sums over all atoms and all atomic orbitals, and therefore may face the task of integrating and summing $\sim 10^{24}$ terms. To make the calculations feasible, we introduce a series of approximations. First, we restrict the general four-center integrals to two-center only, that is, we include only the terms with $\mathbf{R}_{a1} = \mathbf{R}_{a4}$ and $\mathbf{R}_{a2} = \mathbf{R}_{a3}$. This reduces our integration and summation to only $\sim 10^{12}$ terms. Further, for a given coordinate $\mathbf{R}_{a1}$ we single out three cases: (1) when $\mathbf{R}_{a2} = \mathbf{R}_{a1}$ (terms involving the same atom), (2) when coordinates $\mathbf{R}_{a1}$ and $\mathbf{R}_{a2}$ define the nearest neighbors, and (3) where $\mathbf{R}_{a1}$ and $\mathbf{R}_{a2}$ are more remote (a long-range element). In the third case, we replace the coordinate difference $|\mathbf{r}_1 - \mathbf{r}_2|$ in the denominator by the difference $|\mathbf{R}_{a1} - \mathbf{R}_{a2}|$ of the atomic coordinates, i.e., for remote neighbors we treat the partial charges put on the atoms as point charges. In such case the double integral turns to a trivial orthonormalization relation. With all these simplifications, our Coulomb matrix element takes the form:

$$\langle ij|V_{ee}|kl\rangle = \langle ij|V_{ee}^{(O)}|kl\rangle + \langle ij|V_{ee}^{(NN)}|kl\rangle + \langle ij|V_{ee}^{(R)}|kl\rangle, \qquad (5.42)$$

with

$$\langle ij|V_{ee}^{(O)}|kl\rangle = \frac{e^2}{4\pi\varepsilon_0}\sum_{a1}\sum_{\alpha\beta\gamma\delta}F^*(i,a_1,\alpha)F^*(j,a_1,\beta)F(k,a_1,\gamma)F(l,a_1,\delta)$$

$$\times \int d^3r_1 \int d^3r_2 \frac{\phi_\alpha^*(\mathbf{r}_1)\phi_\beta^*(\mathbf{r}_2)\phi_\gamma(\mathbf{r}_2)\phi_\delta(\mathbf{r}_1)}{|\mathbf{r}_1-\mathbf{r}_2|},$$

$$\langle ij|V_{ee}^{(NN)}|kl\rangle = \frac{e^2}{4\pi\varepsilon_0\varepsilon_{NN}}\sum_{a1}\sum_{a2\in NN}\sum_{\alpha\beta\gamma\delta}F^*(i,a_1,\alpha)F^*(j,a_2,\beta)F(k,a_2,\gamma)F(l,a_1,\delta)$$

$$\times \int d^3r_1 \int d^3r_2 \frac{\phi_\alpha^*(\mathbf{r}_1-\mathbf{R}_{a1})\phi_\beta^*(\mathbf{r}_2-\mathbf{R}_{a2})\phi_\gamma(\mathbf{r}_2-\mathbf{R}_{a2})\phi_\delta(\mathbf{r}_1-\mathbf{R}_{a1})}{|\mathbf{r}_1-\mathbf{r}_2|},$$

and

$$\langle ij|V_{ee}^{(LR)}|kl\rangle = \frac{e^2}{4\pi\varepsilon_0\varepsilon}\sum_{a1}\sum_{a2\notin NN}\sum_{\alpha\beta}F^*(i,a_1,\alpha)F^*(j,a_2,\beta)F(k,a_2,\beta)F(l,a_1,\alpha)$$

$$\times \frac{1}{|\mathbf{R}_{a1}-\mathbf{R}_{a2}|}.$$

Note that while separating the terms according to interatomic distance, we have also introduced a distance-dependent dielectric function: for the onsite term $\varepsilon = 1$, the long-range term is scaled by the full bulk constant ε, and the nearest-neighbor constant is typically $\varepsilon_{NN} = \varepsilon/2$. This treatment is chosen to approximate more sophisticated approaches [197–201]; improving our approach in this aspect is one of our next steps.

Out of those terms, the third one is tractable in a straightforward manner, as we know the LCAO wave function coefficients $F(i,a_1,\alpha)$ and the atomic positions needed to calculate the interatomic distances. However, special algorithmic techniques are needed to calculate these terms efficiently. The interested reader will find a description of these techniques in [122]. In order to compute the remaining two terms, we need to have the atomistic orbitals $\phi_\alpha(\mathbf{r})$. To date, we only assumed their angular symmetry (as appropriate for the $sp^3d^5s^*$ basis set), but we have never defined their radial component. In principle we did not need it for our tight-binding calculation, as we have fitted the Hamiltonian matrix elements directly in that model (having the radial component of the wave functions we might have calculated them). However, at present we have to assume them in some form. The typical approach is to choose the Slater formulas appropriate for the multielectron atoms [178, 202]:

$$R_\alpha(r) = Nr^a e^{-br}, \tag{5.43}$$

Table 5.3 Slater orbital coefficients for atoms making up the most important epitaxial and colloidal quantum dots

Orbital	a	b $(1/a_B)$
In s, p	3.0	1.2500
In d	3.0	0.2500
In s^*	3.2	0.3095
As s, p	2.7	1.7027
As d	2.7	0.2703
As s^*	3.0	0.4000
Ga s, p	2.7	1.3514
Ga d	2.7	0.2703
Ga s^*	3.0	0.3250
P s, p	2.0	1.6000
P d	2.0	0.3333
P s^*	2.7	0.4324
Cd s, p	3.0	1.0875
Cd d	3.0	0.2500
Cd s^*	3.2	0.2738
Se s, p	2.7	1.8784
Se d	2.7	0.2703
Se s^*	3.0	0.4375

where N is a normalization constant, and a and b are the Slater coefficients dependent upon the chemical identity of the atom and the orbital. In Table 5.3 we give the values of the coefficients a and b for the semiconductor materials used in this work. It is possible to obtain semi-analytical formulas for the inner integrals present in onsite Coulomb terms (the radial integration has to be carried out numerically). However, the nearest-neighbor terms are more difficult and we compute them using the Fourier transform technique on the grid.

Inclusion of the nearest-neighbor matrix elements is viable only for the very small nanostructures, composed of several hundred atoms at most. In larger nanostructures the computation of the full three-term matrix elements takes a prohibitively long time and we restrict ourselves to the onsite and long-range components only.

To conclude this section, we have computed the matrix element for the electron–electron interaction $\langle i = 1, i = 1 | V_{ee} | i = 1, i = 1 \rangle$ in an InAs disk embedded in a GaAs barrier material. The disk diameter was 25.4 nm, its height was 2.3 nm, and the disk was positioned on a wetting layer of the thickness of 1 monolayer. We will look closer at such systems in the third part of this chapter. The optimization of atomic positions was carried out in a domain of $4 \cdot 10^7$ atoms, while the single-particle states were computed in a domain of $4.5 \cdot 10^5$ atoms. In this system we found that the full Coulomb matrix element $\langle i = 1, i = 1 | V_{ee} | i = 1, i = 1 \rangle$ had a value of 21.32 meV, in which the contribution of the long-range term was 21.13 meV. Thus, the influence of the onsite term here is very small. For the hole–hole elements (full element 25.41 meV, long-range only 25.03 meV) this influence is slightly larger, and for the electron–hole direct elements (full element

23.05 meV, long-range only 22.81 meV) it is of the same order as that in the electron–electron term. However, the onsite element appears to be much more important in the electron–hole exchange: we obtained the value of the full element of 0.148 meV, of which the long-range component is 0.119 meV. In absolute numbers, the contribution of the onsite term is similar to the other three, but relative to the very small value of the element itself, it makes up for about 20% change. For this reason we typically retain the onsite elements in our calculations.

5.2.6 Emission and Absorption Spectra

In the two previous sections we have fully defined the Hamiltonian describing the system of interacting electrons and holes, confined in our nanostructure. By diagonalizing this Hamiltonian in the basis of configurations with the same number of electron–hole pairs, we obtain the eigenstates of these exciton complexes. For instance, the n-th state of the exciton, with energy $E_{X,n}$, can be written as the linear combination

$$|X_n\rangle = \sum_{ip} A_{ip}^{(n)} c_i^+ h_p^+ |v\rangle, \tag{5.44}$$

with the energy $E_{X,n}$ being the eigenvalue, and the vector of coefficients $A_{ip}^{(n)}$—the eigenvector of the Hamiltonian. Similarly, the m-th biexciton state, with energy $E_{XX,m}$, can be written as the linear combination

$$|XX_m\rangle = \sum_{ijpq} B_{ijpq}^{(m)} c_i^+ c_j^+ h_p^+ h_q^+ |v\rangle. \tag{5.45}$$

With these eigenstates we can now compute the emission and absorption spectra of the system. We apply the Fermi's Golden Rule, in which the emission spectra for a biexciton can be calculated as

$$I(\omega, \varepsilon) = \sum_f |\langle X_f | \hat{P}(\varepsilon) | XX_i \rangle|^2 \delta(E_{XX,i} - E_{X,f} - \hbar\omega). \tag{5.46}$$

An analogous formula can be written for any excitonic complex and will connect the two complexes whose numbers of electron–hole excitations differ by one. Here, one of the electron–hole pairs making up the biexciton recombines radiatively, leaving behind the other electron–hole pair. The energy of the outgoing photon, and thus the position of the emission peak, is equal to the difference between the energies of the biexciton and the exciton. Note that for a given initial biexciton state $|XX_i\rangle$ there can be many final exciton states $|X_f\rangle$ and thus many emission maxima. The amplitude of each of these maxima is defined by the matrix element of the interband polarization operator $\widehat{P}(\varepsilon)$, where ε denotes the polarization of light (linear or circular). This operator is defined as

$$\widehat{P}(\varepsilon) = \sum_{ip} d_{ip}(\varepsilon) c_i h_p \tag{5.47}$$

Here, $d_{ip}(\varepsilon)$ is the dipole matrix element connecting the single-quasiparticle states i (electron) and p (hole). We shall discuss this element in the next section.

The calculation of the emission amplitude typically involves executing a long summation of different dipole matrix elements, connecting each of the two-pair configurations to the single-pair configurations. These sums may vanish even if the constituent terms are nonzero. This quantum interference effect has been captured in [194]. We shall discuss the complexity involved here on the examples in the third part of this chapter.

In computing the emission spectra we choose the initial biexciton state (typically it is the ground state of the biexciton) and examine its radiative recombination to all exciton states (ground and excited). If we were to include the temperature effects in our calculation, we would allow several biexciton states to be thermally occupied, with appropriate energy and temperature dependent coefficients; the total emission spectrum would be a superposition of the spectra from each of these biexciton states. In absorption one applies the inverse procedure: one chooses one initial state of a smaller number of electron–hole pairs (e.g., exciton), and examines its optical coupling to the ground and excited states of the higher complex (e.g., the biexciton). The thermal effects can be included here in a manner analogous to that in emission.

5.2.7 Dipole Matrix Elements

The last component of the QNANO package is the dipole matrix element needed to compute the optical response functions, as discussed in the previous section. We define this element as

$$d_{iq}(\varepsilon) = \int d^3 r \Psi_q^*(\mathbf{r}) \varepsilon \cdot \mathbf{r} \Psi_i(\mathbf{r}), \tag{5.48}$$

where we employ the real-space forms of the single-particle wave functions, and account for the photon polarization ε. By substituting the LCAO forms of our single-particle states from Eqs. (5.36) and (5.37), we obtain

$$d_{iq}(\varepsilon) = \sum_{a1}\sum_{\alpha}F(q,a_1,\alpha)\sum_{a2}\sum_{\beta}F(i,a_2,\beta)\int d^3r\phi_\alpha^*(\mathbf{r}-\mathbf{R}_{a1})\varepsilon\cdot\mathbf{r}\phi_\beta(\mathbf{r}-\mathbf{R}_{a2}).$$

$$(5.49)$$

Note that care must be taken to account for the hole character of the orbital q, involving the characteristic lack of conjugacy of the coefficients F.

Similar to the Coulomb matrix element, this element is also evaluated in a series of approximations. First, we limit the range of our two-center integrals by rejecting all terms in which the distance $|\mathbf{R}_{a1}-\mathbf{R}_{a2}|$ is greater than the nearest neighbor distance. Furthermore, in the integral in which $\mathbf{R}_{a1}=\mathbf{R}_{a2}$ we perform the shift of the coordinate system and use the orthogonality of the atomistic orbitals. As a result we obtain

$$\begin{aligned}d_{iq}(\varepsilon) = &\sum_{a1}\sum_{\alpha}F(q,a_1,\alpha)F(i,a_1,\alpha)\varepsilon\cdot\mathbf{R}_{a1}\\&+\sum_{a1}\sum_{\alpha\beta}F(q,a_1,\alpha)F(i,a_1,\beta)\int d^3r\phi_\alpha^*(\mathbf{r})\varepsilon\cdot\mathbf{r}\phi_\beta(\mathbf{r})\\&+\sum_{a1}\sum_{\alpha}F(q,a_1,\alpha)\sum_{a2\in NN}\sum_{\beta}F(i,a_2,\beta)\int d^3r\phi_\alpha^*(\mathbf{r})\varepsilon\cdot\mathbf{r}\phi_\beta[\mathbf{r}-(\mathbf{R}_{a2}-\mathbf{R}_{a1})].\end{aligned}$$

Thus, we deal with three terms: the first one, in which the partial electronic occupations of the atomistic orbitals are treated as point charges, the second one, describing couplings between different orbitals of the same atom, and the third one, describing the nearest-neighbor coupling [203, 204].

In obtaining the numerical values of the above dipole element we utilize a procedure similar to that discussed in Sect. 5.2.5, that is, whenever necessary, we use Slater orbitals to approximate the atomistic wave functions. For illustration, let us give numerical values of the matrix element connecting the lowest-energy electron and hole single-particle state in the example nanostructure discussed in Sect. 5.2.5. We compute the dipole element for the polarization x, that is, $\varepsilon\cdot\mathbf{r}=x$. If we account for the first, "point charge" term only, the absolute value of this element is 0.5407 nm, adding the second term decreases it to 0.4999 nm (about 10%), and adding the third term decreases it further by about 2%, to 0.4886 nm. The nearest-neighbor term entails a somewhat longer computational time and its contribution is typically only several percent, and therefore typically it can be safely neglected. However, the computational complexity introduced by it is nowhere near the requirements of the Coulomb matrix elements.

At this point we have presented the theoretical framework of the QNANO package in its entirety. In the next sections we shall apply this package to example nanostructures and demonstrate how the results of QNANO are interpreted in terms which allow for their comparison to experiments.

5.3   Applications of the QNANO Package

In this section we shall demonstrate the different aspects of the QNANO package on examples of epitaxial and colloidal quantum dots. As is evident from the discussion in the previous section, QNANO gives a large amount of information at its different stages. Below we shall show how this information is to be interpreted in terms of physical properties of the systems.

5.3.1   Engineering of Quasiparticle Degeneracies
in Epitaxial Dots

As the first example we discuss the structure shown in Fig. 5.11a. It is a lens-shaped InAs dot positioned on a monolayer-thick wetting layer. The diameter of the dot is 25 nm and its height is 3.5 nm above the wetting layer. The dot is embedded in the GaAs barrier material. We chose this system in order to test our model against that of the empirical pseudopotentials, whose results have been reported in [205]. Our results closely resemble those from the pseudopotential approach. In our approach we defined the computational domain of $5 \cdot 10^7$ atoms for the calculation of strain, and narrowed it down to $5.61 \cdot 10^5$ atoms for the calculation of single-particle states.

In Fig. 5.11b we show the energies and in-plane probability densities of a few confined electron and hole states. We see that the electron states (blue) form discrete shells: the s shell composed of one Kramers-degenerate state, the p shell containing two closely-lying Kramers doublets, and the d shell composed of three Kramers doublets. The labels of these shells become clear as we examine the probability densities corresponding to the respective states. The s shell has a quasi-circular symmetry (without nodes), in analogy to the s shell of the atom. The two p-type states have a very clear quasi-atomic p-type character. However, there are only two such states (not three as in the atom) because our system is quasi-two-dimensional. Further, the d shell consists of three states with characteristic d-like nodal planes.

For holes (drawn in red) the situation is different: we show the six Kramers doublet levels which do not seem to be organized in any distinctive structures. Their symmetries are also rather indistinct: except for the state closest to the bandgap (i.e., the ground state of the hole), which exhibits the s symmetry, other states have ring-like probability densities.

In Fig. 5.11c we show the joint optical density of states (JDOS), formed by calculating the dipole matrix elements for various pairs of electron–hole states, and plotting them against the energy gaps characteristic for each such pair. In spite of the apparent differences in the wave functions of electron and hole states, the JDOS maxima are grouped in the shell-like structures, labeled in the figure as s, p, and d. We find that the large values of the dipole matrix elements occur between states

Fig. 5.11 (a) Schematic view of a lens-shaped InAs quantum dot placed on a monolayer-thick wetting layer and embedded in the GaAs barrier (barrier atoms not shown). The dot diameter is 25 nm, and its height is 3.5 nm above the wetting layer. (b) The energies of the single-particle states (*center*) and the probability densities of electron (*right*) and hole states (*left*). (c) The joint optical density of states derived from the single-particle spectrum

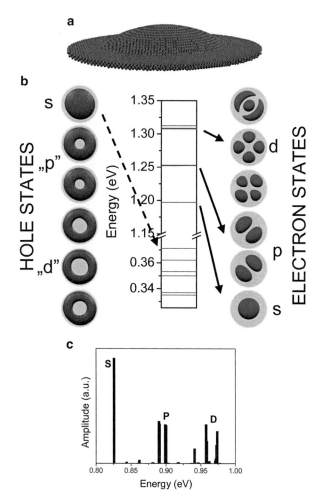

grouped in the same "shells", that is s electron state is strongly connected optically to the s hole state and so on. The cross terms, although nonzero, appear to be small. Their existence is the direct consequence of the complicated character of the hole single-particle states. A more detailed discussion of this system can be found in [180].

Let us now discuss a different structure, shown in Fig. 5.12a. This dot is of the shape of a truncated lens, which can be achieved experimentally with the indium-flush technique mentioned earlier in this chapter [6]. As previously, it is an InAs dot embedded in GaAs, however its diameter is 12.5 nm and its height is 2 nm above the wetting layer. The smaller dot size allowed us to define smaller computational domains: $3.2 \cdot 10^7$ atoms for strain and $3 \cdot 10^5$ atoms for the single-particle states. Figure 5.12b shows the single-particle energies and orbitals. Comparing these results with those for the previous structure, we find that the electron states are

Fig. 5.12 (**a**) Schematic view of an InAs quantum dot with the shape of a truncated lens, placed on a monolayer-thick wetting layer and embedded in the GaAs barrier (barrier atoms not shown). The dot diameter is 12.5 nm, and its height is 2 nm above the wetting layer. (**b**) The energies of the single-particle states (*center*) and the probability densities of electron (*right*) and hole states (*left*)

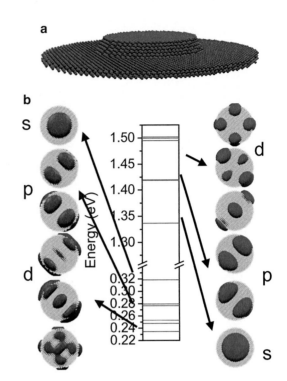

similarly grouped in shells, albeit with larger intershell energy gaps due to the smaller size of the structure. However, there is a qualitative difference in the hole spectrum: now, the hole states are also grouped in shells, and their probability densities (shown in red) also exhibit the characteristic *s*, *p* and *d* symmetries. Therefore, changing the shape of the dot (truncating the lens) can be used to engineer the single-particle spectra of the system. We wish to emphasize that the observed behavior is a result of changing the shape of the dot and not its size. We demonstrate it in [179], where we compare the spectra of our truncated dot to the lens-shaped system of the same diameter.

This qualitative change in the hole spectrum can be understood using simple arguments grounded in the $\mathbf{k} \cdot \mathbf{p}$ theory [114, 115, 119]. In this theory, one typically employs a two-step approach: one first solves for the single-particle states and wave functions in the potential well defined by the nanostructure—as if the electron and the hole were simply particles with an effective mass. In the second step, one accounts for the subband mixing by employing the Luttinger–Kohn Hamiltonian. If we restrict that Hamiltonian to the basis of the two heavy- and two light-hole subbands only, this Hamiltonian reads [114, 115, 119]:

$$\widehat{H}_{LK} = \begin{pmatrix} -P_+ & R & -S & 0 \\ R^* & -P_- & 0 & S \\ -S^* & 0 & -P_- & R \\ 0 & S^* & R^* & -P_+ \end{pmatrix}. \tag{5.50}$$

As we have mentioned, this Hamiltonian is spanned in the hole basis with total angular momentum $J = 3/2$, with the projections $J_z = 3/2$, $-1/2$, $1/2$, $-3/2$, respectively (heavy-, light-, light-, and heavy-hole state). The operators in this Hamiltonian are defined as:

$$P_+ = \frac{\hbar^2}{2m_0}[(\gamma_1 + \gamma_2)k^2 + (\gamma_1 - 2\gamma_2)k_z^2], \tag{5.51}$$

$$P_- = \frac{\hbar^2}{2m_0}[(\gamma_1 - \gamma_2)k^2 + (\gamma_1 + 2\gamma_2)k_z^2], \tag{5.52}$$

$$R = \frac{\hbar^2}{2m_0}(-\sqrt{3})\gamma_{23}k_-^2, \tag{5.53}$$

$$S = \frac{\hbar^2}{2m_0}(2\sqrt{3})\gamma_3 k_- k_z. \tag{5.54}$$

In the above formulas we have the parameters γ_1, γ_2 and γ_3—the Luttinger material parameters, the parameter $\gamma_{23} = (\gamma_2 + \gamma_3)/2$ (axial approximation), and $k_\pm = k_x \pm i k_y$. The momentum operator $\mathbf{k} = -i\nabla$, with $k_\rho^2 = k_x^2 + k_y^2$.

Let us now follow this general approach. We assume that we have solved the simple, single-band Schroedinger equations for the electron and the hole in our system. These single-subband states can be written as

$$\Psi_{i\sigma}(\mathbf{r}) = \Phi_i(x, y, z)u_\sigma(\mathbf{r}), \tag{5.55}$$

where the index i will enumerate envelope functions of the states shown in Figs. 5.11b and 5.12, while σ is the subband index.

Due to the rotational symmetry of the confining potential, and the strong confinement in the vertical direction, we may approximate the envelope functions as $\Phi_i(x, y, z) = \phi_{n,m}(\rho, \varphi) \cos(l\pi z/H)$, where ρ and φ are the radial and angular in-plane coordinates. The in-plane function $\phi_{n,m}(\rho, \varphi)$ will have a well-defined angular momentum m and nodal quantum number n. The vertical component is scaled by the integer quantum number l, defining the vertical subbands, with H being the height of the quantum dot.

As for the Bloch subband functions, the electron functions are $u_{s,1/2}(\mathbf{r}) = |S\rangle|\uparrow\rangle$ and $u_{s,-1/2}(\mathbf{r}) = |S\rangle|\downarrow\rangle$. Here the function $|S\rangle$ accounts for the fact that the electron states are built with the s-type atomic orbitals, and the arrow denotes the electron spin. For the heavy and light holes we have

$$u_{3/2}(\mathbf{r}) = \langle r|3/2, 3/2\rangle = \frac{1}{\sqrt{2}}(|X\rangle + i|Y\rangle)|\uparrow\rangle, \tag{5.56}$$

$$u_{-1/2}(\mathbf{r}) = \langle r|3/2, -1/2\rangle = -\frac{1}{\sqrt{6}}(|X\rangle - i|Y\rangle)|\uparrow\rangle - \sqrt{\frac{2}{3}}|Z\rangle|\downarrow\rangle, \qquad (5.57)$$

$$u_{1/2}(\mathbf{r}) = \langle r|3/2, 1/2\rangle = -\frac{1}{\sqrt{6}}(|X\rangle + i|Y\rangle)|\downarrow\rangle + \sqrt{\frac{2}{3}}|Z\rangle|\uparrow\rangle, \qquad (5.58)$$

$$u_{-3/2}(\mathbf{r}) = \langle r|3/2, -3/2\rangle = \frac{1}{\sqrt{2}}(|X\rangle - i|Y\rangle)|\downarrow\rangle. \qquad (5.59)$$

Here the index of the function u is simply the projection J_z of the microscopic angular momentum of the hole. In the above equations, the functions $|X\rangle, |Y\rangle, |Z\rangle$ are the microscopic orbitals. We note that the hole functions are linear combinations of the p-type orbitals, a fact which we can readily reproduce in our atomistic tight-binding approach.

Since assume that the electron subband does not mix with the hole subbands, the full electron states remain in the form given in the Eq. (5.55). On the other hand, as demonstrated in [119], the cylindrical symmetry of the system allows to isolate two orthogonal subspaces for the mixed hole states (hole spinors) and identify them by the chirality quantum number $\chi = \Uparrow$ or $\chi = \Downarrow$. The chirality $\Uparrow$ states have the form

$$|L, \Uparrow\rangle = \begin{pmatrix} A^+\phi_{n,m}\cos(q_z z)|3/2, 3/2\rangle \\ B^+\phi_{n,m+2}\cos(q_z z)|3/2, -1/2\rangle \\ C^+\phi_{n,m+1}\sin(2q_z z)|3/2, 1/2\rangle \\ D^+\phi_{n,m+3}\sin(2q_z z)|3/2, -3/2\rangle \end{pmatrix}, \qquad (5.60)$$

while the chirality $\Downarrow$ states are

$$|L, \Downarrow\rangle = \begin{pmatrix} A^-\phi_{n,m-3}\sin(2q_z z)|3/2, 3/2\rangle \\ B^-\phi_{n,m-1}\sin(2q_z z)|3/2, -1/2\rangle \\ C^-\phi_{n,m-2}\cos(q_z z)|3/2, 1/2\rangle \\ D^-\phi_{n,m}\cos(q_z z)|3/2, -3/2\rangle \end{pmatrix}. \qquad (5.61)$$

The coefficients $A^\pm, B^\pm, C^\pm, D^\pm$ are eigenvectors of the Hamiltonian (5.50). The coefficient $q_z = \pi/H$. Note that spinors of both types are mixtures of one pair of the heavy- and light-hole state from the lower subband of the vertical motion (the cosine envelope component) and another pair of the higher vertical subband (the sine envelope component). This is due to the element S in the Hamiltonian, which requires the change of symmetry of the vertical envelope component. On the other hand, the in-plane functions have different angular momenta, as both the R and S elements require the changes of their symmetry. The state of each chirality can be, however, characterized by the total angular momentum quantum number L, which is defined as the sum of projections of the envelope and Bloch angular momenta. For the spinor $|L, \Uparrow\rangle$ we have $L = m + 3/2$, while for the other spinor $L = m - 3/2$.

In our discussion above we have stated that the hole shell structure is sensitive not to the lateral dimensions of the dot, but rather its shape along its height (lens versus truncated length). The truncated dot is thinner, i.e., the confinement in the vertical direction is stronger. Because of that, the gaps between states with different vertical character (built on different vertical subbands) are larger. In particular, the heavy-hole states built upon the lowest vertical subband are separated by a larger gap from the light-hole states built upon the second vertical subband. Therefore, the admixture of the light-hole components, defined by the coefficient $C^{\pm}$, in the truncated dot is smaller than that in the lens-shaped dot. Thus, in the truncated dot we see the hole spectra much more as if they were just simple, single-subband objects, similar to electrons, while in the lens-shaped dot this simple picture is not valid. This is precisely the reason for the emergence of the shell structure of hole states in the truncated dot sample.

5.3.2   Radiative Recombination of the Dark Exciton

We can gain even more insight into the nature of the hole single-particle states by examining the emission spectra of an exciton confined in the dot. An extensive discussion of these spectra based on atomistic calculations is presented in [120]. Here we shall summarize its most important points.

In the previous section, we have shown that the single-hole states are mixtures of heavy- and light subbands, and that this mixing can be tuned by choosing the shape of the dot. The question we shall address here is what consequences this mixing has for optical spectra. We return to the spectra of the large lens-shaped dot presented in Fig. 5.11. We present these spectra once again in Fig. 5.13a. Let us now populate the dot with one electron and one hole, and assume for the moment that these carriers do not interact. If we require that these carriers are placed on the respective s shells, as in Fig. 5.13a, we can create four possible configurations owing to the Kramers doublet character of the states. In the full atomistic model it is not straightforward to distinguish between these Kramers doublets. However, if we continue to analyze the system in terms of the $\mathbf{k} \cdot \mathbf{p}$ model, the two electron states are simply spin-up and spin-down, while the two hole states can be distinguished by their chirality, $|\Uparrow\rangle$ and $|\Downarrow\rangle$. The four possible configurations are shown schematically in Fig. 5.13b.

We identify two "bright" configurations, $|B_1\rangle$ and $|B_2\rangle$, in which the carriers have opposite spins, and two "dark" configurations, $|D_1\rangle$ and $|D_2\rangle$, in which the carriers are spin-polarized. The meaning of the labels of these states becomes clear if we analyze the dipole matrix elements connecting them. Let us start with the state $|B_1\rangle$, in which the electron has spin-down, and the hole—chirality up. Thus, the complete single-electron state can be written as $|\Psi_e^{(1)}\rangle = |0,0,e\rangle|S\rangle|\downarrow\rangle$. Here, the first

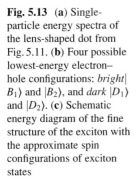

Fig. 5.13 (**a**) Single-particle energy spectra of the lens-shaped dot from Fig. 5.11. (**b**) Four possible lowest-energy electron–hole configurations: *bright* $|B_1\rangle$ and $|B_2\rangle$, and *dark* $|D_1\rangle$ and $|D_2\rangle$. (**c**) Schematic energy diagram of the fine structure of the exciton with the approximate spin configurations of exciton states

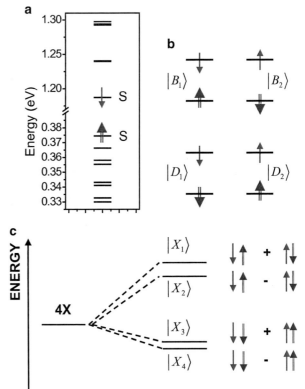

component denotes the envelope function: it has zero angular momentum, the nodal quantum number is also zero, and it is even with respect to the vertical motion (denoted by the letter e). Next we have the Bloch orbital character of the function, which is $|S\rangle$, as appropriate for the conduction band states. Finally, we have the spin state, in this case spin-down. Similarly, the hole state $|\Psi_h^{(\Uparrow)}\rangle = A^+|0,0,e\rangle|3/2,3/2\rangle + B^+|0,2,e\rangle|3/2,-1/2\rangle + C^+|0,1,o\rangle|3/2,1/2\rangle + D^+|0,3,o\rangle|3/2,-3/2\rangle$. It is a mixture of states, whose envelopes have different angular momenta and different parity in the vertical direction (o denotes the odd parity), as well as different microscopic Bloch components.

Let us now calculate the in-plane dipole element $d_{\downarrow,\Uparrow} = \langle\Psi_h^{(\Uparrow)}|x|\Psi_e^{(\downarrow)}\rangle$, corresponding to the polarization x. Since the electron Bloch state is of the s symmetry, while the hole Bloch state—of the p symmetry, the symmetry flip introduced by the position operator has to be "spent" on connecting the two Bloch functions. As a result, the envelope components of the electron and hole functions must be the same and the spins must be opposite, as this is the optical selection rule for the electron–hole systems. Indeed, we find such possibility: the hole Bloch function $|3/2,3/2\rangle$ has spin-up, and it stands next to the envelope function of symmetry properties analogous to that of the electron. We conclude that the dipole element in the x polarization, connecting these two states, is large. Let us now try to perform a similar analysis for the z polarization. Here, the connection can only be established with the light hole state $|3/2, 1/2\rangle$, as only this state has admixtures of the microscopic $|Z\rangle|\uparrow\rangle$ orbital. However, in this case the envelope functions are mismatched—both in the in-plane and the vertical part, and the dipole matrix element in the z polarization must be zero. A similar analysis, with the same conclusions, can be established for the other "bright" state, $|B_2\rangle$. We see that the label "bright" indicates a sizeable dipole matrix element in the in-plane polarization.

Let us now flip the chirality of the hole, i.e., consider the electron–hole configuration denoted as $|D_1\rangle$. In this hole state the order of the vertical parities is reversed. Therefore, the in-plane dipole matrix element must be zero, as the spin-up component $|3/2, 3/2\rangle$ of the hole accompanies the vertically antisymmetric envelope function, incompatible with the symmetric electron function. On the other hand, the dipole element for the z polarization can connect our electron state with the light-hole state $|3/2, 1/2\rangle$, which has the correct spin and vertical symmetry. The only obstacle is the incompatibility of the envelope angular momenta: the electron state has $m = 0$, while the light-hole state has $m = -2$. Thus, in the system with ideal cylindrical symmetry this dipole element would be zero. However, our system has an additional underlying symmetry of the zinc-blende lattice, which breaks the overall rotational symmetry and may lead to a finite, albeit small value of the dipole element in z polarization. Thus, our "dark", spin-polarized configuration is optically inactive in the in-plane polarization, and weakly active in the polarization z. Similar conclusion can be drawn for the other dark configuration $|D_2\rangle$.

We set out to confirm this analysis in our atomistic calculation. However, up to now we have only considered simple electron–hole configurations. As we discussed in Sect. 5.2.4, these configurations are not actual exciton states. We have to form the matrix of the Hamiltonian (5.40) in the basis of our four configurations and diagonalize it numerically. The issues of the fine structure of the exciton are beyond the scope of this chapter; the interested reader may refer to [120, 196] for details. Here let us only state that the resulting energy spectrum consists of four levels and is depicted schematically in Fig. 5.13c. The levels are grouped in two pairs. The pair at a lower energy (states $|X_3\rangle$ and $|X_4\rangle$) is split by a small energy gap (of order of several μeV) and consists predominantly of the "dark" configurations $|D_1\rangle$ and $|D_2\rangle$. The pair at the higher energy (states $|X_1\rangle$ and $|X_2\rangle$) is split by tens of μ eV and is made up from the "bright" configurations $|B_1\rangle$ and $|B_2\rangle$. The two pairs are split by the fine-structure gap of order of hundreds of μ eV.

We compute the emission amplitudes from each of these states according to the prescription discussed in Sect. 5.2.6. The results are presented in Fig. 5.14. In Fig. 5.14 we focus on the bright exciton states, $|X_1\rangle$ and $|X_2\rangle$. The diagrams show the polarization-resolved spectra for the in-plane polarization. We see that the emission is linearly polarized, and the polarization planes are aligned with the (110) and $(1 - 10)$ crystallographic direction due to the underlying crystal lattice.

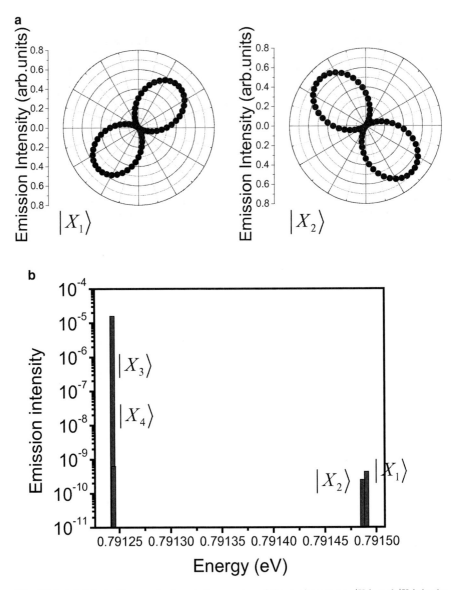

Fig. 5.14 (**a**) Polarization-resolved emission spectra of the exciton states $|X_1\rangle$ and $|X_2\rangle$ in the in-plane polarization. (**b**) Emission amplitudes of the four exciton states in the z polarization [120]

Note that in the rotationally-symmetric $\mathbf{k} \cdot \mathbf{p}$ model the emission would be circularly polarized. We do not report the emission from the dark states in this polarization, as it is negligibly small, in accordance with expectations. The light polarized in-plane propagates upward (along the height of the dot) and therefore can be easily detected.

Figure 5.14b shows the amplitudes of emission in z polarization from all four exciton states. The positions of the emission peaks correspond directly to the energies of the exciton states, as the final state in this emission is vacuum. This allows us to observe directly the fine structure of the exciton levels. Moreover, as expected, the emission amplitude of the "dark" exciton states $|X_3\rangle$ and $|X_4\rangle$, although small, is nonetheless nonzero. At the same time, the emission amplitude from the bright states $|X_1\rangle$ and $|X_2\rangle$ in the z polarization is several orders of magnitude smaller (the "bright" states are "dark" in this polarization). The light in z polarization propagates in the plane of the quantum dot, and its detection is more challenging. However, experimental studies have clearly shown the in-plane emission from the dark exciton [206].

5.3.3 Emission Spectra of Highly Excited Quantum Dots

As the last element of the discussion concerning epitaxial quantum dots, let us describe the emission spectra under high excitation power. In this case the system is pumped by many electrons and holes at a rate which exceeds the radiative recombination rate. In such case the carriers will accumulate and we are able to examine the emission spectra of many-exciton complexes.

We have calculated the emission spectra of complexes consisting of up to eight electron–hole pairs for the large lens-shaped quantum dot from Sect. 5.3.1 [180]. These spectra are presented in Fig. 5.15 as a function of the number of pairs. In the

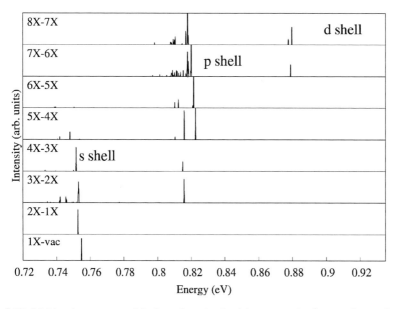

Fig. 5.15 Multiexciton spectra of the lens-shaped epitaxial quantum dot for complexes of up to eight excitons

calculation we assumed the occupation of only ground optically active state, and we assume the in-plane polarization. As we can see, the character of the spectra closely reflects the organization of single-particle states into shells, with the gaps dominated by the splittings of electron states. For the system of up to two electron–hole pairs we see the emission from the lowest shell, since the single-particle s shell for the electron and the hole is composed only of one Kramers doublet each. The third electron–hole pair must be placed on the p shell due to the Pauli exclusion principle, hence the higher-energy emission peak for the three-exciton system. The p shell can be populated altogether by four electron–hole pairs, therefore the seven-exciton system is the first one to exhibit the emission peak from the d shell.

Another property apparent from these spectra is the existence of the "hidden symmetry" [134, 194, 207, 209], that is, the insensitivity of the positions of emission peaks to the number of excitons in the shell. The emission maxima are not exactly aligned from one spectrum to another, which is due to the details of exchange and scattering, as well as the lack of clear shell structure of the hole states.

5.3.4 Exciton-Biexciton Mixing in CdSe Nanocrystals

Let us now apply our QNANO package to a selected set of optical properties of the colloidal nanocrystal. For this discussion we choose the CdSe nanocrystal, whose underlying crystal lattice is of the wurtzite type. A more detailed description of our analysis of this system can be found in [111, 208].

Figure 5.16a shows the schematic view of the atomistic structure of a small colloidal CdSe nanocrystal of diameter of 3 nm. Such a structure is composed of about 300 atoms. The overall shape of the system is chosen to be spherical, however it is clear that the details of the underlying crystal lattice break this symmetry. As we apply the QNANO procedure, we do not calculate the strain. This is an approximation, however, in the realistic structure we deal with surface reconstruction and the presence of surface ligands. At this point QNANO does not treat these mechanisms satisfactorily. Here we only apply the surface passivation scheme as described in Sect. 5.2.3.4.

The results of the calculations of single-particle electron levels in the tight-binding approach are shown in Fig. 5.16b. Even though, as we mentioned above, our system is not an ideal spherical quantum well, we find that the electron levels are grouped into shells. The lowest state (a single Kramers doublet) has a distinct, atom-like s symmetry. This state is separated from the next group of states by a large gap (of the order of 400 meV). The three higher-lying orbitals are close together and exhibit clear p-type overall symmetries. These levels are not degenerate, however. One of them is split off by about 100 meV, a clear consequence of the non-spherical character of the wurtzite structure.

The single-particle hole levels are discussed in Fig. 5.17. This spectrum is very different from that of the electron. We find four closely lying states close to the band gap edge (i.e., ground and three low-lying hole states), separated from the rest

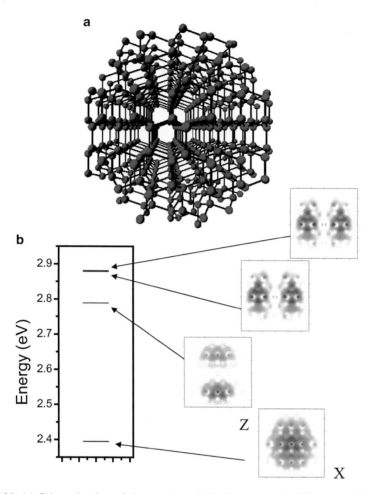

Fig. 5.16 (**a**) Schematic view of the wurtzite spherical nanocrystal of diameter of 3 nm. (**b**) Single-particle electron states and their probability densities for a CdSe nanocrystal with diameter of 3 nm

of the spectrum by a large gap (of about 320 meV). To understand it, we invoke the crystal field splitting present in the wurtzite material, and discussed by us briefly in Sect. 5.2.3.2. The gap separating the parts of the spectrum is size dependent and decreases as the dot size is increased [111]. In the limit of a very large nanocrystal it appears to converge to the characteristic crystal-field gap in the bulk CdSe band structure. The complexity of the hole states is additionally reflected in the lack of apparent symmetry of the four lowest hole states (i.e., the highest-energy states in Fig. 5.17a) We show the cross-sections of their probability density in Fig. 5.17b in the *XY* (horizontal) plane, and in Fig. 5.17c in the *XZ* (vertical) plane. These orbitals do not exhibit any clear similarity with the electron states discussed above.

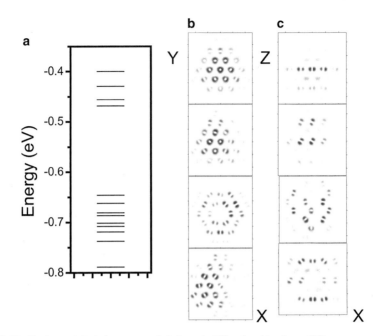

Fig. 5.17 Single-particle hole states and their probability densities for a CdSe nanocrystal with diameter of 3 nm. Panel (**a**) shows the hole energies, while panels (**b**) and (**c**) show the cross-sections of the probability density of the top four hole states in the XY and XZ planes, respectively

In Sect. 5.3.2 we have constructed the low-lying exciton states by considering only the *s*-shell Kramers doublet of the electron and the hole. This was possible because in epitaxial dots the *s* and *p* shells were separated by large energy gaps. Here this approach will be appropriate for the electron, but not for the hole. In constructing the low-energy exciton and biexciton states, we have to consider all four hole Kramers doublets. This is because the Coulomb interaction elements in this system are much larger than the splitting between these levels due to the small size of the nanocrystal.

In the remainder of this section, we shall discuss a different issue, related to the multiexciton generation for solar cells. We have mentioned this phenomenon already in the introduction to this chapter. Since the solar spectrum is very broad in energy, one typically chooses to build the optically active region of the solar cell out of a narrow-bandgap semiconductor. However, the incoming photons, on average, create highly excited electron–hole pairs, which typically relax to the ground exciton state by emitting phonons. Therefore, a large part of the incoming solar energy is converted to heat. In the multiexciton generation process [69] the high-energy carriers give the excess of their energy not to heat, but to electrons in the valence band, promoting them into the conduction band and thereby creating additional electron–hole pairs. Clearly, this process competes with the phonon relaxation, and therefore in order to give estimates of the multiexciton generation efficiency we have to include such relaxation mechanisms in our model. This is the

Fig. 5.18 Schematic diagram of the coupling of exciton and biexciton configurations. The biexciton configuration (*left*) can be coupled to the exciton configuration with an excited electron (**a**) or hole (**b**) [208]

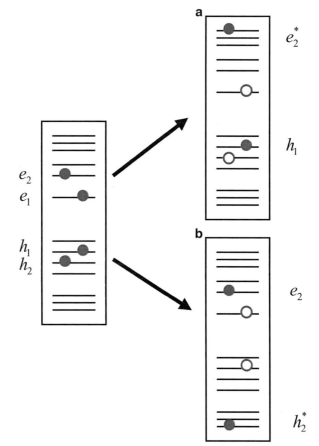

next step in the development of QNANO. At this point, however, we will address a different, but closely related question, concerning the lifetime of the biexciton state.

In Fig. 5.18 (left panel) we show a low-energy two electron–hole pair configuration, which will approximate the biexciton state. This complex can recombine radiatively emitting a photon. However, there exists another biexciton relaxation channel, by the process opposite to the multiexciton generation. Indeed, one constituent electron–hole pair can collapse, and the excess energy can be given to the remaining electron (Fig. 5.18a) or hole (Fig. 5.18b) creating a highly excited exciton. Of course, the overall energy in this system has to be conserved, therefore the excited exciton states must be close in energy to the ground biexciton state. We can describe this coupling in terms of Coulomb interaction. Indeed, even though the exciton and biexciton differ in the number of excitations (one versus two), the number of electrons is the same, therefore the conversion process is simply a Coulomb scattering. We can write explicitly the matrix element of such Coulomb scattering connecting the biexciton configuration and the exciton one with an excited electron in the following form:

$$\langle XX, e_1 e_2 h_1 h_2 | V_{ee} | X, e_1^* h_3 \rangle = (\langle e_1 e_2 | V_{ee} | h_1 e_1^* \rangle - \langle e_1 e_2 | V_{ee} | e_1^* h_1 \rangle) \delta_{h_3 h_2}$$
$$+ (\langle e_1 e_2 | V_{ee} | e_1^* h_2 \rangle - \langle e_1 e_2 | V_{ee} | h_2 e_1^* \rangle) \delta_{h_3 h_1}. \tag{5.62}$$

Here, indices e_1 and e_2 track the electrons belonging to the biexciton configuration, while e_1^* tracks the exciton's electron. Similarly, h_1 and h_2 indicate the occupation of the two holes of the biexciton, and h_3 denotes the hole of the exciton. This notation corresponds to that in Fig. 5.18. In this particular process, one of the electrons annihilates one of the holes, and the second electron is promoted. One hole has to remain unchanged. As it can be accomplished in two ways, there are two terms in the above equation. As is evident, the Coulomb matrix elements responsible for this coupling involve three electron orbitals, and only one hole orbital, and therefore are outside of the electron–hole formulation given in Sect. 5.2.4. However, these terms are naturally present in the approach in which we keep track of all the electrons present in the system.

Compared to the characteristic energy gaps of the system (of order of tens to hundreds of meV), the Coulomb matrix elements connecting the exciton and biexciton configurations are typically very weak (0. 1 meV or less). Therefore, the coupling between these excitations will occur only if their energies are close to resonance. In Fig. 5.19a we show the consequences of this coupling.

With the gray lines we denote the energies of the exciton states, and with the red lines—the biexciton levels obtained with the coupling turned off. We see that the biexciton states are immersed in a dense spectrum of exciton levels. As the coupling is turned on, these levels do not represent the eigenstates of the system any more. The wave functions of actual, mixed levels can be expressed as

$$|n\rangle = A_{XX}^{(n)} |XX\rangle + \sum_i B_{X,i}^{(n)} |X\rangle_i, \tag{5.63}$$

where the state $|XX\rangle$ is here the ground biexciton state before mixing, and the states $|X\rangle_i$ are the exciton states nearby in energy. The index n enumerates the new, mixed eigenstates, while the numbers $A_{XX}^{(n)}$ and $B_{X,i}^{(n)}$ define admixtures of various configurations. The blue bars in Fig. 5.19a give the value $|A_{XX}^{(n)}|^2$ for the complete eigenstates $|n\rangle$, i.e., the value of the spectral function of the biexciton. As expected, we find strong admixtures of the biexciton in the states closest in energy to the original $|XX\rangle$. These admixtures drop off very fast as we move away from that energy, and become negligible already about 1 meV away, owing to the very small value of the coupling matrix elements.

We can also describe this coupling in terms of the time evolution of the system. To this end, we express the time-dependent state of the system as the linear combination of all mixed exciton and biexciton configurations:

Fig. 5.19 (a) Spectral function of the biexciton in the CdSe nanocrystal. The *gray lines* represent the exciton energy levels, and the *red lines*—the biexciton energy levels without coupling. The *blue lines* represent the contribution of the biexciton state to the coupled eigenstates. (b) Time evolution of the probability of finding a biexciton in its ground state in the coupled X-XX system confined in the nanocrystal [208]

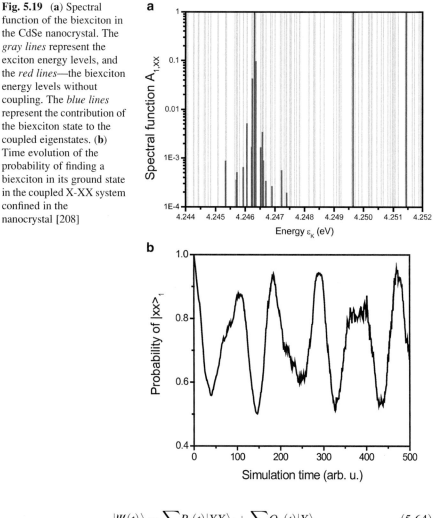

$$|\Psi(t)\rangle = \sum_i P_i(t)|XX\rangle_i + \sum_j Q_j(t)|X\rangle_j. \tag{5.64}$$

In what follows, we assume that at $t = 0$ the system was initialized in the ground biexciton configuration $|XX\rangle_1$. As the time progresses, this state will evolve in time. We now ask how much of the initial ground exciton remains in the state of the system at any time t, i.e., we calculate the quantity

$$|_1\langle XX|\Psi(t)\rangle|^2 = |P_1(t)|^2. \tag{5.65}$$

We can compute this quantity by applying the time propagation operator, including the Hamiltonian $\hat{H}_{TOT}$ with full coupling, to the state at $t = 0$, i.e., to the state $|XX\rangle_1$. We have

$$|\Psi(t)\rangle = \exp\left(-\frac{i}{\hbar}\hat{H}_{\text{TOT}}t\right)|XX\rangle_1. \tag{5.66}$$

To be able to proceed, we have to write our initial biexciton state in terms of the eigenstates of the fully coupled system. To this end we need to perform the inverse rotation to that shown in Eq. (5.63). It is easy to see that

$$|XX\rangle_1 = \sum_n (A_{XX1}^{(n)})^* |n\rangle. \tag{5.67}$$

Therefore,

$$|\Psi(t)\rangle = \sum_n \exp\left(-\frac{i}{\hbar}E_n t\right)(A_{XX1}^{(n)})^* |n\rangle, \tag{5.68}$$

with E_n being the eigenenergies of the full coupled system, and the sought probability of finding our coupled system in the biexciton state $|XX\rangle_1$ is

$$|_1\langle XX| \Psi(t)\rangle|^2 = \left|\sum_n \exp\left(-\frac{i}{\hbar}E_n t\right)|A_{XX1}^{(n)}|^2\right|^2. \tag{5.69}$$

Therefore, the probability of finding the system in the biexciton state at time t is simply expressed as the Fourier transform of the spectral function of the biexciton in the energy domain, shown in Fig. 5.19a. In Fig. 5.19b we show this evolution for our system. We see that the probability $|P_1(t)|^2$ oscillates: the original biexciton converts partially to the exciton states close to it in energy, but then it recovers in a quasi-periodic manner. This coherent time evolution is not regular, because the states which "receive" the spectral weight from the exciton are not distributed symmetrically around the biexciton energy. Instead, we see a complex pattern of beats as in Fig. 5.19b.

As already mentioned, in our model we do not include phonon-assisted relaxation mechanisms. Although the biexciton $|XX\rangle_1$, being the lowest-energy two electron–hole pair state, cannot relax in that manner, the highly excited exciton states coupled to it can. This will take the excitonic component off the energy resonance with the biexciton $|XX\rangle_1$, and thus prohibit its recovery of the spectral weight. Thus, the biexciton will convert to the excitons, and these will irreversibly relax towards the ground exciton state, establishing a finite lifetime for $|XX\rangle_1$. This relaxation happens even before the first oscillation is complete, transforming the oscillating curve from Fig. 5.19b into an exponential decay [39, 71, 210, 211]. Computing the characteristic lifetime defined by this decay is the next major developmental goal of the QNANO package.

5.4 Conclusion

In conclusion, we have presented a tutorial introduction to the atomistic calculations of electronic and optical properties of nanostructures. We have discussed the QNANO package, which takes the atomistic definition of the nanostructure as input. The calculation starts with finding the optimal (relaxed) positions of atoms in the system, and continues with the calculations of single-particle electron and hole states in the strained systems using the empirical tight-binding model. Further, we populate the single-particle states with electrons and/or holes and calculate the many-body eigenstates of this interacting system. The Coulomb scattering matrix elements describing the carrier interactions are computed atomistically based on the single-particle tight-binding wave functions. With the eigenstates of the electron–hole system we then proceed to calculating the optical spectra. We have illustrated this methodology on the examples of epitaxial dots, for which we have discussed the possibility of engineering of single-particle spectra with the dot shape and analyzed the emission spectra of the exciton and multiexciton complexes. Also, we have discussed the coupling between exciton and biexciton states in colloidal nanocrystals.

Acknowledgements The author is grateful to P. Hawrylak, M. Zielinski, O. Voznyy, and E. Kadantsev for discussions.

References

1. F. Henneberger, O. Benson (eds.), *Semiconductor Quantum Bits* (Pan Stanford Publishing, Singapore, 2009)
2. L. Jacak, P. Hawrylak, A. Wojs, *Quantum Dots* (Springer, Berlin 1998)
3. P. Michler (ed.), *Single Quantum Dots, Fundamentals, Applications, and New Concepts.* Topics in Applied Physics, vol. 90 (Springer, Berlin 2003)
4. R.L. Williams, G.C. Aers, J. Lefebvre, P.J. Poole, D. Chithrani, Phys. E **13**, 1200 (2002)
5. R.L. Williams, G.C. Aers, P.J. Poole, J. Lefebvre, D. Chithrani, B. Lamontagne, J. Cryst. Growth **223**, 321 (2001)
6. Z.R. Wasilewski, S. Fafard, J.P. McCaffrey, J. Cryst. Growth **201**, 1131 (1999)
7. M. Bayer, P. Hawrylak, K. Hinzer, S. Fafard, M. Korkusinski, Z.R. Wasilewski, O. Stern, A. Forchel, Science **291**, 451 (2001)
8. J.I. Climente, M. Korkusinski, G. Goldoni, P. Hawrylak, Phys. Rev. B **78**, 115323 (2008)
9. M.F. Doty, J.I. Climente, M. Korkusinski, M. Scheibner, A.S. Bracker, P. Hawrylak, D. Gammon, Phys. Rev. Lett. **102**, 047401 (2009)
10. M. Korkusinski, P. Hawrylak, Phys. Rev. B **63**, 195311 (2001)
11. H.J. Krenner, S. Stufler, M. Sabathil, E.C. Clark, P. Ester, M. Bichler, G. Abstreiter, J.J. Finley, New. J. Phys. **7**, 184 (2005)
12. E.A. Stinaff, M. Scheibner, A.S. Bracker, I.V. Ponomarev, V.L. Korenev, M.E. Ware, M.F. Doty, T.L. Reinecke, D. Gammon, Science **311**, 636 (2006)
13. J. Liu, Z. Lu, S. Raymond, P.J. Poole, P.J. Barrios, D. Poitras, Opt. Lett. **33**, 1702 (2008)
14. P.J. Poole, K. Kaminska, P. Barrios, Z. Lu, J. Liu, J. Cryst. Growth **311**, 1482 (2009)
15. N. Gisin, G. Ribordy, W. Tittel, H. Zbinden, Rev. Mod. Phys. **74**, 145 (2002)

16. N. Gisin, R. Thew, Nat. Photon. **1**, 165 (2007)
17. C. Santori, D. Fattal, J. Vuckovic, G.S. Solomon, Y. Yamamoto, Nature **419**, 594 (2002)
18. A.J. Shields, Nat. Photon. **1**, 215 (2007)
19. N. Akopian, N.H. Lindner, E. Poem, Y. Berlatzky, J. Avron, D. Gershoni, B.D. Gerardot, P.M. Petroff, Phys. Rev. Lett. **96**, 130501 (2006)
20. A. Greilich, M. Schwab, T. Berstermann, T. Auer, R. Oulton, D.R. Yakovlev, M. Bayer, V. Stavarache, D. Reuter, A. Wieck, Phys. Rev. B **73**, 045323 (2006)
21. M. Korkusinski, M.E. Reimer, R.L. Williams, P. Hawrylak, Phys. Rev. B **79**, 035309 (2009)
22. R.M. Stevenson, R.J. Young, P. Atkinson, K. Cooper, D.A. Ritchie, A.J. Shields, Nature **439**, 179 (2006)
23. N. Bonadeo, J. Erland, D. Gammon, D. Park, D. Katzer, D. Steel, Science **282**, 1473 (1998)
24. X. Li, Y. Wu, D. Steel, D. Gammon, T. Stievater, D. Katzer, D. Park, C. Piermarocchi, L.J. Sham, Science **301**, 809 (2003)
25. T.H. Stievater, X. Li, D.G. Steel, D. Gammon, D.S. Katzer, D. Park, C. Piermarocchi, L.J. Sham, Phys. Rev. Lett. **87**, 133603 (2001)
26. S. Stufler, P. Ester, A. Zrenner, M. Bichler, Phys. Rev. Lett. **96**, 037402 (2006)
27. F. Troiani, U. Hohenester, E. Molinari, Phys. Rev. B **62**, R2263 (2000)
28. T. Unold, K. Mueller, C. Lienau, T. Elsaesser, A.D. Wieck, Phys. Rev. Lett. **94**, 137404 (2005)
29. M. Zecherle, C. Ruppert, E.C. Clark, G. Abstreiter, J.J. Finley, M. Betz, Phys. Rev. B **82**, 125314 (2010)
30. A. Zrenner, E. Beham, S. Stufler, F. Findeis, M. Bichler, G. Abstreiter, Nature (London) **418**, 612 (2002)
31. K.J. Vahala, Nature **424**, 839 (2003)
32. E. Yablonovitch, Phys. Rev. Lett. **58**, 2059 (1997)
33. W.-H. Chang, W.-Y. Chen, H.-S. Chang, T.-P. Hsiech, J.-I. Chyi, T.-M. Hsu, Phys. Rev. Lett. **96**, 117401 (2006)
34. D. Englund, D. Fattal, E. Waks, G. Solomon, B. Zhang, T. Nakaoka, Y. Arakawa, Y. Yamamoto, J.J. Vuckovic, Phys. Rev. Lett. **95**, 013904 (2005)
35. J.M. Gerard, Top. Appl. Phys. **90**, 269 (2003)
36. D. Dalacu, K. Mnaymneh, V. Sazonova, P.J. Poole, G.C. Aers, J. Lapointe, R. Cheriton, A.J. SpringThorpe, R.L. Williams, Phys. Rev. B **82**, 033301 (2010)
37. L.E. Brus, Nano Lett. **10**, 363 (2010)
38. A.I. Ekimov, F. Hache, M.C. Schanne-Klein, D. Ricard, C. Flytzanis, I.A. Kudryavtsev, T.V. Yazeva, A.V. Rodina, Al.L. Efros, J. Opt. Soc. Am. B **10**, 100 (1993)
39. R.J. Ellingson, M.C. Beard, J.C. Johnson, P. Yu, O.I. Micic, A.J. Nozik, A. Shabaev, A.L. Efros, Nano Lett. **5**, 865 (2005)
40. D.E. Gomez, M. Califano, P. Mulvaney, Phys. Chem. Chem. Phys. **8**, 4989 (2006)
41. I. Gur, N.A. Fromer, M.L. Geier, A.P. Alivisatos, Science **310**, 462 (2005)
42. V.I. Klimov, Ann. Rev. Phys. Chem. **58**, 635 (2007)
43. G. Nair, S.M. Geyer, L.-Y. Chang, M.G. Bawendi, Phys. Rev. B **78**, 125325 (2008)
44. M. Nirmal, B.O. Dabbousi, M.G. Bawendi, J.J. Macklin, J.K. Trautman, T.D. Harris, L.E. Brus, Nature **383**, 802 (1996)
45. A. Pandey, P. Guyot-Sionnest, J. Chem. Phys. **127**, 111104 (2007)
46. G. Scholes, Adv. Funct. Mater. **18**, 115 (2008)
47. G. Scholes, G. Rumbles, Nat. Mater. **5**, 683 (2006)
48. Y. Yin, A.P. Alivisatos, Nature **437**, 664 (2005)
49. D. Yu, C. Wang, P. Guyot-Sionnest, Science **300**, 1277 (2003)
50. A.P. Alivisatos, Nat. Biotech. **22**, 47 (2004)
51. W.C.W. Chan, S.M. Nie, Science **281**, 2016 (1998)
52. A.H. Fu, W.W. Gu, C. Larabell, A.P. Alivisatos, Curr. Opin. Neurobiol. **15**, 568 (2005)
53. S. Chanyawadee, P.G. Lagoudakis, R.T. Harley, M.D.B. Chariton, D.V. Talapin, H.W. Huang, C.-H. Lin, Adv. Mater. **22**, 602 (2009)

54. V.L. Colvin, M.C. Schlamp, A.P. Alivisatos, Nature **370**, 354 (1994)
55. B.O. Dabbousi, M.G. Bawendi, O. Onitsuka, M.F. Rubner, Appl. Phys. Lett. **66**, 1316 (1995)
56. G. Konstantatos, E.H. Sargent, Nat. Nanotechnol. **5**, 391 (2010)
57. V. Sukhovatkin, S. Hinds, L. Brzozowski, E.H. Sargent, Science **324**, 1542 (2009)
58. A.G. Pattantyus-Abraham, H. Qiao, J. Shan, K.A. Abel, T.-S. Wang, F.C.J.M. van Veggel, J.F. Young, Nano Lett. **9**, 2849 (2009)
59. V.I. Klimov, A.A. Mikhaelovsky, S. Xu, A. Mlko, J.A. Hollingsworth, Science **290**, 314 (2000)
60. C. Weisbuch, J. Cryst. Growth **138**, 776 (1994)
61. H.W. Hillhouse, M.C. Beard, Curr. Opin. Colloid Interface Sci. **14**, 245 (2009)
62. C. Wadia, A.P. Alivisatos, D.M. Kammen, Environ. Sci. Technol. **43**, 2072 (2009)
63. J.M. Luther, M. Law, M.C. Beard, Q. Song, M.O. Reese, R.J. Ellingson, A.J. Nozik, Nano Lett. **8**, 3488 (2009)
64. J.J. Choi, Y. Lim, M.B. Santiago-Berrios, M. Oh, B. Hyun, L. Sun, A.C. Bartnik, A. Goedhart, G.G. Malliaras, H.D. Abruna, F.W. Wise, T. Hanrath, Nano Lett. **9**, 3479 (2009)
65. K.W. Johnston, A.G. Pattantyus-Abraham, J.P. Clifford, S.H. Myrskog, D.D. McNeil, L. Levina, E.H. Sargent, Appl. Phys. Lett. **92**, 151115 (2008)
66. A. Franceschetti, J.M. An, A. Zunger, Nano Lett. **6**, 2191 (2006)
67. G. Nair, M.G. Bawendi, Phys. Rev. B **76**, 081304 (2007)
68. A.J. Nozik, Ann. Rev. Phys. Chem. **52**, 193 (2001)
69. A.J. Nozik, Phys. E **14**, 115 (2002)
70. E. Rabani, R. Baer, Nano Lett. **8**, 4488 (2008)
71. R.D. Schaller, J.M. Pietryga, V.I. Klimov, Nano Lett. **7**, 3469 (2007)
72. A.J. Nozik, Chem. Phys. Lett. **457**, 3 (2008)
73. R.D. Schaller, V.I. Klimov, Phys. Rev. Lett. **92**, 186601 (2004)
74. J.A. McGuire, M. Sykora, J. Joo, J.M. Pietryga, V.I. Klimov, Nano Lett. **10**, 2049 (2010)
75. R.D. Schaller, M. Sykora, J.M. Pietryga, V.I. Klimov, Nano Lett. **6**, 424 (2006)
76. K.S. Novoselov, A.K. Geim, S.V. Morozov, D. Jiang, M.I. Katsnelson, I.V. Grigorieva, S.V. Dubonos, A.A. Firsov, Nature **438**, 197 (2005)
77. K.S. Novoselov, A.K. Geim, S.V. Morozov, D. Yiang, Y. Zhang, S.V. Dubonos, I.V. Grigorieva, A.A. Firsov, Science **306**, 666 (2004)
78. M.L. Sadowski, G. Martinez, M. Potemski, C. Berger, W.A. de Heer, Phys. Rev. Lett. **97**, 266405 (2006)
79. Y. Zhang, Y.W. Tan, H.L. Stormer, P. Kim, Nature **438**, 201 (2005)
80. A.H. Castro Neto, F. Guinea, N.M.R. Peres, K.S. Novoselov, A.K. Geim, Rev. Mod. Phys. **81**, 109 (2009)
81. S.D. Sarma, S. Adam, E.H. Hwang, E. Rossi, Rev. Mod. Phys. **83**, 407 (2011)
82. A.D. Guclu, P. Potasz, M. Korkusinski, P. Hawrylak, *Graphene Quantum Dots* (Springer, Berlin, 2014)
83. V.N. Kotov, B. Uchoa, V.M. Pereira, F. Guinea, A.H. Castro Neto, Rev. Mod. Phys. **84**, 1067 (2012)
84. M.L. Mueller, X. Yan, J.A. McGuire, L. Li, Nano Lett. **10**, 2679 (2010)
85. X. Yan, X. Cui, B. Li, L. Li, Nano Lett. **10**, 1869 (2010)
86. X. Yan, X. Cui, L. Li, J. Am. Chem. Soc. **132**, 5944 (2010)
87. X. Yan, B. Li, L. Li, Acc. Chem. Res. **46**, 2254 (2013)
88. I. Ozfidan, M. Korkusinski, A.D. Guclu, J.A. McGuire, P. Hawrylak, Phys. Rev. B **89**, 085310 (2014)
89. A.D. Guclu, P. Potasz, P. Hawrylak, Phys. Rev. B **82**, 155445 (2010)
90. J. Akola, H.P. Heiskanen, M. Manninen, Phys. Rev. B **77**, 193410 (2008)
91. P. Potasz, A.D. Guclu, P. Hawrylak, Phys. Rev. B **81**, 033403 (2010)
92. T. Yamamoto, T. Noguchi, K. Watanabe, Phys. Rev. B **74**, 121409 (2006)
93. Z.Z. Zhang, K. Chang, F.M. Peeters, Phys. Rev. B **77**, 235411 (2008)
94. M. Ezawa, Phys. Rev. B **76**, 245415 (2007)

95. J. Fernandez-Rossier, J.J. Palacios, Phys. Rev. Lett. **99**, 177204 (2007)
96. A.D. Guclu, P. Potasz, O. Voznyy, M. Korkusinski, P. Hawrylak, Phys. Rev. Lett. **103**, 246805 (2009)
97. J. Jung, A.H. MacDonald, Phys. Rev. B **79**, 235433 (2009)
98. P. Potasz, A.D. Guclu, A. Wojs, P. Hawrylak, Phys. Rev. B **85**, 075431 (2012)
99. O. Voznyy, A.D. Guclu, P. Potasz, P. Hawrylak, Phys. Rev. B **83**, 165417 (2011)
100. W.L. Wang, S. Meng, E. Kaxiras, Nano Lett. **8**, 241 (2008)
101. A.D. Guclu, P. Hawrylak, Phys. Rev. B **87**, 035425 (2013)
102. J. Dukelsky, S. Pittel, Rep. Prog. Phys. **67**, 513 (2004)
103. A.E. Feiguin, E. Rezayi, C. Nayak, S. Das Sarma, Phys. Rev. Lett. **100**, 166803 (2008)
104. S.R. White, Phys. Rev. B **72**, 180403 (2005)
105. J.R. Chelikovsky, S.G. Louie, *Quantum Theory of Real Materials* (Kluwer, Norwell, 1996)
106. J.M. Soler, E. Artacho, J.D. Gale, A. Garcia, J. Junquera, P. Ordejon, D. Sanchez-Portal, J. Phys. Condens. Matter **14**, 2745 (2002); the package available at http://departments.icmab.es/leem/siesta
107. VASP package and documentation are available at http://www.vasp.at
108. Quantum ESPRESSO package and documentation are available at http://www.quantum-espresso.org
109. J. Deslippe, G. Samsonidze, D.A. Strubbe, M. Jain, M.L. Cohen, S.G. Louie, Comput. Phys. Commun. **183**, 1269 (2012); BerkeleyGW code and documentation are available at http://www.berkeleygw.org
110. A. Marini, C. Hogan, M. Gruening, D. Varsano, Comput. Phys. Commun. **180**, 1392 (2009); YAMBO package and documetation are available at http://www.yambo-code.org
111. M. Korkusinski, O. Voznyy, P. Hawrylak, Phys. Rev. B **82**, 245304 (2010)
112. O. Voznyy, J. Phys. Chem. C **115**, 15927 (2011)
113. A.H. Ip, S.M. Thon, S. Hoogland, O. Voznyy, D. Zhitomirsky, R. Debnath, L. Levina, L.R. Rollny, G.H. Carey, A. Fisher, K.W. Kemp, I.J. Kramer, Z. Ning, A.J. Labelle, K.W. Chou, A. Amassian, E.H. Sargent, Nat. Nanotechnol. **7**, 577 (2012)
114. J.M. Luttinger, Phys. Rev. **102**, 1030 (1956)
115. J.M. Luttinger, W. Kohn, Phys. Rev. **97**, 869 (1955)
116. I. Puerto Gimenez, M. Korkusinski, P. Hawrylak, Phys. Rev. B **76**, 075336 (2007)
117. A. Wojs, P. Hawrylak, S. Fafard, L. Jacak, Phys. Rev. B **54**, 5604 (1996)
118. C.-Y. Hsieh, R. Cheriton, M. Korkusinski, P. Hawrylak, Phys. Rev. B **80**, 235320 (2009)
119. L.G.C. Rego, P. Hawrylak, J.A. Brum, A. Wojs, Phys. Rev. B **55**, 15694 (1997)
120. M. Korkusinski, P. Hawrylak, Phys. Rev. B **87**, 115310 (2013)
121. M. Korkusinski, P. Hawrylak, Sci. Rep. **4**, 4903 (2014)
122. W. Sheng, S.-J. Cheng, P. Hawrylak, Phys. Rev. B **71**, 035316 (2005)
123. A.L. Efros, M. Rosen, Phys. Rev. B **58**, 7120 (1998)
124. M. Grundmann, O. Stier, D. Bimberg, Phys. Rev. B **52**, 11969 (1995)
125. C. Pryor, Phys. Rev. B **57**, 7190 (1998)
126. O. Stier, M. Grundmann, D. Bimberg, Phys. Rev. B **59**, 5688 (1999)
127. F. Guffarth, R. Heitz, A. Schliwa, O. Stier, N.N. Ledentsov, A.R. Kovsh, V.M. Ustinov, D. Bimberg, Phys. Rev. B **64**, 085305 (2001)
128. O. Stier, D. Bimberg, Phys. Rev. B **55**, 7726 (1997)
129. P. Nagpal, V.I. Klimov, Nat. Commun. **2**, 486 (2011)
130. G. Bester, A. Zunger, Phys. Rev. B **71**, 045318 (2005)
131. R. Singh, G. Bester, Phys. Rev. Lett. **104**, 196803 (2010)
132. G. Klimeck, F. Oyafuso, T.B. Boykin, R.C. Bowen, P. von Allmen, Comput. Model. Eng. Sci. **3**, 601 (2002)
133. M. Bayer, G. Ortner, O Stern, A. Kuther, A.A. Gorbunov, A. Forchel, P. Hawrylak, S. Fafard, K. Hinzer, T.L. Reinecke, S.N. Walck, J.P. Reithmayer, F. Klopf, F. Schafer, Phys. Rev. B **65**, 195315 (2002)
134. M. Bayer, O. Stern, P. Hawrylak, S. Fafard, A. Forchel, Nature **405**, 923 (2000)

135. The package QNANO is currently under development. Please contact the Author of this Chapter directly for more detailed information about distribution.
136. A.-L. Barabasi, Appl. Phys. Lett. **70**, 2565 (1997)
137. K.E. Khor, S.D. Sarma, Phys. Rev. B **62**, 16657 (2000)
138. C.-H. Lam, C.-K. Lee, L. Sander, Phys. Rev. Lett. **89**, 216102 (2002)
139. M. Meixner, E. Scholl, V.A. Shchukin, D. Bimberg, Phys. Rev. Lett. **87**, 236101 (2002)
140. G. Russo, P. Smereka, J. Comput. Phys. **214**, 809 (2006)
141. J.G. Keizer, P.M. Koenraad, P. Smereka, J.M. Ulloa, A. Guzman, A. Hierro, Phys. Rev. B **85**, 155326 (2012)
142. V. Jovanov, T. Eissfeller, S. Kapfinger, E.C. Clark, F. Klotz, M. Bichler, J.G. Keizer, P.M. Koenraad, M.S. Brandt, G. Abstreiter, J.J. Finley, Phys. Rev. B **85**, 165433 (2012)
143. J.M. Ulloa, M. Bozkurt, P.M. Koenraad, Solid State Commun. **149**, 1410 (2009)
144. A. Mikkelsen, N. Skold, L. Ouattara, M. Borgstrom, J.N. Andersen, L. Samuelson, W. Seifert, E. Lundgren, Nat. Mater. **3**, 519 (2004)
145. A. Wierts, J.M. Ulloa, C. Celebi, P.M. Koenraad, H. Boukari, L. Maingault, R. Andre, H. Mariette, Appl. Phys. Lett. **91**, 161907 (2007)
146. C. Cornet, A. Schliwa, J. Even, F. Dore, C. Celebi, A. Letoublon, E. Mace, C. Paranthoen, A. Simon, P.M. Koenraad, N. Bertru, D. Bimberg, S. Loualiche, Phys. Rev. B **74**, 035312 (2006)
147. A.D. Giddings, J.G. Keizer, M. Hara, G.J. Hamhuis, H. Yuasa, H. Fukuzawa, P.M. Koenraad, Phys. Rev. B **83**, 205308 (2011)
148. D.M. Bruls, J.W.A.M. Vugs, P.M. Koenraad, H.W.M. Salemink, J.H. Wolter, M. Hopkinson, M.S. Skolnick, F. Long, S.P.A. Gill, Appl. Phys. Lett. **81**, 1708 (2002)
149. P.N. Keating, Phys. Rev. **145**, 637 (1966)
150. R.M. Martin, Phys. Rev. B **1**, 4005 (1970)
151. A.J. Williamson, L.W. Wang, A. Zunger, Phys. Rev. B **62**, 12963 (2000)
152. L.D. Landau, L.P. Pitaevskii, A.M. Kosevich, E.M. Lifshitz, *Theory of Elasticity* (Butterworth-Heinemann, Oxford, 1986)
153. T.B. Bahder, Phys. Rev. B **45**, 1629 (1992)
154. C. Pryor, J. Kim, L.W. Wang, A.J. Williamson, A. Zunger, J. Appl. Phys. **83**, 2548 (1998)
155. N.W. Ashcroft, N.D. Mermin, *Solid State Physics* (Saunders College, Philadelphia, 1976)
156. G.W. Bryant, M. Zielinski, N. Malkova, J. Sims, W. Jaskolski, J. Aizpurua, Phys. Rev. Lett. **105**, 067404 (2010)
157. G.W. Bryant, M. Zielinski, N. Malkova, J. Sims, W. Jaskolski, J. Aizpurua, Phys. Rev. B **84**, 235412 (2011)
158. Y.H. Huo, B.J. Witek, S. Kumar, J.R. Cardenas, J.X. Zhang, N. Akopian, R. Singh, E. Zallo, R. Grifone, D. Kriegner, R. Trotta, F. Ding, J. Stangl, V. Zwiller, G. Bester, A. Rastelli, O.G. Schmidt, Nat. Phys. **10**, 46 (2013)
159. H. Kosaka, T. Inagaki, R. Hitomi, F. Izawa, Y. Rikitake, H. Imamura, Y. Mitsumori, K. Edamatsu, Phys. Rev. A **85**, 042304 (2012)
160. R. Vrijen, E. Yablonovitch, Phys. E **10**, 569 (2001)
161. J.R. Chelikovsky, M.L. Cohen, Phys. Rev. B **14**, 556 (1976)
162. M.L. Cohen, T.K. Bergstresser, Phys. Rev. **141**, 789 (1966)
163. M.S. Hybertsen, S.G. Louie, Phys. Rev. B **34**, 2920 (1986)
164. Y. Kurokawa, S. Nomura, T. Takemori, Y. Aoyagi, Phys. Rev. B **61**, 12616 (2000)
165. J.R. Cardenas, G. Bester, Phys. Rev. B **86**, 115332 (2012)
166. M. Gong, K. Duan, C.-F. Li, R. Magri, G.A. Narvaez, L. He, Phys. Rev. B **77**, 045326 (2008)
167. L.-W. Wang, J. Kim, A. Zunger, Phys. Rev. B **59**, 5678 (1999)
168. L.-W. Wang, A. Zunger, Phys. Rev. B **59**, 15806 (1999)
169. A.J. Williamson, A. Zunger, Phys. Rev. B **59**, 15819 (1999)
170. A.J. Williamson, A. Zunger, Phys. Rev. B **61**, 1978 (2000)
171. M. Haugk, J. Elsner, T. Frauenheim, T.E.M. Staab, C.D. Latham, R. Jones, H.S. Leipner, T. Heine, G. Seifert, M. Sternberg, Phys. Status. Solidi. B **217**, 473 (2000)

172. R. Santoprete, B. Koiler, R.B. Capaz, P. Kratzer, Q.K.K. Liu, M. Scheffler, Phys. Rev. B **68**, 235311 (2003)
173. D.J. Chadi, Phys. Rev. B **16**, 790 (1977)
174. J.C. Slater, G.F. Koster, Phys. Rev. **94**, 1498 (1954)
175. J.-M. Jancu, R. Scholz, F. Bertram, F. Bassani, Phys. Rev. B **57**, 6493 (1998)
176. G. Klimeck, R.C. Bowen, T.B. Boykin, C. Salazar-Lazaro, T.A. Cwik, A. Stoica, Superlattice. Microst. **27**, 77 (2000)
177. T.B. Boykin, G. Klimeck, F. Oyafuso, Phys. Rev. B **69**, 115201 (2004)
178. J.G. Diaz, G.W. Bryant, Phys. Rev. B **73**, 075329 (2006)
179. M. Korkusinski, M. zielinski, P. Hawrylak, J. Appl. Phys. **105**, 122406 (2009)
180. M. Zielinski, M. Korkusinski, P. Hawrylak, Phys. Rev. B **81**, 085301 (2010)
181. K. Leung, S. Pokrant, K.B. Whaley, Phys. Rev. B **57**, 12291 (1998)
182. T.B. Boykin, G. Klimeck, R.C. Bowen, F. Oyafuso, Phys. Rev. B **66**, 125207 (2002)
183. A. Janotti, C.G. van de Walle, Phys. Rev. B **75**, 121201 (2007)
184. E.S. Kadantsev, M. Zielinski, M. Korkusinski, P. Hawrylak, J. Appl. Phys. **107**, 104315 (2010)
185. Y.-H. Li, X.G. Gong, S.-H. Wei, Phys. Rev. B **73**, 245206 (2006)
186. S.-H. Wei, A. Zunger, Phys. Rev. B **60**, 5404 (1999)
187. E.S. Kadantsev, P. Hawrylak, J. Appl. Phys. **98**, 023108 (2011)
188. P.R.C. Kent, G.L.W. Hart, A. Zunger, Appl. Phys. Lett. **81**, 4377 (2002).
189. T.B. Boykin, M. Luisier, M. Salmani-Jelodar, G. Klimeck, Phys. Rev. B **81**, 125202 (2010)
190. Y.M. Niquet, D. Rideau, C. Tavernier, H. Jaouen, X. Blase, Phys. Rev. B **79**, 245201 (2009)
191. M. Zielinski, Phys. Rev. B **86**, 115424 (2012)
192. S. Lee, F. Oyafuso, P. von Allmen, G. Klimeck, Phys. Rev. B **69**, 045316 (2004)
193. The complete information about the NEMO3D package can be found on the webpage of the Nanohub nanoelectronic modeling group under the address https://engineering.purdue.edu/gekcogrp/
194. P. Hawrylak, Phys. Rev. B **60**, 5597 (2000)
195. E.S. Kadantsev, P. Hawrylak, Phys. Rev. B **81**, 045311 (2010)
196. A.H. Trojnar, M. Korkusinski, E.S. Kadantsev, P. Hawrylak, Phys. Rev. B **84**, 245314 (2011)
197. C. Delerue, M. Lannoo, G. Allan, Phys. Rev. B **68**, 115411 (2003)
198. A. Franceschetti, H. Fu, L.W. Wang, A. Zunger, Phys. Rev. B **60**, 1819 (1999)
199. I. Moreels, G. Allan, B. de Geyter, L. Wirtz, C. Delerue, Z. Hens, Phys. Rev. B **81**, 235319 (2010)
200. S. Ogut, R. Burdick, Y. Saad, J.R. Chelikovsky, Phys. Rev. Lett. **90**, 127401 (2003)
201. L.W. Wang, M. Califano, A. Zunger, A. Franceschetti, Phys. Rev. Lett. **91**, 056404 (2003)
202. J.C. Slater, Phys. Rev. **36**, 57 (1930)
203. G.W. Bryant, W. Jaskolski, Phys. Rev. B **67**, 205320 (2003)
204. S. Schulz, S. Schumacher, G. Czycholl, Phys. Rev. B **73**, 245327 (2006)
205. L. He, A. Zunger, Phys. Rev. B **73**, 115324 (2006)
206. T. Smolenski, T. Kazimierczuk, M. Goryca, T. Jakubczyk, L. Klopotowski, L. Cywinski, P. Wojnar, A. Golnik, P. Kossacki, Phys. Rev. B **86**, 241305 (2012)
207. P. Hawrylak, Solid State Commun. **127**, 753 (2003)
208. M. Korkusinski, O. Voznyy, P. Hawrylak, Phys. Rev. B **84**, 155327 (2011)
209. A. Wojs, P. Hawrylak, Solid State Commun. **100**, 487 (1996)
210. M. Califano, ACS Nano **3**, 2706 (2009)
211. R.D. Schaller, V.M. Agranovich, V.I. Klimov, Nat. Phys. **1**, 189 (2005)

Chapter 6
What Happens When Molecules Meet Nanostructures: The Convergence of Chemistry and Electronics at the Nanoscale

Stuart Lindsay

Abstract A number of solid-state devices yield useful chemical information based on interactions between molecules and molecule-sized features on the device. These include nanopores, nanowire field-effect transistor sensors, nanoscale redox sensors, and tunnel junctions. Here, we consider the electrochemical requirements needed for the stable operation of such devices. We also review the sources of noise and show how the correlations in noise generated by nanoscale sensors can be used to carry out single-molecule spectroscopy.

6.1 Introduction

Solid-state devices with features similar in size to single molecules have become a reality in recent years. Examples include solid-state nanopores [1] (particularly for DNA sequencing [2]), carbon nanotube field-effect transistors (FETs) [3] and semiconductor nanotube FETs [4] for detection of single-molecule interactions, nanoscale detectors of single-molecule redox chemistry [5], and tunnel junctions [6, 7] that sense molecules via the electron tunnel current passing through them. These devices portend a revolution in electronics, combining the computing potential of CMOS with the diversity of chemistry, a combination that has endless potential [8]. To give just one concrete example from our lab, it has recently been shown that tunnel junctions can identify amino acids and peptides [9], opening the way to single-molecule sequencing of proteins. This could be a game changer in medicine,

S. Lindsay (✉)
Department of Physics, Biodesign Institute, Arizona State University, Tempe, AZ 85287, USA

Department of Chemistry and Biochemistry, Biodesign Institute, Arizona State University, Tempe, AZ 85287, USA
e-mail: Stuart.Lindsay@asu.edu

© Springer International Publishing Switzerland 2015
A. Korkin et al. (eds.), *Nanoscale Materials and Devices for Electronics, Photonics and Solar Energy*, Nanostructure Science and Technology, DOI 10.1007/978-3-319-18633-7_6

where currently DNA sequencing dominates the growth of personalized medicine. However, proteins are the real indicator of health status. But proteins are difficult to analyze and often present only in tiny concentrations. There is no equivalent of the polymerase chain reaction for DNA (which amplifies even single molecules) for proteins. So single-molecule protein analysis could open a whole new approach to personalized diagnostics.

The purpose of this chapter is not to review this new field, though we will describe some devices by way of examples. Rather, the aim is to introduce the physical chemistry aspects of these devices that may not be familiar to the general semiconductor physics audience, and to outline some important, and thus far, underappreciated aspects, of fluctuations at the nanoscale. (A primer on electrochemistry aimed at physical scientists can be found in Chap. 8 of *Introduction to Nanoscience* [10].)

The most important lessons from physical chemistry are that chemical reactions *will* occur at a metal–solution interface in all but the most special of circumstances and that their effect is most significant when electrodes are very small.

The most important statistical lessons are that signals are always stochastic, but that even the "noise" can contain useful chemical information if the system is small enough. In fact, in certain circumstances, statistical analysis of signals *is not required*; the "noise" contains enough information that single signal events can be interpreted in nanoscale single-molecule sensing devices.

After surveying some types of device, we turn to the electrochemical requirements for operating them. These include chemical stability and interfacial noise. We end with a discussion of statistical aspects of single-molecule devices, including the practical importance of nonrandom correlations in signals.

6.2 Examples of Single-Molecule Solid-State Sensing Devices

6.2.1 Nanopores

Nanopores are tiny Coulter counters, consisting of an orifice of a few nm in diameter connecting two reservoirs of electrolyte. Their operation is illustrated in Fig. 6.1a. The pore penetrates a membrane, controlling the amount of ion current that passes between two reference electrodes. The reference electrodes are large area wires in equilibrium with their electrolysis products (e.g., silver wire and AgCl coating and Ag^+ and Cl^- in solution). They maintain a constant potential drop with respect to the electrolyte, because of the Faradaic processes that take place at their surface to maintain the current flow (they are referred to as *ideal nonpolarizable electrodes*, INPE).

The first implementations of nanopore devices utilized biological protein pores suspended in an insulating lipid membrane [11]. Such pores can be remarkably

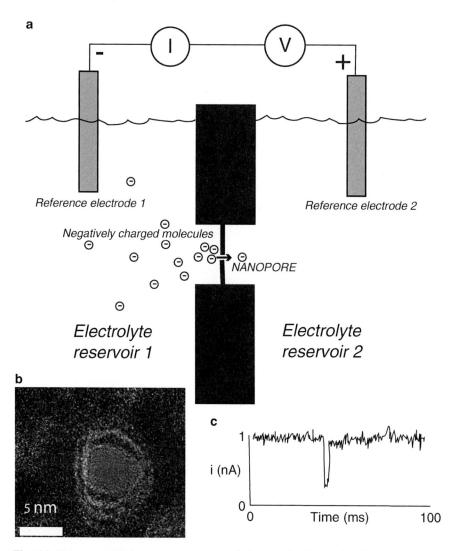

Fig. 6.1 Nanopore with ion-current detection of single-molecule translocations. (**a**) The ion current passing through the pore is measured using reference electrodes inserted into electrolyte reservoirs on each side of the pore. (**b**) TEM image of a pore drilled with an electron beam in a SiN membrane (scale bar is 5 nm). (**c**) Current-time trace showing a single blockade event

robust, the weak link usually being the lipid membrane support. Solid-state pores were introduced by the Golovchenko group [12] to overcome this limitation. These pores are fabricated by drilling holes in thin (ca. 50 nm) SiN membranes using ion or electron beams. A TEM image of a pore drilled in our lab is shown in Fig. 6.1b (the scale bar is 5 nm). The current through the pore is dominated by the resistance of the solution in the pore for long pores, the current being given by [13].

$$I_0 = V[n_+\mu_+ + n_-\mu_-]e\left(\frac{4h}{\pi d^2} + \frac{1}{d}\right)^{-1} + \frac{V\mu_\otimes \pi d\sigma}{h}. \qquad (6.1)$$

Here, the first term, $\left(\frac{4h}{\pi d^2}\right)^{-1}$, corresponds to the solution resistance in the tube formed by the pore. The second term, $\left(\frac{1}{d}\right)^{-1}$, is the access (Sharvin) resistance and the third term gives the electroosmotic current. At low salt concentration, the current is dominated by electroosmotic flow if the surface of the nanopore is charged. In Eq. 6.1, d is the nanopore diameter and h its thickness. μ_+, μ_-, and $\mu_\otimes$ are the cation, anion, and electroosmotic mobilities and σ is the surface charge density in the pore. For thick pores, the pore resistance dominates, and a 5 nm pore of 50 nm thickness in 1 M 1:1 electrolyte is about $10^9\,\Omega$, so a bias of a volt across the reference electrodes produces a current of about 1 nA. A large (e.g., 4 nm in diameter) molecule passing through the pore blocks most of the current, reducing the current almost to zero (see the current-time trace in Fig. 6.1c). Clearly, the current response depends strongly on the material properties of the pore, and solid-state pores have yet to attain the reproducibility shown by protein pores (where every atom is in a reasonably well-defined position). Protein pores are reproducible enough that a whole genome has recently been sequenced using them [14].

6.2.2 CNT FET Sensors for Single-Molecule Dynamics

There have been several remarkable demonstrations of nanowire FETs as single-molecule sensors, including the incorporation of a semiconductor nanowire FET into a nanopore for the sensing of individual molecule translocations through the pore [4]. Here, we focus on measurements of binding kinetics using carbon nanotube FETs, in a discussion that largely follows that published elsewhere [15]. Carbon nanotubes can act as single-molecule sensors as a consequence of strong scattering at a defect site [16], in effect inserting a large series resistance into an otherwise low-resistance circuit. Thus, despite the fact that only a small part of the carbon nanotube is modified by a single molecule, large changes in conductance are possible. Chemically modified carbon nanotubes have found applications as biosensors [17], gas sensors [18], and electrochemical sensors [19]. The sensitivity of CNT FETs is a function of their chirality, metallic tubes being insensitive to gating. (For this reason, semiconductor nanowires have an advantage as sensors, because they can be prepared with a controlled band structure [20].)

A single-molecule sensing FET is illustrated in Fig. 6.2. A single-walled CNT is contacted by source and drain electrodes and immersed in an electrolyte solution. The tube is polarized by means of a gate electrode, consisting of a highly doped p^{++} region underneath the device. Application of a bias to the gate with respect to the CNT results in the formation of electric double layers on the gate and the CNT. A large molecule attached to the CNT displaces ions in the double layer and creates a

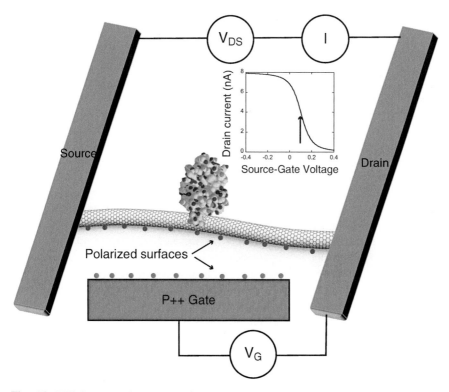

Fig. 6.2 FET for measuring enzyme fluctuations. A CNT is contacted by source and drain electrodes and functionalized with an enzyme (here lysozyme) modified with a pyrene ring that pi stacks on the CNT. The system is submerged in electrolyte and polarized by a gate electrode under the device with the device poised at the point of maximum gate sensitivity (*arrow* on inset). Figure is reproduced from Lindsay, Single-Molecule Nanoelectronics [15]

localized charge defect that causes a change in the conductance of the CNT [21]. The thickness of the double layer is given by the Debye formula as

$$L_D = \frac{0.3\,nm}{\sqrt{M}} \tag{6.2}$$

where M is the molarity of a 1:1 salt. Thus in 10 mM salt, the Debye length, L_D, is 3 nm, so a molecule would have to be about this dimension to displace a significant amount of charge. Importantly, the CNT was connected to a reference electrode for noise reduction [3] presumably for the reasons discussed in Sect. 6.3.1 below.

Devices like this have been used to detect binding events on CNTs [22] and semiconductor nanowires [23]. More recently, CNT FETs have been used to study enzyme motion at the single-molecule level [3, 24], a remarkable feat that we describe further here. These measurements rely on tethering an enzyme to the CNT using a modified amino acid residue. The modification is made on a part of the protein that undergoes large functional motions and consists of coupling an inserted cysteine residue to a pyrene ring that is absorbed strongly onto the surface of the CNT. Importantly, this region also contains at least one charged residue, so that, as the protein fluctuates, the charged residue moves in and out of the Debye layer where the enzyme is tethered, modulating the local electric field and hence the conductance of the CNT. The measured fluctuation time scales are found to be in agreement with single-molecule Förster resonant energy transfer (FRET) experiments. However, the electronic measurements have the advantage of giving additional data points at very short time scales (shorter than the acquisition times for single-molecule optical signals) and out to very long time scales (longer than the photobleaching times in FRET experiments).

Choi et al. [3] used a cysteine-modified lysozyme, with the modification chosen to lie in a part of the enzyme that undergoes substantial motion over the catalytic cycle. By controlling the concentration of the enzyme and the incubation time, it proved relatively straightforward to attach a single molecule to the CNT FET, as illustrated in Fig. 6.2. Lysozyme is an enzyme secreted in tears, breast milk, saliva, and mucus, and its function is to destroy bacterial cell walls by hydrolyzing peptidoglycans on their surface. Figure 6.3 shows a current-time trace recorded from a CNT FET functionalized with a single lysozyme molecule. The FET was biased at the point of maximum transconductance (dI_{DS}/dV_G—inset in Fig. 6.2) and the current trace was filtered to remove slow $1/f$ noise fluctuations. When the peptidoglycan substrate was added (at $t = 0$ in Fig. 6.3), rapid fluctuations in current were immediately observed. An expanded trace (inset) shows that these fluctuations are random telegraph noise (RTS), consisting of jumps between two current levels.

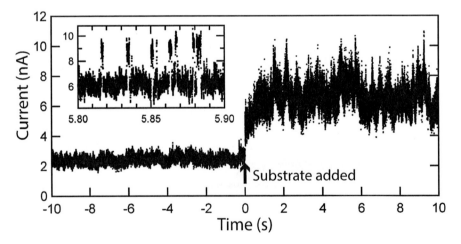

Fig. 6.3 Showing the onset of current fluctuations ($t = 0$) as the peptidoglycan substrate is added to the CNT FET device shown in Fig. 6.2, inducing the structural fluctuations associated with enzyme motion. The *inset* is an expanded trace showing the random telegraph signals that result from the two-state motion of the enzyme between its resting and catalytic states with bound substrate (a $1/f$ noise slow background has been subtracted). The enzyme also displays much more rapid telegraph noise associated with unproductive bindings of the substrate. Data is courtesy of P.G. Collins. Figure is reproduced from Lindsay, Single-Molecule Nanoelectronics [15]

Two families of RTS were observed: rapid (ms) fluctuations that were shown to be a consequence of abortive substrate-binding events and slower (10 ms) fluctuations that reflect completed hydrolysis of the peptidoglycan (these are the fluctuations illustrated in the inset in Fig. 6.3). Each type of event produced an exponential distribution of "on" times for the RTS, a result expected for a Poisson occupation of a simple two-state system. The Poisson statistics were confirmed by showing that the ratio of the variance, σ^2, of the distribution to the sum of the squares of "on" times, $\sum_i t_i^2$, was unity, i.e., $\sigma^2 = \sum_i t_i^2$. The current modulation measured in each experiment varied from CNT to CNT, but, using measured values of transconductance, the fluctuations in the current, ΔI_{DS}, were translated into fluctuations in gate voltage, ΔV_G. An almost constant value was found for all the devices of $\Delta V_G \sim 0.2$ V. This demonstrates the electrostatic origin of the current modulation. The magnitude of the field change is consistent with the known hinge motion of the enzyme and the consequent motion of charged residues near the attachment point to the CNT.

The quality of the recordings and the long recording times mean that rare events can be captured. A similar device was used by Sims et al. [24] to characterize a more complex process in which a substrate is modified in an ATP-dependent manner, i.e., the phosphorylation of a peptide target by cAMP-dependent protein kinase A (PKA).

6.2.3 Single-Molecule Redox Detectors

When a molecule is oxidized or reduced at an electrode, a charge of one, two, or sometimes more electrons passes into (oxidation) or out of (reduction) the electrode. This current can be large simply because the number of redox-active molecules in solution is high. However, another way to achieve a measurable signal is to have two electrodes so close that a single molecule can shuttle between them very rapidly, being oxidized on the positive electrode and reduced on the negative electrode. For this to work, the molecule has to have very fast electron transfer kinetics once it reaches the electrode. The time required for a molecule to transfer from one electrode to the other is on the order of

$$t \sim \frac{d^2}{2D} \tag{6.3}$$

where d is the gap size and D is the diffusion constant. Taking $D = 5 \times 10^{-6}$ cm^2/s and $d = 10$ nm leads to $t \sim 10^{-7}$ s or a current of a picoampere. Electron transfer rates will be different from this diffusion-limited rate when the molecule is fixed in place (i.e., in an attached monolayer), but tunneling rates into and out of the molecule can be high if the coupling is significant. Single-molecule electrochemical currents generated by this charge shuffling were demonstrated by Bard [25] using a scanning probe microscope. The key point is that these currents are only observable when both electrodes are poised near the formal potential (the point where redox currents are maximum) decaying very rapidly as the electrodes are moved away from the formal potential. (The assumption here is that the sensing bias is small and that the electrodes are fixed in potential relative to a reference electrode in contact with the solution.)

A practical solid-state device for detecting single molecules via their redox chemistry has not yet been demonstrated, though concentration fluctuations owing to the finite number of redox-active molecules in a small cavity have been detected [5]. A device for sequencing DNA molecules based on the attachment of labels with different formal potentials has been proposed by inventors at Intel Corporation [26].

6.2.4 Single-Molecule Electron-Tunneling Detectors

The electron-tunneling current between two electrodes is sensitive to the electronic states in the gap, as can be seen from the famous Breit–Wigner formula for the conductance of a tunnel junction containing a level E_M relative to the Fermi energy of the electrodes:

$$G = \frac{4e^2}{h} \frac{\Gamma_L \Gamma_R}{(E_M - V)^2 + (\Gamma_L + \Gamma_R)^2} \tag{6.4}$$

where Γ_L and Γ_R are the tunneling rates from the left electrode to the state and from the state to the right electrode. The conductance is a maximum when an applied bias, V, is equal to E_M. Molecular detail is included in more comprehensive calculations of tunnel current by using a Green's function propagator, and Zwolak and Di Ventra used such a calculation to predict that DNA could be sequenced by reading the tunnel current passing through a tunnel junction as each of the DNA bases passes through it [27]. A single-molecule sensing device that identifies molecules from their electron-tunneling current signatures requires the fabrication of sub-5 nm-sized gaps between electrodes. Previous attempts to build tunnel sensors have used gaps in wires arranged colinearly. Such junctions have been made by cutting gold [28, 29] or carbon [30] nanowires with an electron beam or by using electromigration [6] or mechanical stress [31] to break a gold wire or electron-beam-induced deposition of opposed pairs of wires [32, 33]. It is difficult to scale the fabrication of devices when individual gaps have to be tailored in these ways. Furthermore, the chemical identification of small molecules has been demonstrated only in devices with very small (0.8 nm) gap sizes [34]. Scalable fabrication based on shadow evaporation in a channel [35] or electron-beam lithography [36] has yet to produce gaps less than about 5 nm, too large for significant tunnel current to flow. We have recently reported a novel "stacked

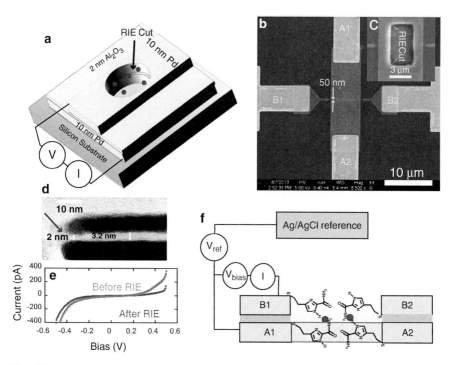

Fig. 6.4 (**a**) The "stacked" junction—the tunnel gap is defined by a 2–3 nm layer of Al_2O_3 exposed by reactive ion etching (RIE). (**b**) SEM image of a device. (**c**) A 3×5 μm cut. (**d**) A TEM projection image showing the 2 nm tunnel gap at the edge of the device. (**e**) Tunnel characteristics before and after the RIE cut. (**f**) The biasing arrangements and the functionalization with reader molecules

junction" geometry that is readily scalable and generates current signals that are characteristic of individual analyte molecules in solution [7].

The layout of a stacked junction is shown in Fig. 6.4a. In this device, the tunnel gap is determined by a 2 nm-thick layer of Al_2O_3 deposited by atomic layer deposition (ALD). This layer separates two planar palladium electrodes, and the edges of the electrodes are exposed in a controlled manner by etching through the stack using reactive ion etching (RIE). A key feature of our tunneling devices is the use of specially designed recognition molecules, strongly attached to the metal electrodes, but forming weak, non-covalent bonds with the analyte molecules, an arrangement we call recognition tunneling (RT) [37]. RT displaces water and ions from the tunneling gap, forming direct contacts with the target by means of highly conductive hydrogen bonds [38]. The chemical attachment of the recognition molecules to the electrodes displaces hydrocarbon contaminants from the tunneling path. A schematic layout of the junction showing a pair of molecules (red dots) trapped by recognition molecules is shown in Fig. 6.4f. Yet another advantage of RT is that the (relatively) conductive recognition molecules allow significant tunneling signals to be collected with the gap being as large as 2 nm, compared to the 0.8 nm required to contact DNA bases without recognition molecules [34]. This both eases fabrication constraints and allows big molecules (like DNA) to pass through the gap.

An SEM image of the top of a device is shown in Fig. 6.4b, with the inset (Fig. 6.4c) showing the center of the device after a large (3 × 5 μm) RIE cut through the region where the 50 nm top wire passes over the larger lower electrode. A side view cross section of the tunnel junction (taken by TEM using a section cut out by an FIB) is shown in Fig. 6.4d. The lack of damage to the electrodes is reflected in the tunnel characteristics (Fig. 6.4e), the current falling after the cut because the junction area is decreased. (Other processes that result in the displacement of the metal cause the current to *increase* after cutting.) The biasing arrangement of the device is shown in Fig. 6.4f. Note the use of a reference electrode, a point to which we will return.

At low concentrations (nM) of analyte, the device gives distinct steps in the current, presumably reflecting the binding and unbinding of single molecules. A scatterplot of the plateau widths vs. the step amplitude for a large number of events recorded as each of the four DNA nucleotides (A,G,C,T) were flowed through the device is shown in Fig. 6.5. Analytes are well separated by step amplitude for the longer lifetime events. Data are also shown for the epigenetically modified 5-methylcytidine (mC) and many of those reads are well separated from the C reads. These data were recorded for analyte concentrations of 10 nM, much smaller than the micromolar levels needed to obtain signals with an STM tunnel junction, presumably reflecting the greater accessibility of the gap.

In order to sequence DNA with a device like this, a nanopore is required (as opposed to the large area RIE opening shown in Fig. 6.4) and this is an object of current research. However, RT will analyze a variety of analytes [9] and the

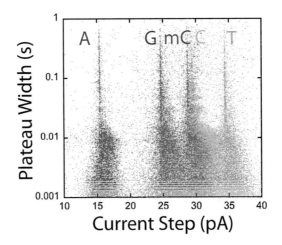

Fig. 6.5 Scatterplot of (vertical axis) the duration of current pulses vs. (horizontal axis) the amplitude of the current step for current fluctuations in the stacked tunnel junction for 10 nM solutions of dAMP (A), dGMP (G), dCMP (C), dTMP (T), and d(5-methylCMP) (mC) showing how individual DNA bases can be identified from the tunnel current signals

capability for single-molecule analysis at low concentrations and very small sample volumes means that this device will find analytical applications even without the direct sequencing capability that a nanopore will bring.

6.3 Electrochemical Aspects of Nanoscale Sensors

As mentioned earlier, any electronic material in contact with an electrolyte will undergo chemical reactions. In the case of the palladium electrodes used in our tunnel junctions, hydrogen evolution is particularly rapid at negative potentials in water. A second issue lies with the accumulation of ions on the electrode surface, resulting in very high interfacial capacitance and, indirectly, significant current noise. These issues are examined here.

6.3.1 Chemistry at the Interface

All electrochemical devices require the presence of a salt in solution to make it conductive. This is because the intrinsic concentration of ions in pure water is very low, giving it a high resistivity (18 MΩ cm). This is a limitation because trace contaminants dictate interfacial potential drops when devices are operated in "pure" water. Operation in salt solutions allows the potential drop at the interface to be controlled. At first sight, it might seem problematical to operate electronic devices in salt solutions, but nanodevices have such a small surface area that electrochemical leakage current is readily minimized.

In a solution containing an electroactive species (i.e., an ion that is readily oxidized at a low interfacial potential drop, such as $Fe^{++} \leftrightarrow Fe^{+++}$) of concentration C molar, the current is

$$\Delta I = 4\pi n F D C r \tag{6.5}$$

where n is the number of electrons transferred in a reaction, D is the ion diffusion constant, F is the Faraday constant, and r is the radius of a disk electrode. A typical value for D is 10^{-5} cm^{2}s^{-1} and, with $F = 9.6 \times 10^4$ C/mol, so even a very high concentration of electroactive molecules (10 mM) results in the current of a few picoampere for electrodes of nm dimension. Clearly, this is only true if the device is passivated at all points except the active region in contact with the fluid.

Once an electrode is place in contact with a conducting electrolyte (which can be chosen to be particularly inert—e.g., sodium perchlorate), then the potential drop across its interface with the solution can be controlled with respect to a reference electrode (an electrode of constant polarization described above). When the potential of the electrode is swept with respect to the reference in, e.g., an aqueous electrolyte, current will rise when the surface is made negative (owing to hydrogen evolution) and again when the surface is made positive (owing to oxygen evolution—with other reactions possible at lower potentials depending on what is dissolved in the electrolyte). In the gap between the potentials that result in these oxidation or reduction Faradaic currents, the current is relatively constant and given by

$$i_{DL} = C_{DL} \frac{dV}{dt} \tag{6.6}$$

Here, C_{DL} is the double-layer capacitance and dV/dt the voltage sweep rate. In this potential range, where the accumulated charge is directly proportional to the potential (no DC current flows into the electrode), the behavior is opposite to that of a reference electrode (an ideal nonpolarizable electrode, INPE), and it is referred to as an *ideal polarizable electrode* (IPE). Outside these ideal limiting regimes (IPE and INPE), both polarization and Faradaic currents will flow. The double-layer capacitance is controlled by the thickness of the ion accumulation layer on the electrode, and this, in turn, is determined by the salt concentration as given by Eq. (6.2). The smallness of the Debye length in typical electrolytes means that this capacitance is very high—on the order of 0.1 F/m^2. Nonetheless, if the exposed electrode area is very small (e.g., 10^{-16} m^2 in our tunneling device described above in Sect. 6.2.4), the capacitance is also very small ($\sim 10^{-17}$ F).

This region of constant current given by Eq. 6.6 defines the potential window of the electrolyte/electrode system. In absolute terms, it will be given as an upper and lower voltage with respect to a particular reference electrode (such as the Ag/AgCl/ Ag$^+$Cl$^-$ reference described above). Clearly, a device must operate within this window if electrochemical reactions are to be avoided. An exception is the redox sensor described in Sect. 6.2.3, though a technical challenge for such sensors lies in stabilizing their surfaces even as electrode material is consumed in the reactions.

To give a concrete example, referring to Fig. 6.4f, the lower electrode (A1, A2) is held at a potential V_{ref} with respect to the Ag/AgCl reference. Since hydrogen evolution starts close to 0 V vs. Ag/AgCl on a Pd surface, V_{ref} is set to be a few tens of millivolt positive. The top electrode is set at $V_{bias} + V_{ref}$ with respect to the Ag/AgCl reference (where V_{bias} must also be positive to avoid hydrogen evolution at the top electrode). Oxidation currents begin to become significant above about 600 mV positive of Ag/AgCl for Pd electrodes, so this limits V_{bias} to a maximum of about 500 mV.

Clearly, such control of the electrodes could not be achieved without the use of a reference electrode. This becomes especially important in the case of nanodevices because of their very small capacitance. Specifically, taking the example of a device with an interfacial capacitance of 10^{-17} F, adsorption of just one electronic charge will change the electrode potential by 1.6×10^{-2} V or 16 mV. Our RT electrodes were designed specifically to capture DNA bases by hydrogen bonding, and since each nucleotide carries a negative charge, the electrode rapidly moves into the hydrogen evolution region as molecules are captured. The result is an oscillating "on–off" behavior as reduction currents discharge the electrodes, only to have them become charged again as more nucleotides are captured. This problem is cured completely by connecting one of the electrodes to a reference electrode, thus

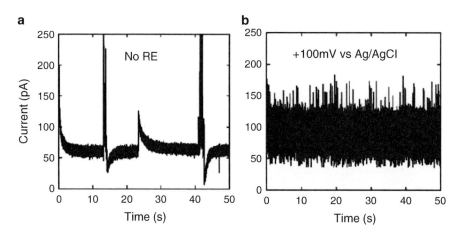

Fig. 6.6 Signals from a tunnel current device operated in a solution of guanosine monophosphate before (**a**) and (**b**) after the connection of one of the electrodes to an Ag/AgCl reference electrodes. The current undergoes violent swings with slow decays between cycles (**a**) when the potential is not controlled

controlling the adsorption of charged species from solution. Figure 6.6 shows the dramatic difference made by connecting a device to a reference electrode (same device, same experimental run).

6.3.2 Noise at the Interface

Noise originates from dissipative processes, and, in the case of an IPE, there is no equilibrium current, so the system is in thermal equilibrium and current fluctuations can be calculated using the fluctuation–dissipation theorem. The resistance in the circuit arises from mobile ions in the double layer and in the bulk. In addition, much (or most) of the polarization is formed by ions trapped at the surface where their finite radius precludes motion, so they contribute only capacitance but no resistance. This inner, immobile layer is called the Stern layer. An equivalent circuit for an IPE device connected to a reference electrode is shown in Fig. 6.7. The current noise is given by the Johnson noise in the resistance of the bulk solution, R_B, and the resistance of the double layer, R_D. In a device like the tunnel junction, operated at constant voltage, the current noise increases as the electrolyte and double-layer resistance decreases. Since this resistance is inversely proportional to the exposed electrode area, minimizing this area is important. In a nanodevice, the area of the interconnecting electrodes is generally many orders of magnitude bigger than that of the device itself, so passivation of the surrounding circuitry is critical in limiting noise. Note too that, in the devices described here, a reference electrode is connected to one of the device electrodes, so current fluctuations owing to the reference electrode will also contribute. These have been analyzed by Hassibi et al. [39]. The reference electrode, an INPE, maintains its potential by means of

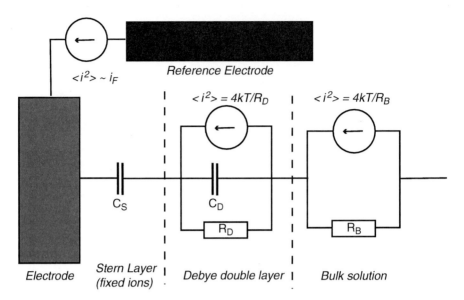

Fig. 6.7 Circuit model for noise sources at an ideal polarizable interface ("electrode"), including a connection to a reference electrode where the noise is generated by Faradaic currents

Faradaic current flow (i_F, in Fig. 6.7). The frequency dependence of these fluctuations can be quite complicated [39] but the mean square current scales with the Faradaic current. In the case of a single-molecule sensor, the current will be given by the adsorption rate of charged molecules on the sensor surface, and it will be very small (much smaller than the tunnel current fluctuations sensed in the device described above).

6.4 Statistics in Nanodevices

Thus far we have talked of these devices in terms of bulk quantities, but fluctuations dominate when the number of molecules in the device is small. Thus, the arrival time of analyte molecules is stochastic, as are times between binding and dissociation events. On the other hand, correlations become even more important as the number of molecules in the system is reduced, an important and underappreciated aspect of nanodevices that we elaborate on below.

6.4.1 Markov (Random) Noise

Noise is regarded as a useless artifact because, in the thermodynamic limit, it conveys no information, each step in a random walk or diffusion process conveying no information about the previous step. Such a process is called Markovian.

The arrival of molecules at a detector by diffusion is an example of a Markovian process, and the distribution of arrival times is given by a simple exponential:

$$N(t) = N_0 e^{-\frac{t}{\tau}}. \tag{6.7}$$

This exponential distribution of intervals between signals means that some signals that follow at very small time intervals are lost owing to the finite frequency response of the detection system, while very long intervals may not be recorded owing to a finite recording time.

6.4.2   Noise Correlations and Chemical Analysis

Equation (6.7) illustrates the lack of information in Markovian processes—the distribution is characterized by just one parameter, τ. However, processes that are highly correlated may have much more complicated noise signatures. An example is shown in Fig. 6.8 where Fig. 6.8a shows data points obtained by averaging thousands of current–voltage measurements take at single bias values on single molecules as a function of surface potential, E_S. The distinct peak in current is a negative differential resistance that results from oxidation of the molecule in the tunnel gap [40]. A peak is also seen in the current when the voltage is swept in a single-molecule device. Figure 6.8b shows four current sweeps for four different

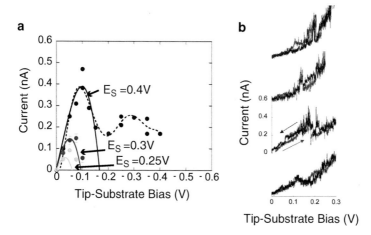

Fig. 6.8 "Noise" in a nanoscale device. (**a**) Averaged current–voltage characteristics (*dots*) taken using many break junction measurements. The peak in current is a consequence of reduction of a molecule trapped in the junction. Different *colored dots* are for different surface potentials, E_S, and the solid lines are fits to a bulk electrochemical model. (**b**) Some I–V traces taken on single molecules. A peak is present in each case, but it is sharp and appears to occur at random values of bias, suggesting that the data are just noise. However, the signals repeat as the voltage sweep direction is changed, showing that the different characteristics reflect different microenvironments

devices. All four show some kind of peak but all four look different and at first glance look like experimental noise. Note, however, that the current on the upsweep is reasonably overlapped by the current on the downsweep. This is not noise but rather reflects microscopic difference in the polarization of each single-molecule device. Thus the "noise" conveys information.

In the example just given, the "noisy" signal was static, but time-dependent stochastic signals can contain useful information too. Figure 6.5 shows how signal spikes in a tunnel junction can be assigned to an analyte with high accuracy using two signal features. Here we give another example taken from recognition tunneling signals obtained from amino acids [9]. Figure 6.9a, b shows the stochastic train of signal spikes obtained from two enantiomers (L- and D-asparagine). These have identical compositions and bonding, save that one is the mirror image of the other. As might be expected, the RT signals produced by the two molecules are very similar, and distributions of signal features are highly overlapped. Figure 6.9c, d shows the distributions of amplitudes of two Fourier components of the signals spikes. If the value of either one of these signal features were to be used to assign an unknown signal based on its most probable value, the outcome would be little better than random (0.53 for the data shown in Fig. 6.9c and 0.57 for the data shown in Fig. 6.9d, as opposed to 0.5 for a random guess). However, when the measured probability is plotted as a function of *both* signal features (Fig. 6.9e), nonlinear correlations result in a separation of the data points. This probability map predicts the identity of an unknown signal spike to 85 % accuracy, using two data sets that, taken by themselves, did little better than random guessing. Using many signal features, it has proved possible to assign RT signals from a number of amino acids to >90 % accuracy [9]. Thus, despite the stochastic nature of the signals, useful chemical information can be extracted. Figure 6.9f shows a signal train obtained from a mixture of L- and D-asparagine, where each cluster of spikes is generated by a single trapped molecule. The colors show the assignment of the signal made by a trained machine learning algorithm. Thus single binding events can be analyzed using stochastic "noise" with a fair degree of accuracy.

6.5 Conclusions

At this time, there are several examples of electronic devices that interact with single molecules based on having feature sizes of molecular dimension. Some, like the nanopore ion-current sensor, are close to or in commercial production. Others face technical obstacles, but can work quite well. The FET single-molecule sensors based on CNTs are limited by the need to select suitable CNTs, though such limitations may not apply to semiconductor nanowire devices. Tunnel junctions can now be made in a scalable way, though they will have to be integrated with nanopores if they are to sequence DNA and proteins. Devices with nanoelectrodes are subject to electrochemical constraints. In particular, their surface potential must be controlled to keep them operating in a region where the electrodes are stable.

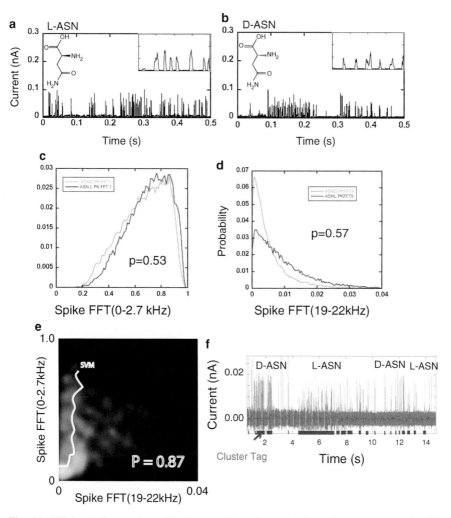

Fig. 6.9 RT signals from amino acids, illustrated here for a pair of enantiomers, L-asparagine (**a**) and D-asparagine (**b**). The *insets* are expanded traces (20 ms) to show the complex shapes of individual spikes. Many features can be used to characterize these spikes: shown here are the normalized amplitudes of the Fourier components in low (0–2.7 kHz, **c**) and high (19–22 kHz, **d**) bands of frequency. The *red curves* are distributions for L-asparagine and the *green curves* are distributions for D-asparagine. The distributions are very similar, so relying on one parameter alone does not separate the data well. For example, assigning all points with high-frequency amplitudes less than about 0.006 as L-asparagine (**d**) yields a correct call 57 % of the time, only just a little better than random (50 %). However, when two such distributions are plotted together (**e**), separation improves dramatically. In the 3D plot, *red* is more probably L-asparagine and *green* is more probably D-asparagine. Assigning all points to the *red* or *yellow* as L-asparagine results in 87 % correct calls. The support vector machine (SVM) makes this assignment by partitioning space to give the best separation (*white line*). (**f**) Machine calling of spikes from a mixed sample (1:1 L-asparagine and D-asparagine; *yellow* is L-asparagine and *purple* is D-asparagine). Note how all the calls in a cluster are the same, consistent with the long trapping time for single molecules in the gap and the notion that each cluster represents a distinct single-molecule binding event (Adapted from Lindsay, Single-Molecule Nanoelectronics [15].)

Noise is both a limiting factor in these devices and a potentially useful source of signal. Remarkably, despite the stochastic origin of single-molecule signals, quite rich chemical information can be extracted from single events. Once ways are found to build these devices in massively parallel arrays, they offer the possibility of complex analysis of proteins, metabolites, and other biomarkers on scales hitherto unthinkable.

Acknowledgments This work was carried out with the support of the National Human Genome Research Institute (Grant number R01 HG006323). Many members of the lab contributed to the work described here including Pei Pang, Brian Ashcroft, Weisi Song, and Yanan Zhao. I am grateful to Phil Collins of UC Irvine for providing the data for the plots used in Fig. 6.3. The assistance of the NNIN facility in the Center for Solid State Electronics Research at ASU and the John Cowley Center for High Resolution Electron Microscopy is gratefully acknowledged.

References

1. B.N. Miles, A.P. Ivanov, K.A. Wilson, F. Dogan, D. Japrung, J.B. Edel, Chem. Soc. Rev. **42** (1), 15 (2013)
2. D. Branton, D. Deamer, A. Marziali, H. Bayley, S.A. Benner, T. Butler, M. Di Ventra, S. Garaj, A. Hibbs, X. Huang, S.B. Jovanovich, P.S. Krstic, S. Lindsay, X.S. Ling, C.H. Mastrangelo, A. Meller, J.S. Oliver, Y.V. Pershin, J.M. Ramsey, R. Riehn, G.V. Soni, V. Tabard-Cossa, M. Wanunu, M. Wiggin, J. Schloss, Nat. Biotechnol. **26**, 1146 (2008)
3. Y. Choi, I.S. Moody, P.C. Sims, S.R. Hunt, B.L. Corso, I. Perez, G.A. Weiss, P.G. Collins, Science **335**, 319 (2012)
4. P. Xie, Q. Xiong, Y. Fang, Q. Qing, C.M. Lieber, Nat. Nanotechnol. **7**(2), 119 (2012)
5. M.A. Zevenbergen, D. Krapf, M.R. Zuiddam, S.G. Lemay, Nano Lett. **7**(2), 384 (2007)
6. M. Tsutsui, S. Rahong, Y. Iizumi, T. Okazaki, M. Taniguchi, T. Kawai, Nat. Sci. Rep. **1**, 46 (2011)
7. P. Pang, B. Ashcroft, W. Song, P. Zhang, S. Biswas, Q. Qing, J. Yang, R.J. Nemanich, J. Bai, J. Smith, K. Reuter, V.S.K. Balagurusamy, Y. Astier, G. Stolovitzky, S. Lindsay, ACS Nano **8**, 11994–12003 (2014)
8. S. Lindsay, J. Phys. Condens. Matter **24**, 164201 (2012)
9. Y. Zhao, B. Ashcroft, P. Zhang, H. Liu, S. Sen, W. Song, J.O. Im, B. Gyarfas, S. Manna, S. Biswas, C. Borges, S. Lindsay, Nat. Nanotechnol. **9**, 466 (2014)
10. S.M. Lindsay, *Introduction to Nanoscience* (Oxford University Press, Oxford, 2009)
11. J.J. Kasianowicz, E. Brandin, D. Branton, D.W. Deamer, Proc. Natl. Acad. Sci. U. S. A. **93**, 13770–13773 (1996)
12. J. Li, D. Stein, D. McCullan, D. Branton, M.J. Aziz, J.A. Golovchenko, Nature **412**, 166 (2001)
13. M. Wanunu, Phys. Life Rev. **9**, 125 (2012)
14. A.H. Laszlo, I.M. Derrington, B.C. Ross, H. Brinkerhoff, A. Adey, I.C. Nova, J.M. Craig, K.W. Langford, J.M. Samson, R. Daza, K. Doering, J. Shendure, J.H. Gundlach, Nat. Biotechnol. **32**, 829 (2014)
15. S. Lindsay, Single-molecule nanoelectronics, in *Nanoelectrochemistry*, ed. by S. Amemiya, M.V. Mirkin (CRC Press, Boca Raton, FL, 2015), pp. 179–202
16. B.R. Goldsmith, J.G. Coroneus, V.R. Khalap, A.A. Kane, G.A. Weiss, P.G. Collins, Science **315**, 77 (2007)
17. B.L. Allen, P.D. Kichambare, A. Star, Adv. Mater. **19**, 1439 (2007)
18. M. Trojanowicz, Trends Anal. Chem. **25**, 480 (2006)
19. C. Gao, Z. Guo, J.-H. Liua, X.-J. Huang, Nanoscale **4**, 1948 (2012)

20. M. Law, J. Goldberger, P. Yang, Annu. Rev. Mater. Res. **34**, 83 (2004)
21. K. Besteman, J.-O. Lee, F.G.M. Wiertz, H.A. Heering, C. Dekker, Nano Lett. **3**, 727 (2003)
22. J.-C. Star, P. Gabriel, K. Bradley, G. Grüner, Nano Lett. **3**, 459 (2003)
23. Y. Cui, Q. Wei, H. Park, C.M. Lieber, Science **293**, 1289 (2001)
24. P.C. Sims, I.S. Moody, Y. Choi, C. Dong, M. Iftikhar, B.L. Corso, O.T. Gul, P.G. Collins, G.A. Weiss, J. Am. Chem. Soc. **135**, 7861 (2013)
25. F.R.F. Fan, A.J. Bard, Science **267**, 871 (1995)
26. X. Su, D. Liu, Patent No. 8372585 B2, (2013).
27. M. Zwolak, M. Di Ventra, Rev. Mod. Phys. **80**, 141 (2008)
28. M.D. Fischbein, M. Drndić, Nano Lett. **7**, 1329 (2007)
29. K. Healy, V. Ray, L.J. Willis, N. Peterman, J. Bartel, M. Drndic, Electrophoresis **23**, 3488 (2012)
30. P.S. Spinney, S.D. Collins, D.G. Howitt, R.L. Smith, Nanotechnology **23**, 135501 (2012)
31. M. Tsutsui, K. Shoji, M. Taniguchi, T. Kawai, Nano Lett. **8**, 345 (2007)
32. A.P. Ivanov, E. Instuli, C.M. McGilvery, G. Baldwin, D.W. McComb, T. Albrecht, J.B. Edel, Nano Lett. **11**, 279 (2011)
33. A.P. Ivanov, K.J. Freedman, M.J. Kim, T. Albrecht, J.B. Edel, ACS Nano **8**, 1940 (2014)
34. T. Ohshiro, K. Matsubara, M. Tsutsui, M. Furuhashi, M. Taniguchi, T. Kawai, Nat. Sci. Rep. **2**, 501 (2012)
35. X. Liang, S.Y. Chou, Nano Lett. **8**, 1472 (2008)
36. A. Fanget, F. Traversi, S. Khlybov, P. Granjon, A. Magrez, L. Forró, A. Radenovic, Nano Lett. **14**, 244 (2013)
37. S. Lindsay, J. He, O. Sankey, P. Hapala, P. Jelinek, P. Zhang, S. Chang, S. Huang, Nanotechnology **21**, 262001 (2010)
38. M.H. Lee, O.F. Sankey, Phys. Rev. E **79**, 051911.1 (2009)
39. A. Hassibi, R. Navid, R.W. Dutton, T.H. Lee, J. App. Phys. **96**, 1074 (2004)
40. F. Chen, J. He, C. Nuckolls, T. Roberts, J. Klare, S.M. Lindsay, Nano Lett. **5**, 503 (2005)

Chapter 7
Terahertz Wave Generation Using Graphene and Compound Semiconductor Nano-Heterostructures

Taiichi Otsuji, Victor Ryzhii, Stephane Boubanga Tombet, Akira Satou, Maxim Ryzhii, Vyacheslav V. Popov, Wojciech Knap, Vladimir Mitin, and Michael Shur

Abstract This paper reviews recent advances in terahertz wave generation using graphene and compound semiconductor nano-heterostructures. The excitation of two-dimensional (2D) plasmons in high-electron mobility transistors (HEMTs) and related semiconductor nano-heterostructures has been used for emission of THz electromagnetic radiation. Plasmons in graphene (which is one or several mono-layers of a honeycomb carbon lattice) have a higher velocity and peculiar transport properties owing to the massless and gapless energy spectrum of graphene. Optical and/or injection pumping of graphene results in a negative-dynamic conductivity in the THz spectral range, which may enable new types of THz lasers. Fundamental physics behind the device operation mechanisms and experimental results are demonstrated including coherent monochromatic THz radiation from InP-based HEMT-type emitters and stimulated emission of THz radiation with a giant gain via excitation of surface plasmon polaritons in optically pumped monolayer intrinsic graphene.

T. Otsuji (✉) • V. Ryzhii • S.B. Tombet • A. Satou
RIEC, Tohoku University, Sendai, Japan
e-mail: otsuji@riec.tohoku.ac.jp

M. Ryzhii
Department of CSC, University of Aizu, Aizu-Wakamatsu, Japan

V.V. Popov
Kotelnikov Institute of Radio Engineering Electronics, Saratov Branch, RAS, Saratov, Russia

W. Knap
LC2-Laboratory, CNRS-Universite Montpellier 2, Montpellier, France

V. Mitin
Department of EE, University at Buffalo, SUNY, Buffalo, NY, USA

M. Shur
Department of ECSC, Rensselaer Polytechnic Institute, Troy, MI, USA

© Springer International Publishing Switzerland 2015
A. Korkin et al. (eds.), *Nanoscale Materials and Devices for Electronics, Photonics and Solar Energy*, Nanostructure Science and Technology, DOI 10.1007/978-3-319-18633-7_7

237

7.1   Introduction

The development of compact, tunable, and coherent sources and detectors operating in the terahertz (THz) regime is one of the key tasks of modern microelectronics [1]. The excitation of two-dimensional (2D) plasmons in high-electron mobility transistors (HEMTs) and related semiconductor nano-heterostructures has been used for detection, modulation, mixing, and emission of THz electromagnetic radiation [2, 3]. Plasmons in graphene (which is one or several monolayers of a honeycomb carbon lattice) have a higher velocity and peculiar transport properties owing to the massless and gapless energy spectrum of graphene [4]. Optical and/or injection pumping of graphene results in a negative-dynamic conductivity in the THz spectral range [5, 6], which may enable new types of THz lasers. This paper reviews recent advances in terahertz wave generation using graphene and compound semiconductor nano-heterostructures.

7.2   2D Plasmons Physics

At THz frequencies, the electron inertia plays an important role in introducing delays between the applied voltage and the resulting electric current. This results in an inductive component of the overall semiconductor impedance. An equivalent circuit of the two-dimensional electrons in the transistor channel could be presented as a distributed LCR transmission line [6–10] (see Fig. 7.1).

When the mean free path of electrons for collisions with impurities and lattice vibrations exceeds the device dimensions, the resistive component in this transmission line becomes small, and the transistor channel supports waves of the electron density referred to as plasma waves. The transistor channel becomes a resonant cavity for these plasma waves. When electron collisions with impurities and lattice vibrations become important or even dominant, the plasma wave decays and, in the limiting case of the collision-dominated regime, becomes overdamped. In this case, the equivalent circuit could be presented as an RC transmission line (see Fig. 7.1). But even in this case, the overdamped electron oscillations would respond to a THz field impinging on the device. Figure 7.2 (from [11]) compares the electron density distributions in the transistor channels for different mobility values.

Fig. 7.1 Equivalent circuit of the two-dimensional electrons in the transistor channel presented as a distributed LCR transmission line

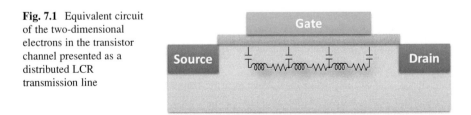

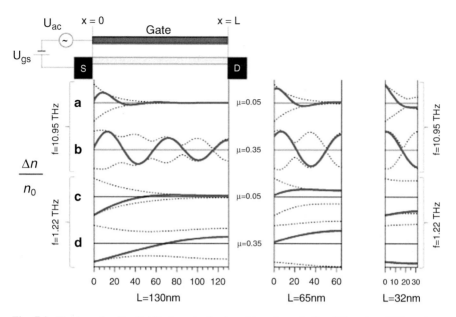

Fig. 7.2 Electron density distributions in the transistor channels for different mobility values (from [11])

The dispersion relations of the plasma waves, ω_p versus k, for the gated and ungated regions of a field effect transistor are given by [12, 13]:

$$\omega_p = \sqrt{\frac{q^2 n_s k}{2 m \varepsilon_1 \varepsilon_{\text{eff}}}}. \tag{7.1}$$

Here, ω_p is the plasma frequency, k is the wave vector, n_s is the electron sheet density, m is the effective mass, ε_1 is the vacuum dielectric permittivity, and ε_{eff} is the effective dielectric constant:

$$\left(\varepsilon_{\text{eff}}\right)_{\text{ungated}} = \frac{1}{2}\left[\varepsilon_2 + \varepsilon_1 \frac{1 + \varepsilon_1 \tanh(kd)}{\varepsilon_1 + \tanh(kd)}\right], \tag{7.2}$$

$$\left(\varepsilon_{\text{eff}}\right)_{\text{gated}} = [\varepsilon_2 + \varepsilon_1 \coth(kd)]/2. \tag{7.3}$$

where ε_1 and ε_2 are the dielectric constants of the barrier layer (with thickness d) between the transistor gate and the channel and the substrate, respectively.

For $kd \ll 1$ and $kd \gg 1$, these equations yield the plasma-wave dispersion relations for the gated and ungated regions, respectively:

$$\left(\omega_p\right)_{\text{gated}} = \sqrt{\frac{q^2 n_s d}{m \varepsilon_0 \varepsilon_2}} k = \sqrt{\frac{q V_{\text{gt}}}{m}} k, \tag{7.4}$$

$$\left(\omega_p\right)_{\text{ungated}} = \sqrt{\frac{q^2 n_s k}{m^* \varepsilon_0 (1 + \varepsilon_2)}}. \tag{7.5}$$

For symmetrical boundary conditions, the wave vectors of the plasma waves in the transistor channel are given by [14]:

$$\left(k_n\right)_{\text{gated}} = (2n - 1)\pi/L_g, \tag{7.6}$$

$$\left(k_n\right)_{\text{ungated}} = (2n - 1)\pi/L_{ug}, \tag{7.7}$$

where $n = 1, 2, 3, \ldots$.

As pointed out in [2], these dispersion relations for the gated regions are identical with those for shallow water waves or for sound waves. The dispersion relations for the ungated regions are identical with those for the deep-water waves.

The plasma-wave phase velocity, s, in the gated region above threshold is $s = \sqrt{\frac{qV_{gt}}{m}}$, which is, typically, much larger than the electron drift velocities (see Fig. 7.3).

A more general expression for the plasma-wave velocity describes the above and below threshold regimes [16]:

$$s = \sqrt{\frac{\eta k_B T}{m} \left[1 + \exp\left(-\frac{qV_{gt}}{\eta k_B T} \right) \right] \ln \left[1 + \exp\left(\frac{qV_{gt}}{\eta k_B T} \right) \right]}. \tag{7.8}$$

Here, η is the subthreshold ideality of the current–voltage characteristic.

In the resonant regime, the plasma waves propagating in the transistor channel with asymmetrical boundary conditions become unstable [2]. Figures 7.4 and 7.5 show early results reporting on the THz emission from the plasmonic FETs. Even though one of such observations establishes the link between the plasma frequency and the radiation frequency (see Fig. 7.5b), the emission mechanism was not

Fig. 7.3 Plasma and transit frequencies for Si, GaAs, and GaN transistor channels [15]

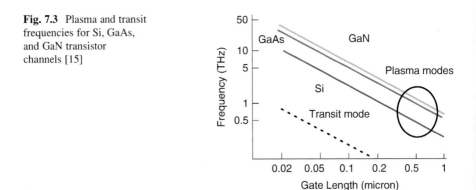

Fig. 7.4 Radiation intensity from a 1.5-μm gate-length GaN-based HFET at 8 K [17]

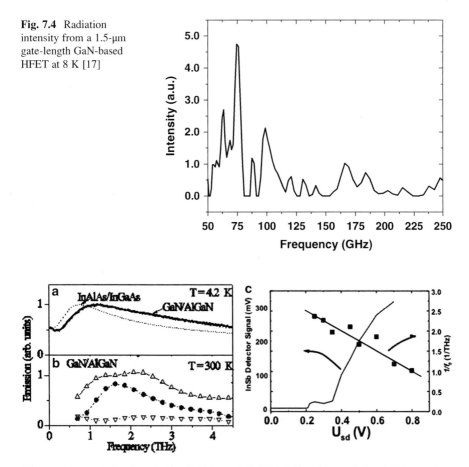

Fig. 7.5 THz emission from InAlAs/InGaAs and GaN/AlGaN HEMTs [18, 19]. (**a**) Comparison of the emission spectra at 4.2 K between InAlAs/InGaAs and GaN/AlGaN HEMTs. (**b**) Emission spectra for GaN/AlGaN HEMTs at 300 K with different gate biases U_{gs}. Δ: $U_{gs} = -4$ V, ●: $U_{gs} = -1.1$ V, ▽: $U_{gs} = 0$ V. (**c**) Dependence of peak emission intensity (*left axis*) and frequency (*right axis*) on drain bias U_{sd} for the InAlAs/InGaAs HEMT

entirely clear because of the broadband nature of this emission [17–19]. Only recently, the Dyakonov-Shur instability was clearly shown to lead to narrowband emission [20]. These results are described in the next section.

7.3 THz Emission Using 2D Plasmons

A big improvement in the plasmonic THz radiation resulted from the development of the 2D plasmon-resonant microchip emitter as a new terahertz light source [20–24]. The structure is based on an HEMT and features interdigitated dual-grating gates (DGGs) (Fig. 7.6).

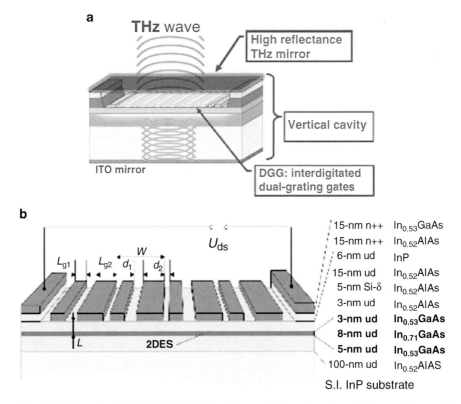

Fig. 7.6 Schematic (**a**) of A-DGG HEMT with a vertical cavity and typical composition and dimensions (**b**) [20]

The dual-grating gates act as a distributive antenna and induce the periodically distributed plasmonic cavities (~100-nm width in microns distance) along the channel [21]. Under pertinent drain–source dc bias conditions, dc electron drift current will lead to the instability owing to (1) periodic modulation of electron drift velocities [25, 26], which we refer to as the Ryzhii-Satou-Shur (R-S-S) instability; (2) Doppler-shift effect at the asymmetric open-short boundaries [2], which we call the Dyakonov-Shur (D-S) instability; and (3) (still to be observed) the instability from periodic transitions from the regions with the plasma velocity is smaller than drift velocity to the regions, where it is higher than the electron drift velocity [15]. Such instabilities result in self-oscillations with the characteristic frequencies in the THz range.

The device shown in Fig. 7.6b was fabricated using InGaP/InGaAs/GaAs and/or InAl/InGaAs/InP material systems [20, 22–24]. So far, a broadband THz emission ranging from 1 to ~6 THz has been obtained reflecting multimode coherent/incoherent plasmons [22]. The excitation of hot plasmons and chirped plasmon modes [27] might contribute to this emission. The DGG HEMT THz emitter could

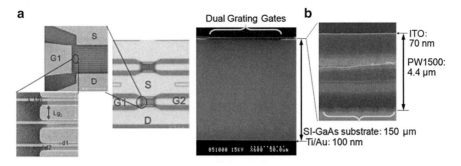

Fig. 7.7 Schematic view and SEM images (**a**) and cross-sectional TEM images (**b**) of an A-DGG HEMT. $L_{g1} = 200$ nm, $L_{g2} = 1,600$ nm, $d_1 = 200$ nm, $d_2 = 400$ nm [20]

find THz spectroscopic and imaging applications as an incoherent broadband terahertz microchip source. The demonstrated applications include fine identification of the water vapor absorptions as well as the fingerprints of sugar groups [23, 24].

The coherent monochromatic THz emission was achieved using an improved resonant-enhanced high-Q vertical cavity structure combined with our original asymmetric DGG (A-DGG) structure (Fig. 7.7) [20]. The A-DGG structure implemented with asymmetric interfinger spaces creates a strong asymmetric field distribution significantly enhancing the D-S instability [2, 20]. We expect that cooperative effects between the R-S-S and D-S instability mechanisms accelerate the injection-locking operation resulting in the so-called giant plasmon instability. As a consequence, the hot plasmon-incoherent spontaneous emission gives rise to stimulated superradiant emission. Numerical simulations of the THz electric field distribution and the resulting photoresponse using a self-consistent electromagnetic approach and hydrodynamic equations in HEMTs under periodic electron density modulation conditions [28] predict a giant enhancement of the responsivity by four orders of magnitude in an asymmetric DGG HEMT under drain-unbiased conditions compared to that for a symmetric DGG HEMT for a dc-drain bias current density of 0.1 A/m (see Fig. 7.8, [20]).

When the asymmetric DGG HEMT is dc-drain biased, the asymmetry of the plasmonic cavity is enhanced even more, resulting in further responsivity increase by an order of magnitude. We expect a similar enhancement for the plasmon instability in the asymmetric DGG HEMT.

AC-DGG HEMTs were designed and fabricated using InAlAs/InGaAs/InP materials (see Fig. 7.6b) [20]. Two grating gates G1 and G2 were formed with 70-nm-thick Ti/Au/Ti by a standard liftoff process. The asymmetric factor, the ratio of the interfinger spaces, d_1/d_2, was fixed to be 0.5. The grating gates G1 with narrower fingers L_{g1}, serving plasmon cavity gates, were designed to be chirped ranging from 215 to 430 nm. After processing the AC-DGG HEMT, a high-Q vertical cavity with a high-quality factor of ~60 was formed using a 4.4-μm-thick transparent PW1500 resist serving as a low-loss buffer layer, a 70-nm-thick ITO

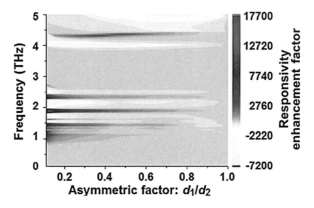

Fig. 7.8 Simulated responsivity enhancement for A-DGG HEMT as a function of the asymmetry factor d_1/d_2 with respect to that for a symmetric DGG HEMT. DC-drain bias current 0.1 A/mm [20]. $L_{g1} = 200$ nm, $L_{g2} = 1.6$ μm, $d_1 + d_2 = 600$ nm, $W = 2.4$ μm. Electron density under the gate G_2 is 2.5×10^{11} cm^{-2}, whereas that is 2.5×10^{12} cm^{-2} under gate G_1

Fig. 7.9 Measured emission spectra at 290 K for different material systems and DGG structures

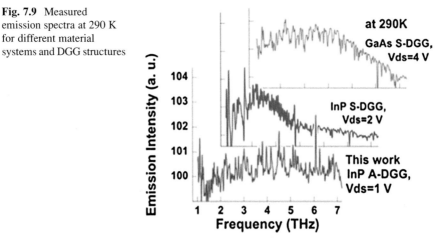

mirror coating on top, and a 100-nm Ti/Au coating on the polished back surface (see Fig. 7.7b). Its fundamental resonant frequency (free spectral range of the Fabry-Perot modes) was designed to be 65 GHz.

The fabricated HEMTs exhibited normal dc transfer characteristics with good pinch-off and gate modulation for both G1 and G2 gates with threshold voltages of -1.1 V and -0.9 V, respectively. The THz spectroscopic measurements used a Fourier-transform far-infrared spectrometer and a 4.2-K-cooled Si composite bolometer. The gate bias for the plasmon cavities V_{g1} was 0 V. Figure 7.9 compares the emission spectra at 290 K with those for the previously reported GaAs-based and InP-based symmetric DGG (S-DGG) HEMTs. The drain bias was equal to the voltage drop along the unit DGG section. The intense backgroundless emission of

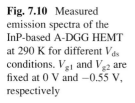

Fig. 7.10 Measured emission spectra of the InP-based A-DGG HEMT at 290 K for different V_{ds} conditions. V_{g1} and V_{g2} are fixed at 0 V and −0.55 V, respectively

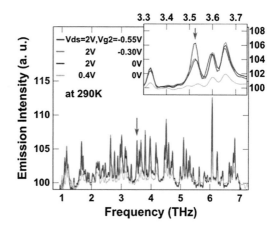

Fabry-Perot modes demonstrates the advantage of the A-DGG structure and InP-based materials with higher electron mobility (~11,000 cm^2/(V s)). The highest peak intensity is observed at ~6.2 THz with a sharp linewidth ~1.23 cm^{-1} (~37 GHz) corresponding to the fifth plasmon mode. Figure 7.10 plots the dependence of the emission spectra on V_{ds} and V_{g2}. By increasing V_{ds} from 0.4 to 2.0 V, the emission intensity of Fabry-Perot modes raises, reflecting the hot plasmon-originated broadband background emission. Furthermore, by applying an appropriate bias for V_{g2} at −0.55 V to make a strong contrast on the electron densities on plasmonic cavities, the peak at ~3.55 THz is enhanced, reflecting the third harmonics of the plasmon resonance driven by the D-S and R-S-S instabilities. Cooling down to 140 K suppressed the hot plasmon residual spurious modes resulting in intense monochromatic emission at the third (at 3.6 THz) plasmon modes matching the frequencies of the adjacent Fabry-Perot modes (see Fig. 7.11). The emission intensity was of the order of a sub-microwatt. The emission peak at the fifth (at 6.1 THz) plasmon mode disappeared, possibly due to the temperature dependence of the 2D electron density in the plasmon cavities detuning the plasmon modes.

7.4 THz Light Amplification by Interband Population Inversion in Graphene

7.4.1 Carrier Dynamics in Optically Pumped Graphene

Figure 7.12 illustrates the carrier relaxation dynamics in optically pumped graphene at relatively high temperatures. When the photogenerated electrons and holes are heated, the collective excitations due to the carrier–carrier scattering (e.g., intraband plasmons) lead to the ultrafast carrier quasi-equilibration (Fig. 7.12b), [29]. Then carriers at the high-energy tails of the distribution are cooled via optical

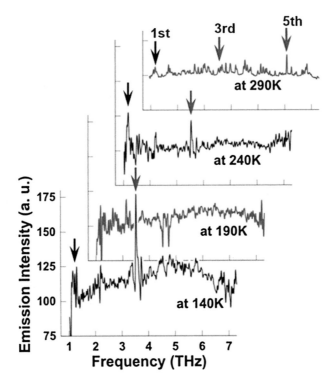

Fig. 7.11 Temperature dependence of the emission spectra of the InP-based A-DGG HEMT under optimal biased conditions ($V_{ds} = 1$ V, $V_{g1} = 0$ V, $V_{g2} = -0.55$ V)

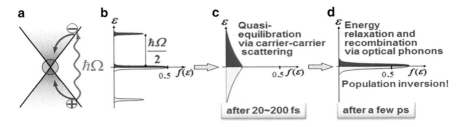

Fig. 7.12 Carrier dynamics in optically pumped graphene

phonon emission and accumulate around the Dirac points (Fig. 7.12c). We numerically simulated the temporal evolution of the quasi-Fermi energy and carrier temperature after pulsed pumping with 0.8-eV photon energy [30, 31]. As shown in Fig. 7.13, due to a fast intraband relaxation (during a ps or less) and relatively slow interband recombination ($\gg 1$ ps) of the photoelectrons and holes, the population inversion is reached under a high pumping intensity $> 10^7$ W/cm^2 [30].

The population inversion is a necessary but not sufficient condition for the gain because of the Drude absorption by free carriers in graphene. The gain at frequency

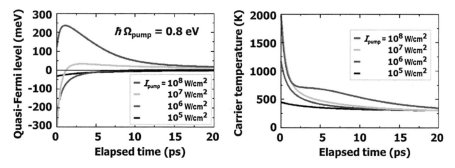

Fig. 7.13 Numerically simulated time evolution of the quasi-Fermi level (*left*) and carrier temperature (*right*) of monolayer graphene after impulsive pumping with a pumping photon energy of 0.8 eV [30]

ω occurs when the real part of the net dynamic conductivity $\mathrm{Re}\sigma_\omega$ (consisting of the intraband $\mathrm{Re}\,\sigma_\omega^{\mathrm{intra}}$ and the interband $\mathrm{Re}\,\sigma_\omega^{\mathrm{inter}}$ contributions) becomes negative. $\mathrm{Re}\sigma_\omega$ is proportional to the rate of the absorption of photons with frequency ω and is given by [5]:

$$\mathrm{Re}\,\sigma_\omega = \mathrm{Re}\,\sigma_\omega^{\mathrm{inter}} + \mathrm{Re}\,\sigma_\omega^{\mathrm{intra}} \approx \frac{e^2}{4\hbar}(1 - 2f_{\hbar\omega}) + \frac{(\ln 2 + \varepsilon_F/2k_BT)e^2}{\pi\hbar}\frac{k_BT\tau}{\hbar(1 + \omega^2\tau^2)},$$

$$(7.9)$$

where e the elementary charge, $\hbar$ is the reduced Planck constant, k_B is the Boltzmann constant, and τ is the momentum relaxation time of carriers. The intraband contribution $\mathrm{Re}\,\sigma_\omega^{\mathrm{intra}}$ corresponds to the Drude-like absorption and is always positive contributing to a loss. In the optically or electrically pumped graphene, the excess electrons and holes occupy the states up to the quasi-Fermi level ε_F in the conduction and valence band, respectively. In this case, Pauli blocking prevents the absorption of photons with energies smaller than $2\varepsilon_F$, and the recombination of electrons and holes at energies corresponding to the photon energy might lead to the stimulated photon emission. In this case, the interband conductivity reaches the negative maximum value of $-e^2/4\hbar$.

Figure 7.14 shows the typical simulated dependences of $\mathrm{Re}\sigma_\omega$ versus time and frequency at a fixed pump intensity 1×10^8 W/cm^2 with the momentum relaxation times of $\tau = 3.3$ ps (left) and $\tau = 10$ ps (right), respectively [31]. To focus on the negative conductivity area, the positive conductivity area is clipped to the null level with red color. It is clearly seen that the gain-spectral bandwidth widens, and a broad terahertz gain bandwidth from ~1.5 to ~10 THz is obtained on the picosecond timescale after pumping for the high-quality graphene with $\tau = 10$ ps.

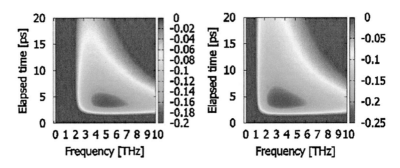

Fig. 7.14 Simulated time evolution of the terahertz dynamic conductivity of graphene after impulsive pumping with a photon energy 0.8 eV at an intensity 10^8 W/cm^2 at 300 K [31]. The carrier momentum relaxation time was assumed to be 3.3 ps (*left*) and 10 ps (*right*), respectively

7.4.2 Observation of Amplified Stimulated THz Emission

Our THz time domain spectroscopy measurements for the femtosecond-laser-pumped graphene samples showed that graphene amplifies an incoming terahertz field [32–34]. An exfoliated monolayer graphene/SiO$_2$/Si sample was placed on the stage. A 0.12-mm-thick CdTe(100) crystal placed on the sample acted as a THz probe pulse emitter as well as an electro-optic sensor [33]. A single 80-fs, 1,550-nm, 4-mW, 20-MHz fiber laser provided the optical pump and probe signals and also generated the THz probe beam. The THz probe pulse double-reflected to stimulate the THz emission in graphene was detected as a THz photon echo signal (marked with number "2" in Fig. 7.15a) [33]. Figure 7.15b shows the measured temporal response. As seen, the secondary pulse, the THz photon echo signal, obtained with graphene (GR) is more intense compared with that obtained without graphene. Figure 7.15c shows a threshold-like behavior when the pumping intensity decreased below 1×10^7 W/cm^2. This is evidence of the light amplification due to the stimulated THz emission. A Lorentzian-like normal dispersion around the gain peak also confirms the amplification mechanism attributed to the stimulated emission of the photocarriers in the inverted states. The Raman spectroscopic measurements of the graphene sample identified the defect density to be quite low with almost no "D" band peak [34]. The room temperature momentum relaxation time τ was estimated from the ratio of the G-band peak to the D-band peak intensity to be 3.3 ps [34]. Figure 7.15d shows the gain-spectral profile calculated using this value of τ in Eq. (7.9). These calculations are in qualitative agreement with the observed results shown in Fig. 7.15c. In an appropriate cavity, the graphene gain medium should lead to a new type of THz lasers [35].

The obtained gain factor exceeded the theoretical limit ($\pi e^2/\hbar c \approx 2.3\%$) given by the quantum conductance $e^2/4\hbar$ (see Sect. 7.4.1) by more than an order of magnitude [33]. We excluded all possible artifacts that might be caused by the experimental setup (such as the gain multiplication due to multiple reflection and

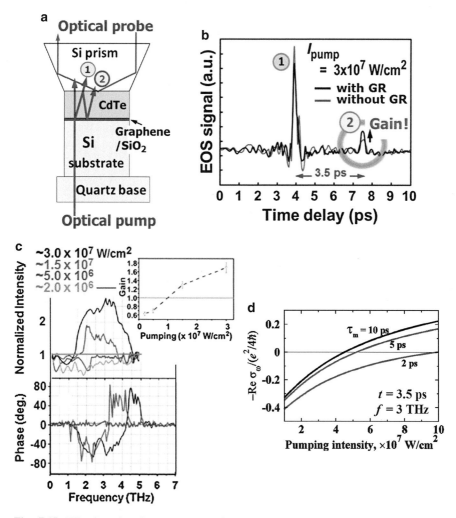

Fig. 7.15 THz time–domain spectroscopy for 80-fs infrared laser-pumped graphene [33]. (**a**) Pump and probe geometry, (**b**) temporal response of the THz probe pulse with the photon echo signal, (**c**) Fourier spectra normalized to the one without graphene and gain vs. pumping intensity, (**d**) calculated gain spectra for $\tau = 3.3$ ps at $t = 3.5$ ps corresponding to the experimental conditions shown in (**c**)

reflective index change due to increase in free carriers.). We concluded that this giant gain originated from the amplified stimulated plasmon emission by the excitation of surface plasmon polaritons (SPPs). This mechanism (which was theoretically revealed and experimentally verified by the authors [36, 37]) will be described in detail in Sect. 7.5.

At a similar time of the authors' observation of terahertz-stimulated emission in optically pumped graphene, Li et al. demonstrated population inversion and stimulated emission at the near-infrared frequencies on very short timescales (~200 fs) [38] (at about the same time as the authors' observation of terahertz-stimulated emission in optically pumped graphene). They intensively pumped graphene with a photon energy of 1.55 eV and a pump fluence two to three orders of magnitude higher than that in our experiments. Before starting the optical phonon emission, they impinged a near-infrared femtosecond pulse with a slightly lower photon energy of 1.16 or 1.33 eV to stimulate the electron–hole recombination. This resulted in near-IR photon emission with the photon energies corresponding to the direct transitions. Generally, as is illustrated in Fig. 7.12, the carrier–carrier scattering lowers the quasi-Fermi energy, preventing the population inversion. However, sufficiently intense pumping can preserve the inverted population, and they observed the stimulated near-IR emission.

7.4.3 Toward the Graphene THz Injection Laser

Figure 7.16 shows the simulated time evolution of the quasi-Fermi levels and electron temperatures for different pumping photon energies when graphene is impulsively pumped at $t = 0$ s. As seen, the optical pumping with a rather high photon energy of the order of an eV significantly heats the carriers, which

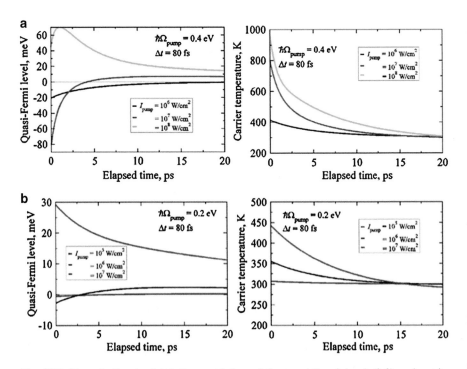

Fig. 7.16 Numerically simulated time evolution of the quasi-Fermi level (*left*) and carrier temperature (*right*) of monolayer graphene after impulsive pumping with a pumping photon energy of (**a**) 0.4 eV and (**b**) 0.2 eV

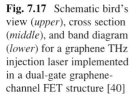

Fig. 7.17 Schematic bird's view (*upper*), cross section (*middle*), and band diagram (*lower*) for a graphene THz injection laser implemented in a dual-gate graphene-channel FET structure [40]

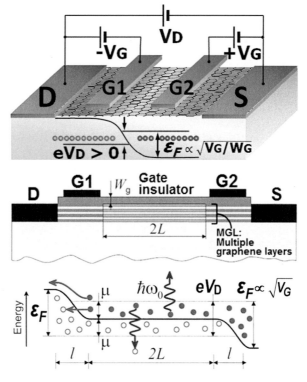

dramatically increases the pumping threshold and prevents the population inversion [39, 40]. Hence, the pumping photon energy should be reduced to obtain a higher THz gain at room temperature. The current injection pumping is the best of solution to cope with this issue because electrical pumping can support the pumping energy on the order of meVs or below in p-i-n junctions [6, 40].

Figure 7.17 shows how a dual-gate structure can implement a p-i-n junction in the graphene channel [6]. Gate biasing controls the injection level, whereas the drain bias controls the lasing gain profiles (photon energy and gain). The distance between the dual-gate electrodes must be sufficiently long to minimize the undesired tunneling current lowering the injection efficiency [6, 40]. The theoretical investigations of the structure- and material-dependent characteristics of the negative-dynamic conductivity in graphene p-i-n structures confirmed their superiority compared to the optical pumping [40].

In a proper terahertz cavity, graphene as a gain media should enable lasing. Due to the pi-electron's orbital emitted terahertz, photons have in-plane components of the electric field intensity. Thus, the selection rule permits the propagation of terahertz photons perpendicular to and/or parallel to the graphene plane. And a vertical cavity [35] as well as waveguide-type cavity structures [41] can support a graphene terahertz laser. A waveguide cavity is superior in terms of a higher quantum efficiency and, hence, lower pumping threshold because of a wider gain-overlapping area for the propagating terahertz photons inside the cavity [41]. Recent developments in the epitaxial graphene technology enable multiple-layer stacking while preserving the monolayer graphene properties. The quantum efficiency increases in proportion to the number of the multiple-graphene layers [41]. Another key issue is decreasing attenuation of the waveguided THz waves.

7.5 Giant THz Gain and Plasmonic Lasing in Population-Inverted Graphene

7.5.1 Giant THz Gain via Excitation of Surface Plasmon Polaritons in Inverted Graphene

The negative THz conductivity of the monolayer graphene originated from the interband population inversion cannot exceed the quantum conductivity ($e^2/4\hbar$). Excitation of the surface plasmon polaritons (SPPs) might help overcome this quantum-conductance limit. The extremely slow wave nature of these excitations (three to four orders of magnitude lower than photon speed) should increase the effective interaction efficiency by three to four orders (on the order of the ratio of the speed of plasmons to that of the photons). This increase results in an order of magnitude increase in the gain coefficient. We predicted the amplification of the surface plasmon polaritons propagating in the graphene-channel waveguide under population inversion [36]. Our analysis of the surface plasmons (SPs) propagating along optically pumped single-graphene layer (SGL) and multiple-graphene layer (MGL) structures showed that the damping of the THz SPs can change to amplification at sufficiently strong optical pumping when the real part of the dynamic conductivity of SGL and MGL structures becomes negative in the THz range of frequencies due to the interband population inversion. The absolute value of the absorption coefficient (SP gain) can be large, substantially exceeding that of optically pumped structures with dielectric waveguides, due to the relatively small SP group velocity. A comparison of SGL and MGL structures shows that the number of graphene layers should be properly chosen in order to maximize the SP gain [36].

Compared with THz lasing due to the stimulated emission of electromagnetic modes (photons), the stimulated plasmon emission by interband transitions in population-inverted graphene can be a much stronger process (see 36, 42, 43). The nonequilibrium plasmons in graphene can be coupled to the TM modes of electromagnetic waves when pertinent structures and/or spatial charge–density distributions lead to the SPP formation and propagation. As shown in Ref. [36], the plasmon gain in the pumped graphene can be very high due to the small plasmon group velocity in graphene and strong confinement of the plasmon field in the vicinity of the graphene layer; see Fig. 7.18a. The propagation index (effective

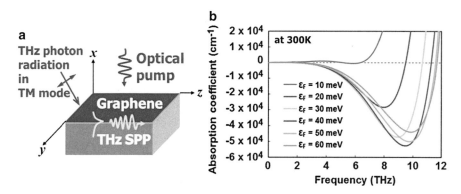

Fig. 7.18 Excitation of SPPs in optically pumped graphene. (**a**) Schematic image, (**b**) simulated frequency dependences of SPP absorption for monolayer population-inverted graphene on SiO_2/Si substrate at 300 K at different levels of population inversion (given by the quasi-Fermi energy $\varepsilon_F = 10, 20, 30, 40, 50, 60$ meV). Carrier momentum relaxation time in graphene is $\tau_m = 3.3$ ps. The results demonstrate giant THz gain (negative values of absorption) of the order of 10^4 cm^{-1} (after Ref. [37])

refractive index for SPPs) ρ of the graphene SPP along the z coordinate is derived from Maxwell's equations [36]:

$$\sqrt{n^2 - \rho^2} + n^2\sqrt{1 - \rho^2} + \frac{4\pi}{c}\sigma_\omega\sqrt{1 - \rho^2}\sqrt{n^2 - \rho^2} = 0 \qquad (7.10)$$

where n is the refractive index of the substrate, c is the speed of light in vacuum, and σ_ω is the conductivity of graphene at frequency ω. When $n = 1$, ρ becomes [36]

$$\rho = \sqrt{1 - \frac{c^2}{4\pi^2\sigma_\omega^2}}. \qquad (7.11)$$

The absorption coefficient α is equal to the imaginary part of the wave vector along the z coordinate: $\alpha = \text{Im}(q_z) = 2\text{Im}(\rho \cdot \omega/c)$. Figure 7.18b plots the simulated values of α for monolayer graphene on a SiO_2/Si substrate ($\text{Im}(n) \sim 3 \times 10^{-4}$) at 300 K [37]. The quasi-Fermi energies are $\varepsilon_F = 10, 20, 30, 40, 50, 60$ meV and the carrier momentum relaxation time $\tau_m = 3.3$ ps. The results demonstrate a giant THz gain (negative values of absorption), on the order of 10^4 cm^{-1}. This giant gain is due to the slow wave nature of the SPPs. Since the absorption coefficients and the resultant gain coefficient (under the negative absorption conditions) are directly linked to the dynamic conductivity σ_ω, as shown in Eq. (7.11), the gain spectra show a similar dependency on the momentum relaxation times and, therefore, on the graphene quality as σ_ω, as discussed in Refs. [5, 6, 29, 31, 44, 45].

We conducted optical pump, THz probe, and optical prove measurements at room temperature for intrinsic monolayer graphene exfoliated from graphite and transferred onto a SiO_2/Si substrate [37]. The experiment used time-resolved near-

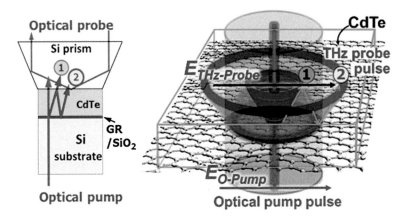

Fig. 7.19 Experimental setup showing the trajectories of the optical pump and THz probe beams. The polarization of the optical pump and the THz probe pulse are depicted with a *red* and *dark-blue arrow* (after Ref. [37])

field reflective electro-optic sampling with fs-IR laser pulse for optical pumping and a synchronously generated THz pulse for probing the THz dynamics of the sample in the THz photon echo regime. Figure 7.19 shows the experimental setup and the pump/probe geometry. A 140-µm-thick CdTe crystal acting as a THz probe pulse emitter as well as an electro-optic sensor was placed on the exfoliated monolayer graphene/SiO$_2$/Si sample. The CdTe can rectify the optical pump pulse to emit the envelope THz probe pulse. The emitted primary THz beam grows along the Cherenkov angle and is detected at the CdTe top surface as the primary pulse (marked with "①" in Fig. 7.19). The THz pulse reflects from the graphene sample. When the sample substrate is conductive, the THz probe pulse transmitting through the graphene reflects back again to the CdTe top surface and is electro-optically detected as the THz photon echo signal (marked with "②" in Fig. 7.19). Therefore, the original temporal response consists of the first forward propagating THz pulse (no interaction with graphene) followed by a photon echo signal (probing the graphene). The delay between these two pulses is given by the total round-trip propagation time of the THz probe pulse through CdTe. The measured waveforms and their corresponding gain spectra (Fig. 7.20) are well reproduced and showed a threshold behavior against the pumping intensity as theoretically predicted in Refs. [5] and [6].

We observe the spatial distribution of the THz probe pulse under the linearly polarized optical pump and THz probe pulse conditions. To measure the in-plane spatial distributions of the THz probe pulse radiation, the optical probe pulse position (at the top surface of the CdTe crystal) was changed step by step by moving the incident point of the optical pump pulse. The pumping intensity $I\Omega$ was fixed at the maximum level. Figure 7.21 shows the observed field distributions for the primary pulse and the secondary pulse intensity. The primary pulse field is situated along the circumference with diameter ~50 µm concentric to the center of

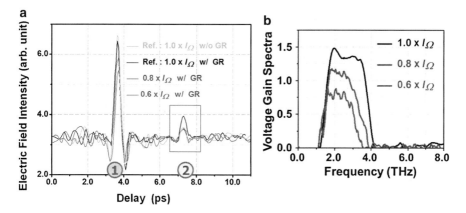

Fig. 7.20 Measured temporal responses (**a**) and corresponding voltage gain spectra (**b**) of the THz photon echo probe pulse (designated with "②") for different pumping intensities $I\Omega$ (3×10^7 W/cm^2), $0.8 \times I\Omega$, and $0.6 \times I\Omega$ (after Ref. [37])

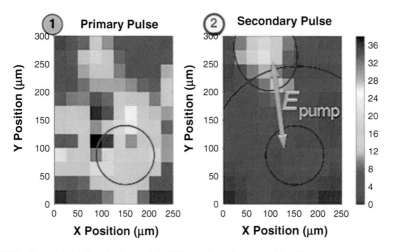

Fig. 7.21 Spatial field distributions of the THz probe pulse intensities. The primary pulse shows nonpolar distribution, whereas the secondary pulse shows a strong localization to the area in which the THz probe pulse is impinged to graphene surface in the TM modes (after Ref. [37])

the optical pumping position. On the other hand, the secondary pulse (THz photon echo) field is concentrated only at the restricted spot area on and out of the concentric circumference with the diameter ~150 μm, where the incoming THz probe pulse excites a TM mode, being capable of supporting the SPPs in graphene [36]. The observed field distribution reproduces the reasonable trajectory of the THz echo pulse propagation in the TM modes inside of the CdTe crystal, as shown in Fig. 7.19 (assuming the Cherenkov angle of 30° in CdTe and the SPP propagation ~10 μm). When the SPPs approach the edge boundary of the optically pumped area, they could mediate the THz emission, which was detected as the secondary THz probe pulse. The observed gain enhancement factor ~50 is in fair agreement with the theoretical calculations presented in Fig. 7.18.

7.5.2 Superradiant THz Plasmon Emission from Inverted Graphene–Metal Micro-Ribbon Arrays

The amplification of THz waves by stimulated generation of resonant plasmons in a planar periodic array of graphene plasmonic microcavities has been theoretically studied in Refs. [42] and [43]. Figure 7.22a shows a typical device structure [43].

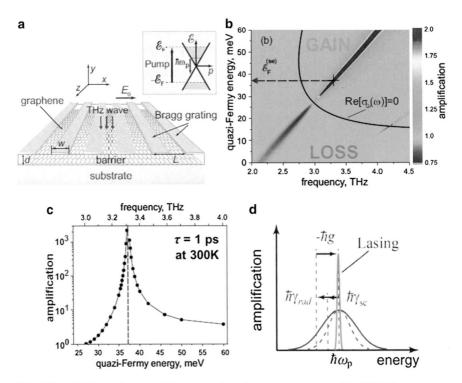

Fig. 7.22 (**a**) Schematic view of the array of graphene micro-/nanocavities [43]. The incoming electromagnetic wave is incident from the top at normal direction to the structure plane with the polarization of the electric field across the metal-grating contacts. The energy band structure of pumped graphene is shown schematically in the *inset*. (**b**) Contour map of the absorbance as a function of the quasi-Fermi energy and the frequency of incoming THz wave for the array of graphene microcavities with period $L = 500$ nm, the length of each graphene microcavity $W = 400$ nm, the gate dielectric layer thickness $d = 50$ nm, and the carrier momentum relaxation time in graphene $\tau_m = 10^{-12}$ s [43]. *Red arrow* marks the quasi-Fermi energy in the plasmonic lasing regime at the fundamental plasmon resonance. (**c**) The variation of the power amplification coefficient along the first plasmon resonance lobe [43]. (**d**) Schematic illustration of the energy rate balance in the plasmon lasing regime [42]

The graphene microcavities are confined periodically underneath the metal-grating gates located on the flat surface of a thin dielectric layer (which can be, e.g., h-BN, SiN, or Al_2O_3). The external THz wave is incident upon the planar array of graphene microcavities at the normal direction with the polarization of the electric field across the metal-grating gates. The graphene is pumped either by optical illumination or by the injection of electrons and holes from the opposite metal gates in each graphene microcavity. In the latter case, the opposite ends of each graphene microcavity adjacent to the metal contacts have to be p- and n-doped. One can easily imagine a biasing scheme for applying dc voltages to successive metal contacts in the interdigital manner [6, 21] in order to ensure the carrier injection into each graphene microcavity. The carrier population and, hence, the dynamic conductivity are characterized by the quasi-Fermi level and carrier temperature (see Eq. (7.9)).

Figure 7.22b shows the contour map of the calculated absorbance as a function of the quasi-Fermi energy (which corresponds to the pumping strength) and the THz wave frequency for the array of the graphene microcavities with period $L = 500$ nm, the length of each microcavity $W = 400$ nm, and the gate dielectric layer thickness $d = 50$ nm. In the amplification regime, the negative value of the absorbance yields the amplification coefficient. The value of $Re[\sigma Gr(\omega)]$ is negative above the solid black line in Fig. 7.22b, corresponding to $Re[\sigma Gr(\omega)] = 0$ (i.e., to transparent graphene). Above this boundary line, negative absorption (i.e., amplification) takes place at all frequencies and pumping strengths. The plasmon absorption resonances below the $Re[\sigma Gr(\omega)] = 0$ line give way to the amplification resonances above this line. The plasmon resonances appear at the frequencies $\omega_n = \omega_p(q_n)$ determined by the selection rule for the plasmon wave vector $q_n = (2n - 1)\pi/a_{eff}$, where a_{eff} is the effective length of the graphene micro-/nanocavity. The frequency of the plasmon resonance is determined mainly by the imaginary part of the graphene conductivity, while the real part of the conductivity (1) is responsible for the energy loss (for $Re[\sigma Gr(\omega)] > 0$) or energy gain (for $Re[\sigma Gr(\omega)] < 0$).

With increasing ε_F, the energy gain can balance the energy loss caused by the electron and hole scattering in graphene, so that the net energy loss becomes zero, $Re[\sigma(\omega)] = 0$ corresponding to the graphene transparency. In this case, the plasmon resonance line exhibits a nonsymmetric Fano-like shape because the real part of the graphene conductivity changes its sign across the plasmon resonance. In this case, the plasmon resonance linewidth is given solely by its radiative broadening (because the dissipative damping is close to zero in this case). Above the graphene transparency line $Re[\sigma Gr(\omega)] = 0$, the THz wave amplification at the plasmon resonance frequency is several orders of magnitude stronger than away from the resonances (the latter corresponding to the photon amplification in the population-inverted graphene). Note that at a certain value of the quasi-Femi energy, the amplification coefficient at the plasmon resonance tends toward infinity with the corresponding amplification linewidth shrinking down to zero. The unphysical divergence of the amplification coefficient is a consequence of a linear approximation approach used in Refs. [42] and [43]. This divergence corresponds to the plasmonic lasing in the graphene micro-/nanocavities in the self-excitation regime. Figure 7.22c shows the behavior of the amplification coefficient around the self-excitation regime [43]. The lasing occurs when the plasmon gain balances the electron scattering loss and the radiative loss; see Fig. 7.22d, [42]. The plasmon oscillations are highly coherent in this case, with virtually no dephasing at all. The red arrow in Fig. 7.22b marks the quasi-Fermi energy corresponding to plasmonic lasing in the first plasmon resonance.

The plasmons in different graphene microcavities oscillate in phase (even without the incoming electromagnetic wave) because the metal contacts act as a synchronizing element between adjacent graphene microcavities [42]. Therefore, the plasma oscillations in the array of the graphene microcavities constitute a single collective plasmon mode distributed over the entire area of the array, which leads to the enhanced superradiant electromagnetic emission. Extraordinary properties of a collective mode in the array of synchronized oscillators are well known in optics: the power of the electromagnetic emission from such an array grows as the square of the number of the oscillators in the array [46].

It is important to stress that the giant amplification enhancement at the plasmon resonance is ensured by strong confinement of the plasmons in the graphene microcavities. As mentioned above, an elevated gain in graphene (approaching the negative of the plasmon radiative damping) is required to meet the self-excitation condition. However, the elevated gain would lead to strong dephasing of a plasma wave over quite long distance of its propagation (which corresponds to the nonresonant stimulated generation of plasmons [47, 48]). Therefore, strong plasmon mode confinement in a single-mode plasmonic cavity is required to ensure the resonant stimulated generation of plasmons. The plasmon confinement to a single-mode microcavity also enhances the rate of the spontaneous electromagnetic emission by the plasmon mode due to the Purcell effect [49]. It is expected that the confinement of plasmons in two-dimensional graphene microcavities (arranged in a chessboard array) could enhance the amplification even stronger.

7.6 Conclusion

Recent advances in THz wave generation from 2D electrons in graphene and III–V semiconductor nano-heterostructures were described. Two-dimensional plasmon resonance in HEMT structures as well as ultrafast nonequilibrium dynamics of massless electrons/holes together with surface plasmon polaritons in graphene are promising mechanisms for enabling new types of practical integrated THz emitters and lasers.

References

1. M. Tonouchi, Cutting-edge terahertz technology. Nat. Photon. **1**, 97–105 (2007)
2. M. Dyakonov, M. Shur, Shallow water analogy for a ballistic field effect transistor: new mechanism of plasma wave generation by dc current. Phys. Rev. Lett. **71**, 2465–2468 (1993)
3. M. Dyakonov, M. Shur, Detection, mixing, and frequency multiplication of terahertz radiation by two-dimensional electronic fluid. IEEE Trans. Electron. Dev. **43**, 1640–1645 (1996)

4. K. Geim, K.S. Novoselov, The rise of graphene. Nat. Mater. **6**, 183–191 (2007)
5. V. Ryzhii, M. Ryzhii, T. Otsuji, Negative dynamic conductivity of graphene with optical pumping. J. Appl. Phys. **101**, 083114 (2007)
6. M. Ryzhii, V. Ryzhii, Injection and population inversion in electrically induced p–n junction in graphene with split gates. Jpn. J. Appl. Phys. **46**, L151–L153 (2007)
7. I. Khmyrova, Y. Seijyou, Analysis of plasma oscillations in high-electron mobility transistor like structures: distributed circuit approach. Appl. Phys. Lett. **91**, 143515 (2007)
8. I.G.R. Aizin, G.C. Dyer, Transmission line theory of collective plasma excitations in periodic two-dimensional electron systems: finite plasmonic crystals and Tamm states. Phys. Rev. B **86**, 235316 (2012)
9. A. Gutin, A. Muraviev, M. Shur, V. Kachorovski, Large signal analytical and SPICE model of THz plasmonic FET, *Lester Eastman Conference Proceedings*, pp. 62–63 (2012), IEEE Catalog Number: CFP12COR-ART
10. T. Otsuji, M. Shur, Terahertz plasmonics. Good results and great expectations. IEEE Microw. J. **15**, 43–50 (2014)
11. S. Rudin, G. Rupper, A. Gutin, M. Shur, Theory and measurement of plasmonic terahertz detector response to large signals. J. Appl. Phys. **115**, 064503 (2014)
12. A.V. Chaplik, Possible crystallization of charge carriers in low-density inversion layers. Sov. Phys. JETP **35**, 395–398 (1972)
13. A.V. Chaplik, Absorption and emission of electromagnetic waves by two-dimensional plasmons. Surf. Sci. Rep. **5**, 289–335 (1985)
14. V.V. Popov, A.N. Koudymov, M. Shur, O.V. Polischuk, Tuning of ungated plasmons by a gate in the field-effect transistor with two-dimensional electron channel. J. Appl. Phys. **104**, 024508 (2008)
15. V. Ryzhii, M.S. Shur, *Plasma Wave Electronics Devices, ISDRS Digest, WP7–07-10* (ISDRS, Washington, DC, 2003), pp. 200–201
16. W. Knap, V. Kachorovskii, Y. Deng, S. Rumyantsev, J.-Q. Lu, R. Gaska, M.S. Shur, G. Simin, X. Hu, M. Asif Khan, C.A. Saylor, L.C. Brunel, Nonresonant detection of terahertz radiation in field effect transistors. J. Appl. Phys. **91**, 9346–9353 (2002)
17. Y. Deng, R. Kersting, J. Xu, R. Ascazubi, X.C. Zhang, M.S. Shur, R. Gaska, G.S. Simin, M.A. Khan, V. Ryzhii, Millimeter wave emission from GaN HEMT. Appl. Phys. Lett. **84**, 70–72 (2004)
18. W. Knap, J. Lusakowski, T. Parenty, S. Bollaert, A. Cappy, V. Popov, M.S. Shur, Terahertz emission by plasma waves in 60 nm gate high electron mobility transistors. Appl. Phys. Lett. **84**, 2331–2333 (2004)
19. N. Dyakonova, A. El Fatimy, J. Lusakowski, W. Knap, M.I. Dyakonov, M.-A. Poisson, E. Morvan, S. Bollaert, A. Shchepetov, Y. Roefens, C. Gaquiere, D. Theron, A. Cappy, Room-temperature terahertz emission from nanometer field-effect transistors. Appl. Phys. Lett. **88**, 141906 (2006)
20. T. Watanabe, A. Satou, T. Suemitsu, W. Knap. V.V. Popov, T. Otsuji, Plasmonic terahertz monochromatic coherent emission from an asymmetric chirped dual-grating-gate InP-HEMT with a photonic vertical cavities, *CLEO: Conference on Lasers and Electrooptics Dig., CW3K.7*, San Jose, CA, USA, June 12, 2013.
21. T. Otsuji, M. Hanabe, T. Nishimura, E. Sano, A grating-bicoupled plasma-wave photomixer with resonant-cavity enhanced structure. Opt. Express **14**, 4815–4825 (2006)
22. T. Otsuji, Y.M. Meziani, T. Nishimura, T. Suemitsu, W. Knap, E. Sano, T. Asano, V.V. Popov, Emission of terahertz radiation from dual-grating-gates plasmon-resonant emitters fabricated with InGaP/InGaAs/GaAs material systems. J. Phys. Condens. Matters **20**, 384206 (2008)
23. Y. Tsuda, T. Komori, T. Watanabe, T. Suemitsu, T. Otsuji, Application of plasmonic micro-chip emitters to broadband terahertz spectroscopic measurement. J. Opt. Soc. Am. B **26**, A52–A57 (2009)
24. T. Otsuji, T. Watanabe, S. Boubanga Tombet, A. Satou, W. Knap, V. Popov, M. Ryzhii, V. Ryzhii, Emission and detection of terahertz radiation using two-dimensional electrons in III-V semiconductors and graphene. IEEE Trans. Terahertz Sci. Technol. **3**, 63–72 (2013)

25. V.Y. Kachorovskii, M.S. Shur, Current-induced terahertz oscillations in plasmonic crystal. Appl. Phys. Lett. **100**, 232108 (2012)
26. V. Ryzhii, A. Satou, M. Ryzhii, T. Otsuji, M.S. Shur, Mechanism of self-excitation of terahertz plasma oscillations in periodically double-gated electron channels. J. Phys. Condens. Matters **20**, 384207 (2008)
27. M.I. Dyakonov, Boundary instability of a two-dimensional electron fluid. Semiconductors **42**, 984–988 (2008)
28. V.V. Popov, D.V. Fateev, T. Otsuji, Y.M. Meziani, D. Coquillat, W. Knap, Plasmonic terahertz detection by a double-grating-gate field-effect transistor structure with an asymmetric unit cell. Appl. Phys. Lett. **99**, 243504 (2011)
29. M. Breusing, C. Ropers, T. Elsaesser, Ultrafast carrier dynamics in graphite. Phys. Rev. Lett. **102**, 086809 (2009)
30. A. Satou, T. Otsuji, V. Ryzhii, Theoretical study of population inversion in graphene under pulse excitation. Jpn. J. Appl. Phys. **50**, 070116 (2011)
31. T. Otsuji, S.A. Boubanga Tombet, A. Satou, H. Fukidome, M. Suemitsu, E. Sano, V. Popov, M. Ryzhii, V. Ryzhii, Graphene-based devices in terahertz science and technology. J. Phys. D **45**, 303001 (2012)
32. H. Karasawa, T. Komori, T. Watanabe, A. Satou, H. Fukidome, M. Suemitsu, V. Ryzhii, T. Otsuji, Observation of amplified stimulated terahertz emission from optically pumped heteroepitaxial graphene-on-silicon materials. J. Infrared Millim. Terahertz Waves **32**, 655–665 (2011)
33. S. Boubanga-Tombet, S. Chan, T. Watanabe, A. Satou, V. Ryzhii, T. Otsuji, Ultrafast carrier dynamics and terahertz emission in optically pumped graphene at room temperature. Phys. Rev. B **85**, 035443 (2012)
34. T. Otsuji, S. Boubanga-Tombet, A. Satou, M. Suemitsu, V. Ryzhii, Spectroscopic study on ultrafast carrier dynamics and terahertz amplified stimulated emission in optically pumped graphene. J. Infrared Millim. Terahertz Waves **33**, 825–838 (2012)
35. A.A. Dubinov, V.Y. Aleshkin, M. Ryzhii, T. Otsuji, V. Ryzhii, Terahertz laser with optically pumped graphene layers and Fabry–Perot resonator. Appl. Phys. Express **2**, 092301 (2009)
36. A.A. Dubinov, Y.V. Aleshkin, V. Mitin, T. Otsuji, V. Ryzhii, Terahertz surface plasmons in optically pumped graphene structures. J. Phys. Condens. Matter **23**, 145302 (2011)
37. T. Watanabe, T. Fukushima, Y. Yabe, S. Boubanga Tombet, A. Satou, A.A. Dubinov, V. Ya Aleshkin, V. Mitin, V. Ryzhii, T. Otsuji, Gain enhancement effect of surface plasmon polaritons on terahertz stimulated emission in optically pumped monolayer graphene. New J. Phys. **15**, 075003 (2013)
38. T. Li, L. Luo, M. Hupalo, J. Zhang, M.C. Tringides, J. Schmalian, J. Wang, Femtosecond population inversion and stimulated emission of dense Dirac Fermions in graphene. Phys. Rev. Lett. **108**, 167401 (2012)
39. V. Ryzhii, M. Ryzhii, V. Mitin, A. Satou, T. Otsuji, Effect of heating and cooling of photogenerated electron–hole plasma in optically pumped graphene on population inversion. Jpn. J. Appl. Phys. **50**, 094001 (2011)
40. V. Ryzhii, M. Ryzhii, V. Mitin, T. Otsuji, Toward the creation of terahertz graphene injection laser. J. Appl. Phys. **110**, 094503 (2011)
41. V. Ryzhii, A. Dubinov, T. Otsuji, V. Mitin, M.S. Shur, Terahertz lasers based on optically pumped multiple graphene structures with slot-line and dielectric waveguides. J. Appl. Phys. **107**, 054505 (2010)
42. V.V. Popov, O.V. Polischuk, A.R. Davoyan, V. Ryzhii, T. Otsuji, M.S. Shur, Plasmonic terahertz lasing in an array of graphene nanocavities. Phys. Rev. B **86**, 195437 (2012)
43. V.V. Popov, O.V. Polischuk, S.A. Nikitov, V. Ryzhii, T. Otsuji, M.S. Shur, Amplification and lasing of terahertz radiation by plasmons in graphene with a planar distributed Bragg resonator. J. Opt. **15**, 114009 (2013)

44. T. Otsuji, S. Boubanga-Tombet, A. Satou, M. Ryzhii, V. Ryzhii, Terahertz-wave generation using graphene-toward new types of terahertz lasers. IEEE J. Sel. Top. Quantum Electron. **19**, 8400209 (2013)
45. T. Otsuji, S.A. Boubanga Tombet, A. Satou, H. Fukidome, M. Suemitsu, E. Sano, V. Popov, M. Ryzhii, V. Ryzhii, Graphene materials and devices in terahertz science and technology. MRS Bull. **37**, 1235–1243 (2012)
46. M. Benedict, A. Ermolaev, V. Malyshev, I. Sokolov, E. Trifinov, *Superradiance: multiatomic coherent emission* (IOP, Bristol, 1996)
47. V. Ryzhii, A. Satou, T. Otsuji, Plasma waves in two-dimensional electron-hole system in gated graphene heterostructures. J. Appl. Phys. **101**, 024509 (2007)
48. F. Rana, Graphene terahertz plasmon oscillators. IEEE Trans. Nanotechnol. **7**, 91–99 (2008)
49. E.M. Purcell, Spontaneous emission probabilities at radio frequencies. Phys. Rev. **69**, 681 (1946)

Chapter 8
Optics of Hybrid Nanomaterials in the Strong Coupling Regime

Adam Blake and Maxim Sukharev

8.1 Coupled Harmonic Oscillators

The idea of coupled harmonic oscillators is readily demonstrated for macroscopic systems such as two mass-spring systems with a coupling spring between the masses. Two coupled pendulums also serve as a good example. Remarkably, the analytical machinery of coupled harmonic oscillators can be applied in the microscopic domain to quantum emitters (an atom or molecule that has a well-defined optical transition and can be treated as a two-level atom), charge density waves, and electromagnetic fields. None of these immediately conjures up an image of a mass on a spring. But when emitters and charge density waves are treated as oscillators coupled by the fields that they emit, features predicted by classical coupled oscillator equations, such as strong coupling and avoided crossing, are clearly observed! We therefore have a firm analytical framework for understanding hybrid materials consisting of emitters and metallic structures. Coupling two oscillators and solving for the eigenfrequencies ω_+ and ω_- yields [1]:

$$\omega_+^2 = \omega_c^2 + \Omega^2 \tag{8.1a}$$

$$\omega_-^2 = \omega_c^2 - \Omega^2 \tag{8.1b}$$

A. Blake (✉)
Department of Physics, Arizona State University, Tempe AZ 85281, USA
e-mail: ahblake@asu.edu

M. Sukharev
Science and Mathematics Faculty, College of Letters and Sciences,
Arizona State University, Mesa AZ 85212, USA
e-mail: Maxim.Sukharev@asu.edu

© Springer International Publishing Switzerland 2015
A. Korkin et al. (eds.), *Nanoscale Materials and Devices for Electronics,*
Photonics and Solar Energy, Nanostructure Science and Technology,
DOI 10.1007/978-3-319-18633-7_8

$$\omega_c^2 = \omega^2 + \Omega^2 \tag{8.1c}$$

ω_c is the frequency that one oscillator would have if the other were held in place and ω is the resonant frequency of each uncoupled oscillator. Clearly, the splitting between normal mode frequencies increases as the coupling strength is increased.

In the case of coupling emitters to charge density waves such as surface plasmon polaritons (SPPs; these will be discussed further in Sect. 8.3), we are dealing with an oscillator and a standing wave. Consider the (separate) dispersion relations of two uncoupled systems: a two-level emitter and an SPP. The dispersion relation of the uncoupled SPP is given as follows [2]:

$$k = \frac{\omega}{c} \sqrt{\frac{\varepsilon_1 \varepsilon_2}{\varepsilon_1 + \varepsilon_2}} \tag{8.2}$$

ε_2 is the permittivity of the dielectric, ε_1 is the permittivity of the metal, k is the in-plane wave vector of the SPP, ω is the angular frequency of the SPP, and c is the speed of light in vacuum. The dispersion relations for each uncoupled system as well as the coupled system are shown in Fig. 8.1. Each of the solid lines represents the dispersion relation that would be obtained from each individual, independent oscillator. A complete picture of the coupling is given in [3]. The Hamiltonian of the coupled exciton-SPP system is as follows:

$$\begin{bmatrix} E_m & \Delta \\ \Delta & E_{pl}(k) \end{bmatrix} \tag{8.3}$$

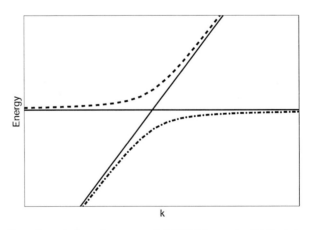

Fig. 8.1 The dispersion relation of an uncoupled SPP (*diagonal solid line*), two-level emitter (*horizontal solid line*), and coupled system consisting of the upper polariton (*curved dashed line*) and lower polariton (*curved dash-dot line*). Note that at larger wave numbers (not shown here), the SPP dispersion curve bends and asymptotically approaches a constant value

$E_{pl}(k)$ is the energy of the uncoupled plasmon energy, E_m is the energy of the uncoupled emitter resonance, and Δ is the coupling energy. This Hamiltonian is diagonalized, yielding the following expression for the energies of the normal modes of the system (referred to as upper and lower polaritons) as a function of wave vector [4, 5]:

$$E_{u,l}(k) = \frac{1}{2}\left\{[E_{pl}(k) + E_m] \pm \sqrt{4\Delta^2 + (E_{pl}(k) + E_m)^2}\right\} \qquad (8.4)$$

$E_{u,l}(k)$ is the energy of the upper or lower polariton as a function of the wave number k and all other quantities are the same as in Eq. 8.3. $E_{pl}(k)$ is the energy of the uncoupled plasmon energy, E_m is the energy of the uncoupled emitter resonance, and Δ is the coupling energy. The minimum energy separation is the Rabi splitting of the system and this value increases as the coupling strength is increased. Where the solid lines in Fig. 8.1 overlap (i.e., both systems driven at resonance) the dispersion of coupled oscillators clearly departs from that of the uncoupled oscillators. Far from resonance, the system response approaches that of uncoupled oscillators.

This leads to the definition of two regimes: weak coupling and strong coupling. When the coupling strength exceeds the upper and lower polaritons' linewidths, the system is in the strong coupling regime. If the damping is strong enough that the splitting is not observable, the system is in the weak coupling regime.

We will look at several examples of weak and strong coupling between light and matter, the first of which is emitters in a reflecting microcavity followed by emitters coupled to SPPs via various geometric structures.

8.2 Coupling Microcavities to Quantum Emitters

The emission characteristics of an emitter depend upon the environment in which it is placed. Modifying the environment in which an emitter is placed can be accomplished in many different ways. Photons can be confined inside of one of various configurations of reflecting microcavities. As will be discussed in the next section, an emitter can be placed in the intense fields of an SPP. Some experiments involve shooting atoms through a cavity one at a time, with each atom interacting with the fields from the atom that preceded it [6]. In this section, the coupling of emitters to microcavities is discussed.

Confining emitters to a microcavity was first demonstrated by Ekimov and Onushchenko [7] and Efros and Efros [8] in which semiconductor nanocrystals were grown in glass. More elaborate structures have since been developed. Micropillars use Bragg mirrors to confine light axially and total internal reflection for radial confinement. Microdisks use total internal reflection for confinement along all directions. Photonic crystals are periodic nanostructures with regions of high and low dielectric constant that can create bandgaps (frequency regions over which propagation is forbidden) for photons. These actually occur naturally as opals. One of the primary differences between these three structures is the size of the emitter (in this case, a quantum dot) that can be placed in the cavity. A larger quantum dot gives a larger dipole moment and therefore stronger coupling. The micropillar and microdisk structures have larger volumes which would lead to weaker coupling were it not for the fact that a larger quantum dot (with a larger dipole moment) can be placed in this larger volume, whereas the photonic crystal offers a smaller volume and higher quality factor Q leading to the largest E-field in the empty cavity [9].

The spontaneous emission rate of a quantum emitter can be controlled by placing it in a reflecting microcavity [6]. Spontaneous emission is suppressed when the size of the cavity is less than the emission wavelength and enhanced when the size of the cavity is equal to the emission wavelength. The reason for this is that spontaneous emission is actually stimulated by vacuum fluctuations and the size of the cavity determines the wavelengths of vacuum fluctuations that are permitted.

The enhancement of spontaneous emission is known as the Purcell effect and is quantified by the Purcell factor:

$$F_p = \frac{3}{4\pi^2} \frac{\lambda_c^3}{n} (Q/V) \tag{8.5}$$

λ_c is the wavelength in the cavity medium, n is the index of refraction of the cavity medium, and V is the mode volume of the cavity.

Quantum information science, which requires photon-on-demand sources, can benefit from more deterministic control of photon emission. It is desirable to know exactly when the photon is going to be emitted, and shorter emission lifetimes accomplish this. The Purcell effect increases the emission rate and therefore decreases the emission time, thereby making the emission of the photon more deterministic. As can be seen from Eq. (8.5), this can be realized by increasing the quality factor and decreasing the effective volume of the cavity. The decreasing volume ultimately restricts the cavity to a single mode [9]. Also, the quantum dephasing rate must be made much smaller than the cavity decay rate. If a quantum emitter/microcavity system is to be used for quantum computing, it will ideally demonstrate antibunching (no more than one photon emitted at a time) and uniform spacing between the emitted photons.

When an emitter is strongly coupled to a microcavity, the states of both systems are altered: one system is coupled to another and the original energy levels of each individual system are modified and become that of a single, hybrid system. The extent to which the original energy levels are modified is determined by the coupling between the two systems. Strong coupling between inorganic quantum emitters and microcavities has yielded Rabi splitting of 3–10 meV, whereas strong coupling between organic quantum emitters (such as J-aggregates) and microcavities shows Rabi splitting of 100–500 meV [10].

The strong coupling regime can be mapped by varying the temperature of the emitter, which sweeps the emitter resonance across that of the microcavity. This is less than ideal, as changing the temperature of the emitter causes its optical properties to change. A less intrusive means of sweeping the emitter resonance involves incrementally condensing Xe onto the interior of the microcavity [11], causing the index of refraction to increase relative to vacuum and the emitter resonance to shift to lower energies.

The dispersion relation for the upper and lower polaritons for a system consisting of an organic semiconductor material inside of a microcavity is derived in [10] and it is equivalent to Eq. 8.4 up to a damping term. This is not surprising; even though one equation deals with SPPs and the other with microcavities, both describe systems of two-level emitters coupled to electromagnetic field modes. The upper and lower polaritons (coherent excitations) are well defined within certain limits of the wave vector k, outside of which (less than k_{min} or larger than k_{max}) wave vector broadening is on the order of the wave vector itself and the excitations become incoherent [10]. The states of a microcavity whose volume contains an organic semiconductor consists of a mixture of a large number of incoherent states and a smaller number of coherent states. Photoluminescence from the former pumps the latter.

It is emphasized in [10] that the excitations, even when not clearly defined, are coupled exciton-cavity excitations. k_{min} can be reduced for inorganic structures because of the lower damping at lower temperatures. At large wave vectors the upper polariton tends toward the cavity curve. These states are coherent, and the upper polariton can decay into an incoherent state, which can decay into the lower polariton. Either of these processes results in an optical phonon.

8.3 Strong Coupling Between SPPs and Quantum Emitters

SPPs are oscillations of charge on a metal-dielectric boundary that are able to produce intense, highly localized evanescent fields. SPPs can be described classically, but the name derives from the more accurate quantum mechanical description. A quantum of surface charge oscillation (a plasmon) couples to a quantum of light (a photon) leading to a system that is a hybrid of the two independent entities: an SPP. The intense, strongly confined fields of SPP oscillations are ideal for strong coupling to quantum emitters. Borrowing a term from cavity quantum electrodynamics (CQED), SPPs are said to have a small effective volume which immediately entails strong coupling with emitters. The largest Rabi splitting between SPPs and emitters observed thus far is 700 meV [12].

For a given frequency, an SPP wave number is larger than the corresponding wave number in vacuum. In order to couple light to an SPP, both the frequency and wave number must be matched and therefore a means of increasing the wave number of light must be introduced. One method is known as the attenuated total reflection (ATR) method which surrounds a metal film by two media with different permittivities. An evanescent wave is generated on the input side of the film (via total internal reflection) whose wave number matches that of the SPP on the output side of the film. The second method, which will be of importance herein, relies on the fact that a periodic metal structure (such as a corrugated grating, as in [13] and

[14]) reduces the SPP wave vector by a factor related to the periodicity of the grating, allowing the radiation to couple to SPP waves [2].

A departure from the usual normal mode splitting occurs when either or both emitter density and emitter dipole moment (increasing either of these increases coupling between SPPs and emitters) are made very large [3] and is observed in simulations as follows. The finite difference time domain (FDTD) method [15] is used to propagate electromagnetic waves. Silver is represented by the Drude model [16] and the emitters have two energy levels, the upper one being degenerate to allow polarization along two directions. The Liouville–von Neumann equation is numerically solved to give the density matrix elements in regions containing emitters [17]. Ordinary normal mode splitting is demonstrated by using lower values of emitter density and dipole moment and varying the period of the slit array (which is experimentally adjustable), causing the SPP resonance to shift. The SPP resonance is swept through the emitter resonance, demonstrating avoided crossing. This however assumes interaction between SPPs and isolated emitters. The interaction between the emitters themselves increases as emitter density and dipole moment increase and in fact, the accuracy of the two-level model begins to suffer as the emitter density and/or dipole moment increases. For the case of ordinary normal mode splitting (no interaction between the emitters), the Rabi splitting increases linearly as $\sqrt{N}$ [1, 18], where N is the emitter density. Simulations from [3] show deviations from this behavior at high emitter densities. Rabi splitting increases linearly with the dipole moment, and deviations are not seen in [3] at higher dipole moments.

In addition to the deviations mentioned, a new peak appears in between the upper and lower polaritons and it is proposed to indicate a collective effect from the entire ensemble of emitters. The progression from ordinary Rabi splitting to the appearance of the collective peak is shown in Fig. 8.2. Experiments verify that this

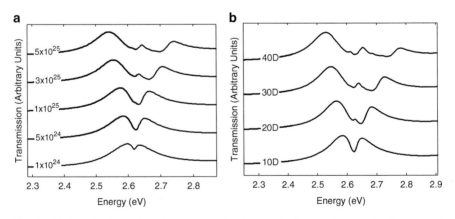

Fig. 8.2 Emitter density and dipole moment affect the coupling between SPPs and emitters. Panel (**a**) shows increasing emitter density (emitters per m^3) and panel (**b**) shows increasing emitter dipole moment. When either the density or dipole moment are low, "ordinary" Rabi splitting is seen. As density or dipole moment is increased, the third "collective" peak emerges and strengthens near the emitter resonance

third peak is present under conditions of very strong coupling [19–21]. It does not deviate appreciably from the emitter resonance even as the SPP resonance is tuned, suggesting that its origin is from the emitters rather than the SPPs. An energy gap is also observed: when the coupling strength is such that the collective mode is not seen, the upper and lower polariton energies asymptotically approach the emitter resonance for large detuning. When the collective peak is present, the asymptotes of the upper and lower polaritons (for large detuning) have a gap between them. The upper polariton demonstrates a larger gap than the lower polariton.

A spacer layer of variable thickness was inserted between the emitter layer and the grating in order to see the dependence of emitter–emitter coupling on the strength of the SPP field. The collective peak is observed to merge with the upper polariton as the spacer thickness increases. Coupling between the emitters themselves is suspected given the fall-off of the collective mode as the SPP field decreases. Furthermore, a simulation was run in which the entire region containing emitters was replaced by a single two-level system (essentially a single quantum emitter with a very large dipole moment), thereby eliminating the spatial variation of the fields. In this case, the collective peak disappears even at extremely high emitter densities.

The emission from quantum dots is essentially isotropic, but results shown in [22] demonstrate that depositing quantum dots onto a periodic slit array causes the dots' emission to become both directional and wavelength-selective. Quantum dots have also been placed upon a nanoscale Yagi–Uda antenna [23].

An aluminum grating on a glass substrate is used, and the emitters are InAs/CdSe core-shell nanocrystal quantum dots (NQDs). The emission spectrum of the NQDs peaks at 1.2 μ m and has a FWHM of about 200 nm with no angular dependence being observed.

A plot of near-field intensity for the NQD/slit system shows intense fields near the slits at zero angle from the slits and much weaker fields at 15°. A plot of emission intensity vs. incident angle shows a FWHM of 3.4° and transmission in the forward direction that is 20 times higher than a sample consisting only of NQDs. This beaming effect is thought to be a result of the NQDs being coupled to the SPP waves.

8.4   Control and Time Dynamics of Strong Coupling

The Rabi splitting for a coupled emitter-SPP system is given by [24]. It depends upon the emitter dipole moment density and the strength of the local SPP electric field:

$$\Omega_R = \int \vec{\mu}_x(r) \cdot \vec{E}_{SPP}(r) dV \qquad (8.6)$$

Ω_R is the Rabi splitting in rad/s, $\vec{\mu}_x$ is the excitonic transition dipole moment density, and $\vec{E}_{SPP}$ is the electric field of the local SPP. The coupling can therefore be manipulated by changing $\vec{\mu}_x$ or $\vec{E}_{SPP}$ and [24] takes advantage of this in order to create an optically switchable metallic mirror. By altering the dipole moment of the emitters (and thus the coupling) on short timescales, the reflection can be reduced by 40 % in parts of the near-infrared. This is a reversible, sub-picosecond effect which is immediately appealing as a switch, especially given that it can be controlled by a single photon.

Results are first presented for J-aggregates on a smooth gold film. At the emitter resonance, the bare film displays a reflectivity of 0.95, which drops to 0.25 when the emitters are added. With a pump pulse, the reflectance (relative to that of the bare film) at the emitter resonance is seen to increase for all values of pump fluence. A system of J-aggregates deposited upon a gold reflection grating is subsequently analyzed. A contour plot of reflectivity as a function of angle of incidence and wavelength shows an avoided crossing which of course indicates strong coupling between the emitters and the SPPs, and calculations show that the SPP fields are localized around the slits. The spectra obtained agree with numerical solutions of Maxwell's equations as applied to the hybrid structure. Non-equilibrium time dynamics is then detailed: a pump pulse is sent in first, followed by a probe pulse 150 fs later. Near the exciton resonance, the reflection behaves as that of the emitters on a smooth film; these excitons are therefore not strongly coupled to the SPP fields. The usual upper and lower polariton features remain, but additional structure (minima in reflections) is seen in these curves. As the pump fluence is increased, these new features separate from the upper and lower polariton curves and begin to resemble those of uncoupled oscillators. This can be understood in terms of a decrease in coupling strength of the excitons that are strongly coupled to the SPP field, as a reduction in the dipole moment occurs due to saturation of the excitonic oscillator strength. Note that the pumping is non-resonant; it is on the tail of the exciton curve. This creates a mixture of uncoupled and coupled excitons.

A pump pulse can therefore switch the system from well-defined upper and lower polaritons to a system with structured upper and lower polariton curves. For an incident angle of 49°, [24] demonstrates that a pump pulse can significantly lower the reflectivity of the system by activating the reflection minima in these structured curves.

Active control of the strong coupling regime is also demonstrated in [25] by a much different means. An ATR setup is used in which light is coupled to SPP waves on the output side of a metal film (which is covered with emitters) via a prism. The oscillator strength of the emitters is controlled by introducing nitrogen dioxide into the system. As the concentration is increased from 0 PPM to 6 PPM, the dispersion changes from an uncoupled SPP wave to that of a hybrid system; avoided crossing is observed with a corresponding Rabi splitting of 130 meV. A coherent energy exchange develops in which "light and matter exchange energy for a certain number of periods before the energy escapes the system" [25]. The system can be uncoupled by heating the sample, causing the reflectance spectrum to display a single ATR minimum instead of two.

Changes in the optical properties of a hybrid material from an ultrashort laser pulse are observed in simulations performed by [26]. The simulations demonstrate

that the coherent energy exchange between emitters and SPPs occurs over femto-
seconds and this time scale can be adjusted via material or laser parameters.

The Rabi oscillations are observed by pumping the emitters and then sending in a
probe pulse after a delay time $\Delta\tau$, with $\Delta\tau = 0$ corresponding to the probe pulse
being sent right as the pump pulse begins. This is clearly different from simply
probing the system, as a much larger number of emitters are excited. A series of
simulations is run, each with a longer $\Delta\tau$. Running this sequence of simulations
maps out the effect of the pump pulse over time; the Rabi oscillations are clearly
seen in a plot of reflectance (R) vs. time.

For a 15 fs pump, a graph of pump delay $\Delta\tau$ versus change in reflectivity R shows
that ΔR increases for the upper and lower polaritons relative to their values without
any pumping. This is thought to be caused by energy transfer between the SPPs and
the emitters: the period of oscillations in the ΔR vs $\Delta\tau$ graph is 3.75 fs and this time
depends only upon the local electric field strength and the emitter dipole moment. It
does not vary appreciably when slit period or emitter density is changed. This
period of oscillation does depend on the peak pump amplitude. For a 30 fs pump
pulse that is resonant with the emitters, ΔR vs. $\Delta\tau$ is plotted for two different peak
pump amplitudes. At 2×10^9 V/m, the period of oscillation is 3 fs and at $4 \times
10^9$ V/m the period of oscillation is 6 fs (see Fig. 8.3). Two surprising results are
also observed: the Rabi period slows down over time and the Rabi frequency is
smaller in these simulations as compared to a simple two-level atom in the same
laser field.

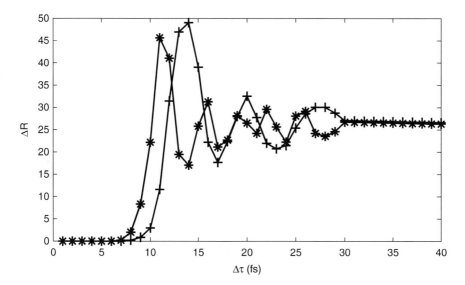

Fig. 8.3 Femtosecond oscillations of ΔR. The Rabi oscillation period depends on the pump
amplitude. *Plus symbols* indicate an amplitude of 2×10^9 V/m (the oscillation period is 6 fs)
and *asterisk symbols* indicate an amplitude of 4×10^9 V/m (the oscillation period is 3 fs)

A longer (180 fs) pump pulse is used to map out the excitation dynamics. The oscillations of ΔR and those of the ground state population are in step with one another, suggesting that the oscillations in the transient spectra are caused by transitions of the emitters between the ground state and the excited state.

Additionally, a pump pulse is applied and then abruptly truncated to zero. This is a promising means of controlling the number of Rabi oscillations. With this in mind, one could conceivably manipulate the time envelope and the incident angle of the pump pulse to control the plasmon energy distribution.

Another instance of optical control of energy distribution in hybrid materials is in [12]. The transmission coefficient is able to be modified by using two different types of emitters each with different resonant energies. These emitters are coupled to a silver sinusoidal grating. Two absorption features are present in the bare grating and they are identified as SPPs by noting that the minima in the transmission spectrum correspond to the maxima in the reflection spectrum. The fields of the SPPs are very inhomogeneous, and plotting the fields reveals the spatial extent of the coupling for each type of emitter.

The SPP resonances are able to be swept across the emitter resonances by varying the amplitude of the grating. When the two types of emitters are added, the lower energy resonance is seen to undergo Rabi splitting as would be expected. The higher energy resonance not only undergoes splitting, but (for an amplitude of 60 nm and greater) exhibits the collective peak that was previously described. The density and dipole moment are not particularly large, but the fields (which of course affect coupling) associated with this higher energy are much larger, giving rise to the collective peak.

A chirped laser pulse (in this case, a pulse whose frequency increases or decreases linearly over time) can be used to selectively excite either one emitter type or the other, which was verified by sending a pulse into a single emitter only and recording the density matrix elements corresponding to the ground and excited states. If the laser pulse starts at the frequency of higher energy emitters and decreases, the lower-energy emitters will be excited and the higher-energy type will not be; and vice versa for the laser pulse starting at the lower emitter frequency and increasing. Given that the two types of emitters couple to different SPP modes, whose fields corresponding to each mode have different spatial distributions, a positive chirp will excite one set of regions and a negative chirp will excite a different set of regions. These chirped pulses effectively pump one type of emitter, causing its response to a subsequent probe pulse to differ from that without the chirped pulse.

As is shown in Fig. 8.4, the transmission at the lower SPP resonance is altered by a negative chirp (but not much by a positive chirp) and transmission at the higher SPP resonance is altered by a positive chirp (but not much by a negative chirp). The simulations show that a negative chirp inverts the lower energy emitters whereas a positive chirp causes the higher energy emitters undergo several Rabi oscillations.

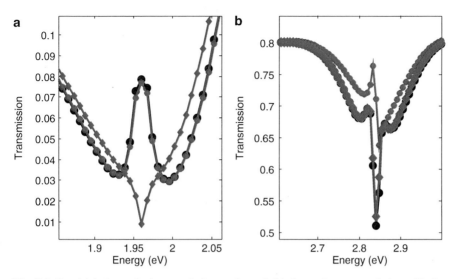

Fig. 8.4 Panel (**a**) shows the lower polariton and panel (**b**) shows the upper polariton. *Circles* indicate no chirp, *diamonds* indicate a negative chirp, and *asterisks* indicate a positive chirp. A negative chirp alters the lower polariton whereas a positive chirp does not. A positive chirp alters the upper polariton whereas a negative chirp has no effect on it

References

1. P. Törmä, W.L. Barnes, Rep. Prog. Phys. **78**, 013901 (2015)
2. H. Raether, *Surface Plasmons on Smooth and Rough Surfaces and on Gratings* (Springer, New York, 1988)
3. A. Salomon, R.J. Gordon, Y. Prior, T. Seideman, M. Sukharev, Phys. Rev. Lett. **109**, 073002 (2012)
4. V.M. Agranovich, M. Litinskaia, D.G. Lidzey, Phys. Rev. B **67**, 085311 (2003)
5. D.G. Lidzey, D.D.C. Bradley, M.S. Skolnick, T. Virgili, S. Walker, D.M. Whittaker, Nature **395**, 53 (1998)
6. S. Haroche, Rev. Mod. Phys. **85**, 1083 (2013)
7. A.I. Ekimov, A.A. Onushchenko, JETP Lett. **34**, 345 (1981)
8. A.L. Efros, A.L. Efros, Sov. Phys. Semicond. **16**, 772 (1982)
9. G. Khitrova, H.M. Gibbs, M. Kira, S.W. Koch, A. Scherer, Nat. Phys. **2**, 81 (2006)
10. V. Agranovich, G. La Rocca, Fundamental optical and quantum effects in condensed matter. Solid State Commun. **135**, 544 (2005)
11. S. Mosor, J. Hendrickson, B. Richards, J. Sweet, G. Khitrova, H. Gibbs, T. Yoshie, A. Scherer, O. Shchekin, D. Deppe, Appl. Phys. Lett. **87**, 141105 (2005)
12. M. Sukharev, J. Chem. Phys. **141**, 084712 (2014)
13. M. Sukharev, P.R. Sievert, T. Seideman, J.B. Ketterson, J. Chem. Phys. **131**, 034708 (2009)
14. I. Avrutsky, Y. Zhao, V. Kochergin, Opt. Lett. **25**, 595 (2000)
15. A. Taflove, S. Hagness, *Computational Electrodynamics: The Finite-Difference Time-Domain Method* (Artech House, Boston, 2005)
16. L. Novotny, S.J. Stranick, Ann. Rev. Phys. Chem. **57**, 303 (2006)
17. M. Sukharev, A. Nitzan, Phys. Rev. A **84**, 043802 (2011)

18. P. Vasa, W. Wang, R. Pomraenke, M. Lammers, M. Maiuri, C. Manzoni, G. Cerullo, C. Lienau, Nat. Photon. **7**, 128 (2013)
19. A. Salomon, C. Genet, T. Ebbesen, Angew. Chem. Int. Ed. **48**, 8748 (2009)
20. Y. Sugawara, T.A. Kelf, J.J. Baumberg, M.E. Abdelsalam, P.N. Bartlett, Phys. Rev. Lett. **97**, 266808 (2006)
21. J.A. Hutchison, D.M. O'Carroll, T. Schwartz, C. Genet, T.W. Ebbesen, Angew. Chem. Int. Ed. **50**, 2085 (2011)
22. N. Livneh, A. Strauss, I. Schwarz, I. Rosenberg, A. Zimran, S. Yochelis, G. Chen, U. Banin, Y. Paltiel, R. Rapaport, Nano Lett. **11**, 1630 (2011)
23. A.G. Curto, G. Volpe, T.H. Taminiau, M.P. Kreuzer, R. Quidant, N.F. van Hulst, Science **329**, 930 (2010)
24. P. Vasa, R. Pomraenke, G. Cirmi, E. De Re, W. Wang, S. Schwieger, D. Leipold, E. Runge, G. Cerullo, C. Lienau, ACS Nano **4**, 7559 (2010)
25. A. Berrier, R. Cools, C. Arnold, P. Offermans, M. Crego-Calama, S.H. Brongersma, J. Gómez-Rivas, ACS Nano **5**, 6226 (2011)
26. M. Sukharev, T. Seideman, R.J. Gordon, A. Salomon, Y. Prior, ACS Nano **8**, 807 (2013)

Index

A

Atomistic simulations
  absorption spectra, 191–192
  CdSe nanocrystal, exciton-biexciton
    atomistic structure, 204, 205
    Coulomb interaction, 207, 208
    Kramers doublets, 206
    multiexciton generation process,
      206–207
    probability densities, 204–206
    single-particle electron states, 204, 205
    single-particle hole states, 205, 206
    solar cells, 206
    spectral function, 208, 209
    time evolution, 209, 210
    wave functions, 208
  computational domain, 157–159
  Coulomb matrix elements
    dielectric function, 189
    interatomic distances, 189
    LCAO form, 188
    Slater orbital coefficients, 190
  dark exciton
    Bloch function, 199, 200
    emission amplitudes, 202, 203
    energy diagram, 200, 201
    Kramers doublet, 199, 200
    polarization-resolved emission spectra,
      201, 202
    single-electron state, 199, 200
  dipole matrix elements, 192–193
  electrons and holes interaction, 185–187
  emission spectra, 191–192
  epitaxial dots
    chirality, 198
    hole shell structure, 199
    joint optical density of states, 194, 195
    Kramers doublets, 194
    lens-shaped InAs dot, 194, 195
    probability densities, 194–196
    single-band Schroedinger
      equations, 197
    single-particle state energies, 194–196
  quantum dots, 156
    colloidal nanocrystal, 151–152
    energy spectrum, 150
    fermionic anticommutation rules, 154
    high excitation power, 203–204
    Schroedinger equation, 155
    self-assembled, 150–152
    single-dot measurements, 156
    tight-binding approach, 156–157
    triangular graphene quantum dot,
      151, 153
  single-quasiparticle levels (*see* Single-
    quasiparticle levels)
  strain
    biaxial strain, 163, 164
    bond bending, 160, 161
    bond stretching, 160, 161
    elastic parameters, 161, 162
    hydrostatic strain, 163, 164
    shear strain, 163, 164
    tensor matrix elements, 162, 163
    VFF approach, 160, 161
    zinc-blende lattice, 159, 160
Attenuated total reflection (ATR)
    method, 267

© Springer International Publishing Switzerland 2015
A. Korkin et al. (eds.), *Nanoscale Materials and Devices for Electronics,
Photonics and Solar Energy*, Nanostructure Science and Technology,
DOI 10.1007/978-3-319-18633-7